A Practical G[illegible]
Quantum Computing

Hands-on approach to quantum computing with Qiskit

Elías F. Combarro
Samuel González-Castillo

<packt>

A Practical Guide to Quantum Computing

Portfolio Director: Kunal Chaudhari
Relationship Lead: Rithika Shetty
Project Manager: Ashwin Dinesh Kharwa
Content Engineer: Divya Anne Selvaraj
Technical Editor: Kushal Sharma
Copy Editor: Safis Editing
Indexer: Manju Arasan
Proofreader: Divya Anne Selvaraj
Production Designer: Ponraj Dhandapani
Growth Lead: Mansi Shah

First published: July 2025

Production reference: 1250725

Published by Packt Publishing Ltd.
Grosvenor House
11 St Paul's Square
Birmingham
B3 1RB, UK.

ISBN 978-1-83588-594-9

www.packtpub.com

Get in touch

Feedback from our readers is always welcome.

General feedback: If you have questions about any aspect of this book, mention the book title in the subject of your message and email us at customercare@packtpub.com.

Errata: Although we have taken every care to ensure the accuracy of our content, mistakes do happen. If you have found a mistake in this book, we would be grateful if you would report this to us. Please visit `www.packtpub.com/support/errata` and fill in the form.

Piracy: If you come across any illegal copies of our works in any form on the Internet, we would be grateful if you would provide us with the location address or website name. Please contact us at copyright@packtpub.com with a link to the material.

If you are interested in becoming an author: If there is a topic that you have expertise in and you are interested in either writing or contributing to a book, please visit `authors.packtpub.com`.

Share your thoughts

Once you've read *A Practical Guide to Quantum Computing*, we'd love to hear your thoughts! Scan the QR code below to go straight to the Amazon review page for this book and share your feedback or leave a review on the site that you purchased it from.

`https://packt.link/r/1835885950`

Your review is important to us and the tech community and will help us make sure we're delivering excellent quality content.

Download a free PDF copy of this book

Thanks for purchasing this book!

Do you like to read on the go but are unable to carry your print books everywhere? Is your eBook purchase not compatible with the device of your choice?

Don't worry, now with every Packt book you get a DRM-free PDF version of that book at no cost.

Read anywhere, any place, on any device. Search, copy, and paste code from your favorite technical books directly into your application.

The perks don't stop there, you can get exclusive access to discounts, newsletters, and great free content in your inbox daily.

Follow these simple steps to get the benefits:

1. Scan the QR code or visit the link below:

   `https://packt.link/free-ebook/9781835885949`

2. Submit your proof of purchase.
3. That's it! We'll send your free PDF and other benefits to your email directly.

Subscribe to Deep Engineering

Join thousands of developers and architects who want to understand how software is changing, deepen their expertise, and build systems that last.

Deep Engineering is a weekly expert-led newsletter for experienced practitioners, featuring original analysis, technical interviews, and curated insights on architecture, system design, and modern programming practice.

Scan the QR or visit the link to subscribe for free.

`https://packt.link/deep-engineering-newsletter`

Join us on Discord!

Read this book alongside other users, developers, experts, and the author himself.

Ask questions, provide solutions to other readers, chat with the authors via Ask Me Anything sessions, and much more. Scan the QR or visit the link to join the community.

https://packt.link/deep-engineering-quantum

Contributors

About the authors

Elías F. Combarro holds degrees in mathematics (1997) and computer science (2002) from the University of Oviedo (Spain). For each of these, he received national awards in recognition of his academic performance—second-highest grades in the country for mathematics, and highest in the country for computer science. He completed several research stays at Novosibirsk State University (Russia) before earning a PhD in mathematics from the University of Oviedo in 2001. His dissertation, supervised by Prof. Andrey Morozov and Prof. Consuelo Martínez, explored properties of computable predicates.

Since 2023, Elías has been a full professor in the Department of Computer Science at the University of Oviedo. He has authored more than 50 research papers in international journals, covering topics such as quantum computing, computability theory, machine learning, fuzzy measures, and computational algebra. His current research focuses on applying quantum computing to problems in algebra, optimization, and machine learning.

He served as a cooperation associate at CERN openlab in 2020 and 2022, and was a visiting scholar at Harvard University in 2024. He represented Spain on the advisory board of the CERN Quantum Technology Initiative from 2021 to 2024. He is also a co-author of *A Practical Guide to Quantum Machine Learning and Quantum Optimization* (Packt, 2023).

To Adela, Paula and Sergio. Always my reason to live.

Samuel González-Castillo holds degrees from the University of Oviedo (Spain) in both mathematics and physics (2021) and a research master's degree in mathematics from the National University of Ireland, Maynooth (2023). He is a mathematics research student at the University of Oviedo, where he works as a graduate teaching assistant. His current research focuses on the application of algebraic techniques to problems in quantum computing.

He completed his physics bachelor thesis under the supervision of Prof. Elías F. Combarro and Prof. Ignacio F. Rúa (University of Oviedo), and Dr. Sofia Vallecorsa (CERN). In it, he worked alongside other researchers from ETH Zürich on the application of quantum machine learning to classification problems in high-energy physics. In 2021, he was a summer student at CERN developing a benchmarking framework for quantum simulators. He has contributed to several conferences on quantum computing, including the Quantum Technology International Conference and the Conference on Computing in High Energy and Nuclear Physics. He is one of the authors of *A Practical Guide to Quantum Machine Learning and Quantum Optimization* (Packt, 2023).

About the reviewers

Francisco Orts is a researcher at the University of Almería (Spain). He holds a PhD in computer science focused on the combined use of quantum computing with classical high-performance computing techniques, winning an extraordinary PhD award. His research interests are multidimensional scaling, quantum computing, and high-performance computing, with currently 23 papers published in high-impact journals and having participated in more than 40 research conferences. Beyond research, he has worked as a computer scientist at construction, stock exchange, and IT services companies, with more than 15 years of experience in the sector. He also teaches quantum computing in master's programs.

Guillermo Botella Juan (Senior Member, IEEE) is full professor at the Department of Computer Architecture and Automation, Complutense University of Madrid (UCM), Spain. He received MSc degrees in physics and electronic engineering, and a PhD from the University of Granada. His research focuses on hardware acceleration, computer arithmetic, and emerging paradigms such as analog and quantum computing. He introduced UCM's first undergraduate course in quantum computing. He has conducted research at UCL, led projects with Banco Santander and IBM, and serves as deputy editor-in-chief of Digital Signal Processing. He actively contributes to sessions on applications of emerging technologies at the DATE conference and has supervised 10 PhD theses.

I would like to thank Conchi and my family for their support and patience during the time I spent reviewing this book. I am also grateful to my APCC students at UCM, as well as those working on their final degree and master's projects in comp. science, physics, and math, and to my PhD students, for their constant inspiration. It has been a privilege to help review a book that brings quantum computing closer to future scientists and engineers.

Table of Contents

Foreword

Quantum computing is no longer a distant dream or the subject of speculative fiction. It's here—not yet fully realized, but unmistakably real—and equipped with all the ingredients to reshape the foundations of modern technology. From chemical simulation and drug discovery to optimization challenges, cybersecurity, and artificial intelligence, the potential of quantum computing is as vast as it is profound. Yet, for many curious minds eager to explore this new computational paradigm, the quantum world often feels opaque—hidden behind walls of complex mathematics and abstract physics.

This is precisely where *A Practical Guide to Quantum Computing* by Elías F. Combarro and Samuel González-Castillo finds its purpose. In a field often dominated by theory and intimidating formalisms, this book offers a clear and accessible path forward. It arrives at a moment when quantum innovation is accelerating at an extraordinary pace—breakthroughs in hardware, error correction, and algorithmic and software development are happening faster than ever before.

Its timing couldn't be more fitting: 2025 has been designated the International Year of Quantum Science and Technology by the United Nations, marking 100 years since the birth of quantum mechanics. This recognition underscores the global importance of quantum technologies and their transition from theoretical concepts to real-world impact.

Combarro and González-Castillo respond to this moment by putting practical tools into the hands of those who will shape the quantum future—whether they're computer scientists, engineers, or passionate tech enthusiasts. With clarity and care, they bridge the often-complicated gap between quantum principles and real-world application. This book does more than explain what a quantum computer is—it invites you to play and learn with it.

The journey unfolds gradually and intentionally. Beginning with the basics—the single qubit and foundational concepts for manipulating it—the authors guide you through increasingly complex systems and algorithms. They culminate the book by exploring the medium- to long-term challenges facing the field. Each step is hands-on, grounded in code and practical exercises, delivering not just knowledge but experience.

On a personal note, I am deeply proud to see how Elías and Samuel have embraced Qiskit—the quantum software development kit I helped create in 2017 alongside Jay Gambetta and Andrew Cross—as a core tool for this guide. It is incredibly rewarding to see Qiskit support such a timely, important, and deeply practical book. With this guide in hand, you're not just learning about quantum computing—you're taking your place in its unfolding story.

Ismael Faro Sertage

VP Quantum + AI, IBM Quantum

Acknowledgements

Feeling gratitude and not expressing it is like wrapping a present and not giving it.

— William Arthur Ward

We would like to thank all the people that helped us, in one way or another, with this book. First and foremost, we need to mention our high school math teacher Tomás Fernández Marcos. Without him, we would have never met, and this book would simply not exist. Every time we write that little *i* representing the imaginary unit, we're reminded that Tomás was the first to introduce us to so many wonderful mathematical concepts.

We would also like to thank Ismael Faro for his generosity in writing the foreword to this book, and for all his efforts in making quantum computers accessible online to students and researchers around the world. Having the opportunity to use real quantum computers for free has truly changed our lives for the better. Thank you so much for your vision!

Many parts of this book originate from courses on quantum computing taught over the years. Those courses would not have been possible without the help and trust of Enrique Arias, Alberto Di Meglio, Melissa Gaillard, Ester Martín Garzón and José Ranilla, among many others.

Many of the insights that we have tried to convey in these pages came from discussions with friends and colleagues all over the world, from whom we have learned lots and lots. There are too many of you to name here, but you know who you are!

We are also thankful for the generous support of the Real Colegio Complutense at Harvard University and the Physics Department at Harvard University, which made it possible for one of us to spend some wonderful time working there. And, of course, a special thanks to Professor Susanne Yelin for her hospitality and kindness, as well as to Cristina Valdés Rodríguez and Daniel Sánchez Mata for all their help.

Ferdous Khan has been a constant source of support over the years. Although we haven't met in person (yet!), he has always offered wise words and encouraged us to do our best. Words can't fully express how much his friendship means.

Of course, we'd also like to thank our team at Packt. It seems they liked what we did with our first book, because they invited us to write another one! Once again, they were incredibly supportive throughout the process—always quick to respond and ready to help with anything we needed.

And what can we say about our technical reviewers? Guillermo Botella and Francisco Orts once again volunteered to read through the entire manuscript, carefully checking for inconsistencies and errors. And what a fantastic job they did! They caught some very subtle "bugs" and were a huge help in improving the book overall. Of course, any remaining mistakes are entirely our own.

We're also deeply grateful to the readers of our first book. We've been truly overwhelmed by the wonderful feedback we've received over the past two years. We never imagined that our unique way of explaining things would resonate so deeply with so many people. That encouragement is what inspired us to do it all over again. We hope you enjoy this one too!

And last but by no means least, we are forever grateful to our friends and families. Their unconditional love and support carried us through the tough moments of the writing process. No combination of words, even in the infinitely vast Hilbert spaces of quantum states, could ever capture how much you mean to us!

Elías F. Combarro, Samuel González-Castillo

Oviedo, April 2025

Preface

Nature isn't classical, dammit, and if you want to make a simulation of nature, you'd better make it quantum mechanical, and by golly it's a wonderful problem, because it doesn't look so easy.

— Richard Feynman

Almost three years ago, we sat down to write a little book (well, at that time, we thought it *would* be little) titled *A Practical Guide to Quantum Machine Learning and Quantum Optimization*. It was a new, exciting and unforgettable adventure, and we had tons and tons of fun because we made it… for us.

It is a cliché to say that you should write the book that you would like to read, but, what can we say? We are all for clichés and that's what we did. We wrote the book that we would've liked to read when we started studying quantum machine learning and quantum optimization. A book with mathematical rigor, because you cannot understand quantum computing without all the formulas, but also with code that you can run to obtain results, because quantum computing should be, first and foremost, about *computing*.

So we wrote the book mainly for us… and we still can't believe how many people found our way of explaining quantum computing useful and even entertaining. Since the book was published, we've received feedback from many readers that liked the book and, to our surprise, wanted more.

At the same time, quantum computing has surged in popularity. In the last few years, more and more people have became interested in learning how to program quantum computers and the number of courses on the topic offered by universities all over the world has grown exponentially. Moreover, with 2025 proclaimed by the United Nations as the International Year of Quantum Science and Technology, this trend will very likely continue and multiply.

For all these reasons, even if we believed that writing *A Practical Guide to Quantum Machine Learning and Quantum Optimization* had been a once-in-a-lifetime adventure, when the wonderful people at Packt reached to us again wondering whether we'd be interested in writing a new book, we thought that it was a brilliant idea. In fact, we still had a thing or two to write about quantum computing, so why not?

In our previous book, we decided to focus on what we call "modern" quantum algorithms: variational algorithms, quantum annealing, and other techniques that try to put the kind of quantum computers that are available today to good use in tasks such as data classification and combinatorial optimization. But, in fact, the development of quantum computing as a field did not start with these kinds of applications in mind. Algorithms such as the ones proposed by Grover, Shor, and other visionaries, are quite different from the ones that we presented in our previous book... but equally important and fascinating.

In *A Practical Guide to Quantum Machine Learning and Quantum Optimization* we decided to skip many wonderful applications of quantum computing to focus on these modern quantum algorithms. And, though we of course love all the new developments that the quantum computing community has come up with in recent years, we've always had a bit of a regret for not having been able to include other, more foundational algorithms and protocols.

Understanding how the quantum Fourier transform works or how quantum teleportation is achieved is not only fascinating in itself, but crucial to comprehend what quantum computers can do and how they do it. It is obviously great to be able to apply variational circuits to learn patters in datasets or to find ground states of Hamiltonians, but if you've never heard of Deutsch's algorithm or you do not know how to implement Grover's search, your grasp of quantum computing will be inevitably incomplete.

With the book that you now hold in your hands, we try to put a remedy to that. Our focus is, now, on the foundations of quantum computing. In some sense, you could think of this new book as a *prequel* to our previous one. If you've already read *A Practical Guide to Quantum Machine Learning and Quantum Optimization*, you will find here some concepts that you already know about. And you will learn much more about them, in the same way that you learned much more about Obi-Wan Kenobi and Anakin Skywalker when you watched The Phantom Menace after being acquainted with them from the original Star Wars trilogy[1]. And the great thing is that, if you've never read our previous book, you will not miss anything![2]

Our approach to explaining the topics covered in this book is, essentially, the same that we followed in *A Practical Guide to Quantum Machine Learning and Quantum Optimization.* We start completely from scratch and we do not assume any knowledge of quantum physics. In fact, we can assure you that quantum physics is even less necessary to understand this book than it was for our previous one. And you don't need to know a lot of math either—just some basic algebraic concepts, such as what a matrix and a vector are, and how to multiply them. By the way, if you need a refresher on these or other mathematical concepts, you can always refer to the Appendices at the end of the book.

Another key characteristic of our approach to introducing concepts is that we firmly believe in the importance of a gradual progression—from simple ideas to more complex ones—allowing you to build confidence and understanding step by step. We start very humbly, focusing on a simple qubit. And we spend a lot of time studying one-qubit systems before moving further. But we think you'll find this approach helpful in developing a deep understanding that serves as a solid foundation for more complex algorithms and techniques.

In *A Practical Guide to Quantum Machine Learning and Quantum Optimization*, most chapters included a mix of mathematical concepts and code. However, in this new book, we have decided to split into different chapters the abstract discussion of the algorithms

[1]We only hope that you enjoy this book better than The Phantom Menace!
[2]But if you like this one, please give the other a try!

and protocols that we introduce, and their practical implementation through code. We feel that this works better in this case, because most algorithms are self-contained and it is better to explore them as a whole before attempting to fully implement them.

The quantum programming framework that we have selected for our implementations is Qiskit, concretely version 2.1. Qiskit is, by far, the most popular quantum computing platform out there and it is especially suitable for the kind of algorithms and protocols that we study in this book. Additionally, over the last few years, Qiskit has reached a much bigger stability and maturity, making it even more appealing to implement foundational quantum algorithms.

The style of our exposition is mainly informal, without following the usual structure of definition-theorem-proof-corollary of many mathematical texts, but without sacrificing rigor at any point in the book. Whenever possible, we give detailed derivations that justify the mathematical properties that we use in our developments and analyses (or, at least, we provide an argument that may be extended to a full proof by just adding some small technical details). In the cases that proving a particular fact is beyond the scope of the book, we provide references in which a full treatment can be found.

Throughout all the text, we propose exercises that will help you understand important concepts and develop practical skills for manipulating formulas and writing your own quantum code. They are intended to be readily solved (we try to give useful hints for those exercises that are a little bit more challenging), but, at the end of the book, we provide full, detailed solutions so that you can check your understanding of the subject.

Quantum computing is a field in constant evolution, so we feel that it is especially important to give pointers to new developments, to variants of the algorithms that we present in the book, and to alternative approaches to solve the kind of problems that we study. We do this by including numerous boxes with the label "To learn more…". You can skip these boxes if you wish, as they are not necessary to follow the main text. However, we strongly recommend reading them, since they help to situate in a wider context the topics under study. Other boxes that we use throughout the book serve to highlight important facts, to give warnings about subtle points, or to remind you of central definitions and formulas. These should not be skipped. They are labeled "Important note" for a reason!

Writing this book was a lot of fun. Again, this is the kind of book that we would have liked to have read when we first delved into the field of quantum computing. We wrote it, mainly, for ourselves. But we hope that you find it useful too. Enjoy the ride!

Who this book is for

This book would be ideal for university-level students in computer science, mathematics, physics, or other STEM fields taking introductory-level courses on quantum computing. It would also suit professionals, researchers, and self-learners with a STEM background. Potential readers of our previous book, *A Practical Guide to Quantum Machine Learning and Quantum Optimization*, will benefit from first building foundational quantum computing skills with this book.

More broadly, this book would also be a good fit for people who are curious about quantum computing and want to understand it from a rigorous and hands-on perspective, exploring the details of the most well-known quantum algorithms.

What this book covers

This book is organized into five parts, an afterword, and some appendices, as follows:

Part 1, One Qubit to Rule Them All: Working with One Qubit

Chapter 1, What Is (and What Is Not) a Quantum Computer, serves as an introduction to the rest of the book, clarifying what a quantum computer is, how it is different from a classical computer, and why quantum algorithms can outperform classical ones on some tasks.

Chapter 2, Qubits, Gates, and Measurements, discusses what single-qubit systems are and how they can be represented, measured, and transformed. This chapter lays down the most basic theoretical foundation for working with quantum algorithms, as qubits are the fundamental unit of quantum information (analogous to bits in classical computing). The remaining chapters in the book build upon this one.

Chapter 3, Applications and Protocols with One Qubit, explores how, while a humble qubit may not seem like much, it already enables a few practical applications. In this chapter, we show how one-qubit systems can be used to implement key distribution schemes and quantum money that is impossible to forge.

Chapter 4, Coding One-Qubit Protocols in Qiskit, introduces the Qiskit framework and we briefly mention and discuss other platforms for quantum computing. We also show how one-qubit protocols can actually be implemented and run using Qiskit.

Part 2, Qubit Meets Qubit: Two Qubits and Entanglement

Chapter 5, How to Work with Two Qubits, takes one step further in generality and introduces two-qubit systems. In a structure analogous to that of *Chapter 2*, we discuss how the state of a two-qubit system can be represented and how these systems can be measured and transformed.

Chapter 6, Applications and Protocols with Two Qubits, looks at how, following the introduction of two-qubit systems, we can exploit new quantum phenomena such as entanglement, and we are in a position to unlock new and very interesting applications; this is what this chapter is devoted to. The applications that we discuss in this chapter are not just appealing in and of themselves, but they also give us an opportunity to discuss how quantum algorithms take advantage of quantum phenomena in the construction of quantum algorithms.

Chapter 7, Coding Two-Qubit Algorithms in Qiskit, succinctly reviews the core principles of the Qiskit framework and introduces how two-qubit systems can be handled in Qiskit. Then we show how the protocols and algorithms discussed in *Chapter 6* can be implemented and run using Qiskit.

Part 3, Working with Many Qubits

Chapter 8, How to Work with Many Qubits, allows us to capitalize on having mastered two-qubit systems, and with a full understanding of how to work with them, we are in a position to take the final step in our inductive journey and unleash the full power of quantum computing, introducing systems with an arbitrarily large number of qubits. This chapter introduces multi-qubit systems as a generalization of two-qubit systems, showing how they can be represented, measured, and transformed.

Chapter 9, The Full Power of Quantum Algorithms, introduces some simple quantum algorithms that fully demonstrate how quantum superposition, entanglement, and interference can be used in practice. These algorithms perfectly illustrate the capabilities and limitations of quantum computing, and they serve as a foundational step towards the more sophisticated algorithms that are introduced in the following chapters.

Chapter 10, Coding with Many Qubits in Qiskit, culminates our inductive introduction to Qiskit by showing how to work with an arbitrary number of qubits in it. As a means of illustrating this and reviewing the content covered in the previous chapter, we show how the Deutsch-Jozsa and Bernstein-Vazirani algorithms can be implemented in Qiskit.

Part 4, The Stars of the Show: Main Quantum Algorithms

Chapter 11, Finding the Period and Factoring Numbers, looks at Shor's factoring algorithm—probably the best-known quantum algorithm. It has been a crucial reason why social interest in quantum computing has risen quickly over the last few decades. In this chapter, we introduce Shor's algorithm, first explaining why it is significant, with a special emphasis on its implications in cybersecurity. We then discuss all the details of the algorithm, taking a rigorous approach, but with plenty of informal and intuitive explanations.

Chapter 12, Searching and Counting with a Quantum Computer, discusses Grover's algorithm, which is a quantum search algorithm. We begin by introducing the problem of searching for elements through an unsorted list, commenting on the computational complexity of this problem for classical computers. This sets the scene for the introduction of Grover's algorithm, which can provide a quadratic speedup over classical methods. We also discuss how the quantum Fourier transform that we covered in the previous chapter can fit with Grover's algorithm in order to allow search results to be counted.

Chapter 13, Coding Shor and Grover's Algorithms in Qiskit, explores how Grover and Shor's algorithm will have far-reaching applications in real-world scenarios once quantum hardware becomes powerful enough to support them. In this chapter, we show how these algorithms can be programmed with Qiskit. We also include a section devoted to the implementation of the quantum Fourier transform in Qiskit.

Part 5, Ad Astra: The Road to Quantum Utility and Advantage

Chapter 14, Quantum Error Correction and Fault Tolerance, introduces quantum error correction, which could pave the road towards fault-tolerant computation and thus may be a cornerstone in the development of useful quantum computers. This chapter begins with a discussion on the necessity for quantum error correction, which is followed by the introduction and implementation of a simple quantum error-correcting code. The chapter finishes with some remarks on fault-tolerant computing and how quantum error correction can help enable it.

Chapter 15, Experiments for Quantum Advantage, is devoted to understanding quantum advantage, specifically recent experiments with random circuit sampling. We unpack the main concepts involved in these experiments and we give some clarifying examples using Qiskit. In addition to this, and as a way of wrapping up the book, we take the chance to sketch some ideas about what lies ahead in the road of quantum computing.

Appendices

Appendix A, Mathematical Tools, provides a refresher on the fundamentals of linear algebra, including vectors and matrices, and important notions such as bases and eigenvalues. In addition, it gives a quick recap of the most relevant properties of complex numbers and how to operate with them, and it even covers some concepts from modular arithmetic.

Appendix B, The Bra-Ket Notation and Other Foundational Notions, explores in depth the details behind the "bra-ket" notation that we use throughout the book and that is ubiquitous in the quantum computing literature. We will also briefly touch upon a very widely used tool to represent one-qubit states: the Bloch sphere.

Appendix C, Measuring the Complexity of Algorithms, serves as a quick introduction to measuring the resources needed to solve problems with algorithms. It defines some important concepts that are used throughout the book, such as the big O notation.

Appendix D, Installing the Tools, guides you through the process of installing the libraries needed in order to run the code included in this book.

Appendix E, Production Notes, gives a glimpse of the process of writing a technical book like this one, including the software used to typeset and prepare the content.

Solutions contains the solutions to all the exercises proposed in the main text.

Conventions used

There are a number of text conventions used throughout this book.

`CodeInText`: Indicates code words in text, filenames, file extensions, URLs, and user input. Here is an example: "We could create a `QuantumCircuit` object and directly use its `draw` method".

A block of code is set as follows:

```
from qiskit import QuantumCircuit
circuit = QuantumCircuit(1)
```

Important ideas are highlighted in boxes like the following:

> **Important note**
>
> I am a box. I feel important. That's because I am important.

We sometimes include material for those of you who want to learn more. We format it as follows:

> **To learn more...**
>
> You don't have to read me if you don't want to.

There are a few exercises in the text, which are displayed as follows:

> **Exercise 0.1**
>
> Prove that every even number greater than two can be written as the sum of two prime numbers.

To get the most out of this book

The concepts explained in this book are better understood by implementing algorithms that solve practical problems and by running them on simulators (or actual quantum computers!). You will learn how to do all that starting from the very beginning of the book, but in order to run the code you will need to install some tools.

We recommend that you download the Jupyter notebooks from the link provided in the following section and that you follow the instructions given in *Appendix D* to get your environment ready to rock!

Download the example code files

The code bundle for the book is also hosted on GitHub at `https://github.com/PacktPublishing/A-Practical-Guide-to-Quantum-Computing`. In case there's an update to the code, it will be uploaded to the existing GitHub repository.

We also have other code bundles from our rich catalog of books and videos available at `https://github.com/PacktPublishing/`. Check them out!

Part 1

One Qubit to Rule Them All: Working with One Qubit

In this part, we begin our journey into the depths of quantum computing. We will start gently, by discussing some generalities about quantum computing and introducing a few general ideas from an intuitive and informal point of view. We shall then explore in detail the most simple of all quantum systems: one-qubit systems. As humble as they may appear to be, you will find out that they have a lot of potential.

This part includes the following chapters:

- *Chapter 1*, What Is (and What Is Not) a Quantum Computer
- *Chapter 2*, Gates, Qubits, and Measurements
- *Chapter 3*, Applications and Protocols with One Qubit
- *Chapter 4*, Coding One-Qubit Algorithms in Qiskit

1

What Is (and What Is Not) a Quantum Computer

It is indifferent to me where I am to begin, for there shall I return again.

— Parmenides

Welcome to the wonderful world of quantum computing! This is the start of a long and exciting journey. A journey that will lead you to learn about mind-blowing algorithms that can efficiently solve problems that are infeasible for mere, classical computers. As we walk together, you will discover the beautiful (but surprisingly simple) mathematics that underlie some of the mysterious physical phenomena that power these quantum systems. And not only that, but you will also learn how to write your own quantum code, and how to run it on actual quantum computers that you can access on the Internet for free. There are many reasons to be excited for the pages ahead.

Soon, you will understand how processing information with quantum technologies could allow you to teleport data, to securely exchange information over communication networks, or even to create money that is impossible to counterfeit (although you might not be able to carry it in your pocket just yet!).

You will also learn how to factor large integers (and, thus, break the security of many of the cryptographic protocols on which the security of the Internet relies) and how to search more efficiently for data on unsorted lists of elements.

As exciting as this all may sound, it is also important to be aware that quantum computers have their own limitations too. Despite the hype that the media may have created, there are problems that quantum computers can't solve faster than their classical counterparts. And there are even problems that they cannot solve at all!

For this reason, we will start by examining the question of what a quantum computer is (and what it is not!), what problems it can solve, and what are the most significant physical principles that it relies on. More concretely, in this chapter, we will cover the following topics:

- Quantum computing: myths and realities
- Superposition, entanglement, interference and other mysterious quantum phenomena
- The shape of things to come

After reading this chapter, you will have a clear image of what a quantum computer is and how it compares to a regular, traditional computer, and you will be familiar with the physical principles that are exploited by quantum algorithms in order to obtain impressive speed-ups over classical methods. You will also know in what kinds of applications quantum computers excel, and what the challenges and limitations of simulating quantum computers with classical ones are.

1.1 Quantum computing: myths and realities

If you are reading this book, chances are that you have already heard quite a lot about the wonderful feats that quantum computers can achieve: greatly speeding up computations that would take longer than the age of the universe to be completed with our regular, classical machines. However, if you really want to understand what quantum computers are capable of, we first need to dispel some very widespread myths about quantum computing and clarify where the power of quantum information processing really comes from.

The first important thing to make clear is that a quantum computer is not a magical device. There will always be computational problems that, even if we had an extremely advanced, fault-tolerant quantum computer (more on this in *Chapter 14*), will be impossible to solve. This is a mathematical fact that not even quantum physics can escape from!

Figure 1.1: The IBM Quantum System One, the first circuit-based commercial quantum computer, introduced by IBM in 2019. It was later succeeded by the IBM Quantum System Two (image courtesy of IBM Research, under a Creative Commons Attribution 2.0 Generic license)

To learn more...

Proving that some problems are impossible to solve—not only with quantum computers, but with any imaginable computing machine—is one of the main topics of a field of mathematics known as **computability theory**. The origins of this theory can be traced back to the first half of the 20th century, when Alan Turing showed that it is impossible to design an algorithm that, in a finite amount of time, can correctly answer the question of whether a machine will eventually stop for a given input [1]. This is known as the **halting problem** and proving that it is unsolvable relies on a simple but beautiful idea. Sadly, discussing this in detail would lead us astray from our path.

Nevertheless, if you want to know more about this fascinating topic, we can recommend a wonderful, non-technical book by David Harel called *Computers Ltd: What They Really Can't Do* [2]. And if you want all the nitty-gritty, technical details, you can check out the books by Sipser [3] and Rich [4].

So quantum computers are not almighty computing devices and they do have their own limitations. Fair enough. But they are *always* faster than classical computers, right? Well, not quite. In fact, nobody knows for certain when it is better to use a quantum computer to solve a problem (this book will teach you a good chunk of what we *do* know about this question!) but we know that, for some problems, you can save your money and just use your good old classical machines. This includes, for example, the problem of sorting a list of elements, where classical algorithms are as good as quantum ones, as proved by Høyer et al. [5].

The good news is that when quantum computers are faster than classical computers… they are much faster! In fact, the difference in those cases cannot be quantified with just one number. You'd probably read in the newspaper headlines such as *"Quantum computer solves problem one million times faster than a supercomputer"*. But that is just a very partial and limited view of the actual gap between quantum and classical computing.

To accurately describe the speed-ups that are achievable through quantum algorithms, we need to take into account that for those problems in which quantum computers have an advantage, the difference between quantum and classical processing times grows larger and larger with the size of the problem. Therefore, for a particular instance of a certain task, the quantum computer can be, say, ten times faster—yet, for a bigger instance of that same task, it could be one thousand times faster!

The right way of measuring the differences in performance of algorithms is by taking into account how the time they take to solve a task grows with the size of the problem (for instance, if searching for a specific record in a database, the size could be the total number of records in the database). This would give us a function that relates the size of the problem with the time a certain algorithm needs to solve it. For example, imagine that you want to find the minimum value in a list that contains n integer numbers. For this, you could write an algorithm that, when run on a certain computer, takes, say, $3n + 17$ microseconds to solve the task. This leads to an analysis of the performance of algorithms that is usually called **asymptotic computational complexity**, and that we cover in more detail in *Appendix C*.

For our discussion here, it suffices to assume that we have a function $f_q(n)$ and a function $f_c(n)$ that give us the time that a quantum algorithm and a classical algorithm, respectively, take to solve an instance of size n of a certain problem that we are interested in. In fact, since the running time will depend on the actual hardware that we are using and may easily decrease if we upgrade the specs of computer, rather than the concrete computation time, we are usually interested in more abstract measures of computational cost (such as the number of steps or instructions that the algorithm needs to run). This is because we want to assess how good an algorithm is, independently of the fact that our computer could use a CPU replacement.

To learn more...

For simplicity, we are assuming that the time that an algorithm takes to solve a problem only depends on its size, but that might not be the case in many situations. For instance, if you are checking whether a list of 100 integers contains any negative number and the first element in the list is -2, you can stop immediately, whereas, if all the 100 numbers were positive, you would need to read the whole list to find out. Hence, size may not be all that matters when analyzing time complexity!

In the case where an algorithm has different running times for instances of a problem with the same size, it is common and sensible to consider a **worst case analysis** and define $f(n)$, the time taken on instances of size n, to be that of the instance of size n that takes the longest. This is the approach that we will adopt in this book, as detailed in *Appendix C*, but other options are possible. For example, one could take $f(n)$ to be the average computational cost over all instances of size n.

In any case, if we compare $f_q(n)$ and $f_c(n)$, we can determine in which situations we may prefer to use the quantum algorithm over the classical one. For example, if $f_q(n) = 10n+30$ and $f_c(n) = n^2 + 1$, although for small values of n the classical algorithm may be preferable, there will be a size of the problem from which the quantum algorithm will always be faster. In fact, it is a very illuminating exercise to compute those critical sizes for different situations, something that we invite you to do in the following exercises (you can find the solutions to all the exercises at the end of the book, but we sincerely encourage you to try them out first on your own and only refer to the solutions if you are truly stuck).

Exercise 1.1

Suppose that you have both a quantum and a classical algorithm for the same problem. The quantum algorithm needs $f_q(n) = 10n + 30$ instructions to solve instances of size n while the classical algorithm needs $f_c(n) = n^2 + 1$. Determine, in each of the following cases, for which sizes n the quantum algorithm is preferable over the classical one:

(a) You have a quantum computer that takes 2 milliseconds per instruction and a classical machine that takes 3 microseconds per instruction.

(b) You have a quantum computer that takes 100 milliseconds per instruction and a classical machine that takes 0.5 microseconds per instruction.

(c) You have a quantum computer that takes 3 seconds per instruction and a classical machine that takes 0.001 microseconds per instruction.

Is there any situation where a classical computer would always be preferable to the quantum one when using these algorithms?

As you will soon find out, there are some important problems for which there exist quantum algorithms whose asymptotic growth is much better than those of any known (or even possible) classical algorithm (for the same task). For instance, in *Chapter 12*, we will learn about **Grover's algorithm**, which solves search problems quadratically faster than any classical algorithm can; more concretely, searching with Grover's algorithm in a list of n elements takes about $\sqrt{n}$ queries to the elements in the list, while a classical algorithm will need roughly n such queries. Another famous quantum algorithm is Shor's algorithm, which is able to find the factors of any large integer L in a time that grows polynomially (slower than n^3, in fact) with n, the number of bits required to write L. By contrast, the time taken by the fastest classical algorithm currently known grows almost exponentially with n (we will learn about all this in *Chapter 11*).

Exercise 1.2

Prove that, for any $k > 0$ and any $c > 1$, the polynomial n^k eventually grows more slowly than the exponential c^n.

But… how is this possible? The reason behind these asymptotical speed-ups comes from the fact that quantum computers are not just faster classical machines. Quantum machines do things very differently from classical computers, allowing them to solve some problems (but not all of them) in new and more efficient ways. In fact, as we are about to explore in more detail in the following section, quantum algorithms explicitly use properties of quantum systems such as **superposition**, **entanglement**, and **interference**, to open new paths in information processing.

1.2 Superposition, entanglement, interference, and other mysterious quantum phenomena

Nobody understands quantum mechanics. Or, at least, that is what Richard Feynman famously said. And Feynman was a certified genius (and a Physics Nobel Award winner, to top it up!) so who are we to disagree? But don't worry, you do not need to know quantum physics to learn how to program a quantum computer. However, it will be very illuminating to explore in some detail the basic quantum phenomena that quantum computers exploit in order to outperform their classical cousins in some important tasks.

Figure 1.2: A picture taken at the Solvay Conference held in 1927. This conference gathered some of the greatest minds of the early 20th century to discuss quantum theory

Among all quantum phenomena, we must highlight superposition, entanglement, and interference, for we will encounter them time and again throughout this book. We will also have special mentions for the **uncertainty principle**, which is exploited by some protocols, or the impossibility of perfectly copying quantum information (formally known as the **no-cloning theorem**). Soon, we will develop the mathematical tools that we will need in order to formalize all of them (they will be very easy to learn, we promise!), but, in this first chapter, we want to give you some intuition on what they are and how they are used in quantum algorithms.

To learn more...

You don't need to understand all the physics behind quantum mechanics in order to understand quantum computing, so we are not going to get very physic-sy here.

However, if you are curious and want to know more, we can recommend some very good books for you to read.

How to Teach Quantum Physics to your Dog, by Chad Orzel [6], not only has a striking title, but is a very readable popular account of quantum mechanics that goes deeper than other books with a similar scope. If you like heavy metal music (we[a] love it!) and even if you don't (nobody is perfect), you should take a look at *When the Uncertainty Principle Goes to 11: Or How to Explain Quantum Physics with Heavy Metal*, by Philip Moriarty [7]. It really rocks!

If you want to get more technical but you do not feel like reading a regular quantum physics treatise, *Quantum Mechanics: the Theoretical Minimum*, by Susskind and Friedman [8], is certainly the best option. And if you want a bonafide textbook, *A First Introduction to Quantum Physics*, by Pieter Kok [9], is a good place to start before you jump into classical books, such as the one by Griffiths and Schroeter [10].

[a]Well, at least one of us!

Let's start by talking about superposition and some really strange cats!

1.2.1 Superposition

Superposition is one of the most famous quantum phenomena. If by any chance you've heard of **Schrödinger's cat**, you already are somewhat familiar with superposition. This thought experiment, proposed by Erwin Schrödinger, one of the fathers of quantum physics, invites us to imagine a poor little feline trapped inside a box with a deadly poison that is to be activated by the decay of a certain subatomic particle. The way of describing this situation in the language of quantum theory implies that, until we open the box and observe the cat, it is not dead or alive but *both* dead and alive at the same time. We usually say that the cat is in a state of superposition.

We can all agree that this is a very weird situation. We don't ever see cats (or any other animals) that are both dead and alive at the same time. Like, come on, what does it even *mean* for a cat to be both alive and dead? Well, as detached as these ideas may be from the physical reality at our scale, at the subatomic level, superposition completely explains the results of many experiments and has been validated in innumerable occasions. And, strange as it may sound, in quantum mechanics, it makes sense to talk about particles that have spin up and spin down at the same time or about electrons that have two different energies at once. At least, until you **measure** them. Then, they **collapse** and have a very definite spin, energy, or position.

Because, yes, that is another strange thing about superposition. When you try to observe the state of a system in superposition… it will stop being in superposition and it will randomly decide to adopt one definite state. This is usually called the **wavefunction collapse** and, according to the standard interpretation of quantum physics, occurs whenever we measure a quantum system. This will make quantum computing an inherently and intrinsically *probabilistic* procedure, in which running the same instructions with the same inputs can lead to different outcomes! In fact, when designing quantum algorithms, it will be of utter importance to be able to manipulate (in our favor, of course) the probabilities of obtaining different results so that we can efficiently solve our problems. And a huge part of this book is devoted to explaining how to achieve that for some important tasks.

We won't blame you if you find this a little bit (or an awful lot) confusing. But before you throw this book out of the window and look for something that makes more sense, let us assure you that, in order to program quantum computers, you won't have to think about particles that are in two different locations at the same time or any other strange such behavior.

In fact, in the next chapter, we will give a very direct mathematical formulation of what superposition is, and it will only involve linear combinations of vectors (if you need a refresher on vectors and linear algebra in general, we have you covered: take a look at *Appendix A*). And for the remainder of this book, we promise, you will only have to think about superposition in those simple mathematical terms. That formulation will also give us a very easy way of determining the probability of each outcome when we perform a measurement, so we will kill two birds with one stone (metaphorically, of course; we love birds, cats, and all animals, despite the colorful examples in this section!).

Alright, but why is superposition relevant for quantum computing? As you will soon learn, most quantum algorithms start with an initial step that puts the system in superposition—a combination of all possible solutions, so to say. This will be helpful to design procedures that work on all those possible solutions at the same time and it is one of the sources of the kind of advantage that we can obtain when processing data with quantum devices.

In fact, quantum computers store information on something called **qubits** (you will learn all about them in *Chapter 2*) and, if your quantum machine has n qubits, it can implicitly store 2^n quantities, exponentially more than a classical computer! We will use superposition in most of the algorithms that we will study in this book, especially from *Chapter 6* on, in order to exploit this massive potential of information storage and processing.

However, this ability of working with many solutions at once is, most of the times, not enough to obtain a speed-up over classical algorithms. For that, we will need other ingredients, such as the spooky entanglement that we introduce in the next section.

1.2.2 Entanglement

Entanglement is no less famous than superposition... and it's probably just as mysterious! The great Albert Einstein had a really hard time accepting it (he called entanglement the "spooky action at a distance") because it seemed to be contrary to every physical intuition we may have. But it has been demonstrated in labs time and again and, in fact, in 2022, Alain Aspect, John F. Clauser, and Anton Zeilinger were awarded the Physics Nobel Prize "for experiments with *entangled photons*, establishing the violation of Bell inequalities and pioneering quantum information science".

Entangled particles show correlations that cannot be explained with classical physics alone, as changes in one of them cause instantaneous changes in the others, regardless of how far apart they may be. This was one of the reasons that led Einstein to claim that entanglement was "spooky", since it seems to act instantaneously at a distance, and break the principle of locality.

Entanglement is central in quantum information processing and quantum computing. Were it not for entanglement, quantum computers could be easily simulated with traditional computers and quantum computing would thus not provide any algorithmic speed-ups. Also, protocols such as **quantum teleportation** or **superdense coding**, which we will study in detail in *Chapter 6*, rely on the use of entangled qubits to carry out tasks that are impossible with classical devices. In more elaborate quantum procedures, we will use entanglement in combination with superposition in order to simultaneously associate to each possible input of a problem its correct output, effectively computing an exponential number of operations in just one go.

Again, the mathematical description of entanglement, which we will introduce in *Chapter 5*, is surprisingly simple, and has to do with the fact that some vectors that describe quantum systems cannot be expressed as the product of two smaller vectors. But in spite of its humble mathematical formulation, entanglement will become one of our most powerful tools in quantum algorithms, especially when combined with interference, which is our next stop in this magical mystery tour of quantum properties.

1.2.3 Interference

Interference is the unsung hero of quantum algorithms. Most popular accounts of quantum computing put superposition and entanglement front and center, and almost completely forget about interference. But truth is that it is the most important part of most quantum algorithms and protocols. In fact, in most cases, getting a system into superposition and entanglement is just the first step in quantum algorithms... and is relatively simple to achieve, at least from a computational point of view. The tricky bit comes when we have to take advantage of interference.

As mentioned in the previous section, superposition and entanglement will allow us to simultaneously pair each possible input together with its correct output. But then the difficult part will be to select, from all of them, the ones that we actually need in order to solve our problem. This is where interference comes to the rescue. As we will very soon learn (in fact, in *Chapter 2*), the coordinates in the vectors that describe qubit states can be negative numbers (and even complex ones; see *Appendix A* for a reminder of these fascinating quantities). In our quantum algorithms, we will use this to our advantage. By performing certain operations, we will be able to use negative interference to cancel the solutions that we are not interested in, while reinforcing the results that we would like to obtain. This will increase the probability of measuring exactly the solution that we need—happy days!

The problem is that using interference effectively is the most intricate element in the design of quantum algorithms, and we only (yet) know how to do it for tasks such as the ones you will learn about in future chapters. Getting unwanted solutions to interfere with each other is the crucial ingredient of successful quantum procedures and we will pay a lot of attention to explaining how this can be conducted, starting with the simplest algorithms (such as Deutsch's method in *Chapter 6*) and building our way up to the most sophisticated ones (Shor's algorithm in *Chapter 11* and Grover's algorithm in *Chapter 12*). We know that this sounds abstract and weird at this moment, but don't worry; throughout the book, we will have ample opportunities to see interference in action with very clarifying examples.

Superposition, entanglement, and interference are some of the main ingredients used in quantum information processing, but by no means are they the only ones. We are certain of it, as you will see in the next section!

1.2.4 The uncertainty principle and the no-cloning theorem

As we will learn in *Chapter 3*, quantum information processing can find some very interesting applications in the area of secure data transmission. In that context, it will be very important to have some assurance that unwanted parties don't have access to more information than we want them to have. For this goal, the **uncertainty principle** and the **no-cloning theorem** will prove to be invaluable tools.

Very roughly speaking, the uncertainty principle states that, with just one measurement, it is impossible to get all information of a quantum system, including whether it is in superposition or not. We will use this in our favor when designing some protocols for **quantum cryptography**. In particular, it will be very useful in order to securely distribute secret keys between users of a telecommunication channel, something that is known as **Quantum Key Distribution (QKD)**.

The uncertainty principle has our backs covered if someone tries to measure a single quantum state to extract information about it. But what if that person makes some copies and tries to perform more measurements on those copies? Well, that is indeed a sneaky attempt, but one that is doomed to fail. As it turns out, without some additional knowledge about what state we are trying to copy, the laws of quantum physics completely forbid making a perfect, independent copy of that state. This is known as the no-cloning theorem, and as strange as it may sound, it is a very simple consequence of the postulates of quantum mechanics, as we will prove in *Chapter 5*.

These are not, by any means, all the strange properties of quantum systems, but they are certainly (no pun intended) the ones that we will most frequently encounter in our study of quantum protocols and algorithms. Let us insist one more time: don't be afraid; we promise that the mathematical formulation of these properties will be surprisingly simple and you won't have to dwell on the their physical implications—unless you want to, of course. If you're up for the adventure, the *meaning* of quantum theory is indeed a profound and entertaining philosophical question.

Okay, so quantum physics has some weird properties. But how does that affect our ability to simulate quantum systems using classical computers? That is a very good question. So good indeed that it will be the focus of our next section!

1.2.5 Simulating quantum computers with classical computers

Despite what you may have heard in the media, classical computers can solve *exactly* the same problems that quantum computers can. As you will learn, starting from the next chapter, what quantum computers do (at least, those based on the quantum circuit model, which are the ones that we will study throughout this book) is to multiply (big) matrices with (big) vectors (refer to *Appendix A* for all the relevant linear algebra concepts). And that is something that, given enough time and memory, a classical computer can always do. The catch is that "enough time and memory" can be, in some cases, *a humongous lot* of time and memory. So much that, in practice, replicating the operations of quantum computers with a traditional one is just unfeasible.

As we have already mentioned, the amount of classical memory needed to store the state of a quantum computer grows exponentially with the number of qubits it has. This means that, even if we used each atom of the universe to store one number, we would be unable to represent even a minuscule fraction of the state of a quantum computer with 1000 qubits. That idea of working with that (huge) vector doesn't sound so appealing now, does it?

But... what if there were some shortcuts? Actually, it is not known whether storing the state vector of a quantum computer and operating on it is the best way of simulating a quantum computer with a classical one (you can check the papers by Young et al. [11] and Xu et al. [12] for some good surveys on different simulations methods). And, for some particular cases of quantum operations, there are *much better* ways of replicating them with classical algorithms that do not scale exponentially. But all the mathematical evidence that we currently have points in the same direction: simulating a quantum computer seems to be, in general, an extremely demanding task.

To learn more...

Not every quantum algorithm is hard to simulate with a classical computer. For example, if there is not a lot of entanglement in a quantum system, then we can represent its state vector much more succinctly, and thus operate on it much more efficiently. Moreover, there are some algorithms that produce highly entangled states but only use a certain subset of all possible quantum operations, which can be simulated quite easily with something that is known as the **stabilizer method** (if you are curious, take a look at *Section 10.5* in the book by Nielsen and Chuang [13]).

> The moral of it all is that some quantum procedures are easy to simulate while others are hard... but nobody knows for sure where the boundary between them lies. In fact, it is a very open, very important, and very difficult question in quantum complexity theory! And the answer, oddly enough, may have to do with something called **magic quantum states** [14].

In any case, simulating quantum computers with classical ones can be very useful in some situations; for example, when you are debugging your code before running it on quantum hardware. You will be limited to working with a reduced number of qubits (it is, in general, highly resource-consuming to go beyond 40-qubit simulations), but it will help you understand how the algorithms work, and it is a tool that we will use extensively starting from *Chapter 4*.

But before that, let us close this chapter by taking a look at the kind of applications that quantum computers excel at and the algorithms that we will study in the remainder of this book.

1.3 The shape of things to come

In this first chapter, we have focused on the properties that are used in quantum algorithms, mostly at a descriptive level. But starting in the next chapter, we will introduce the mathematical concepts that you will need in order to understand quantum algorithms and program quantum computers.

Don't be afraid. The mathematics that you will need (covered in *Appendix A*) are quite simple. And we will start from the very basics, first considering only one qubit (*Part 1*), introducing systems with two qubits after that (in *Part 2*), and working all the way up to algorithms that use many qubits (*Part 3* and *Part 4*). We will then conclude with some concepts from quantum error correction and fault-tolerant quantum computing and some thoughts on the notion of quantum advantage (*Part 5*).

Every part of the book will follow the same structure. We will begin by introducing the mathematical concepts needed to describe quantum systems of increasing complexity. Then, we will put those concepts to good use by applying them in different protocols and algorithms. Finally, we will show how to write code that implement all those methods using Qiskit, one of the most popular and advanced quantum packages out there, so that you can run them on both simulators and actual quantum computers.

Throughout the rest of the book, we will focus mainly on applications of quantum computing and quantum information processing that require some level of error correction (more on this on *Chapter 14*) and that have been mathematically proven to obtain some advantage over classical computing. They range from proposals for quantum money that cannot be forged and completely secure ways of sharing secrets over telecommunication networks (which we will study in *Chapter 3*) to algorithms to efficiently factor large integers (*Chapter 11*) and to search extremely fast (*Chapter 12*).

These examples will not only showcase the kind of problems in which quantum computers will be useful in the future, but they will also help us highlight the role that quantum phenomena such as superposition, entanglement, and interference play in successfully designing practical quantum algorithms. Understanding the use of these phenomena in concrete applications will clarify how they are combined together to obtain speed-ups over classical algorithms and will illuminate their central role in quantum information processing.

To learn more...

Our focus in this book is on quantum algorithms that, when run on fault-tolerant quantum computers, are asymptotically faster than their classical counterparts. However, in recent years, a lot of attention has also been paid to other quantum algorithms whose advantage has not been mathematically proven but that can be run on the kind of quantum computers that are available today. These devices are usually called **noisy-intermediate scale quantum** (or **NISQ**) machines (see the wonderful survey by John Preskill [15] for more on them) and have a limited number

> of qubits that are not fully connected and that are not protected against errors and external noise.
>
> Despite these limitations, there have been many proposals of quantum algorithms that seek to use these early quantum devices for practical applications, including problems in machine learning, optimization, and physics and chemistry simulations. If you want to learn more about them, please take a look at our book *A Practical Guide to Quantum Machine Learning and Quantum Optimization: Hands-on Approach to Modern Quantum Algorithms*, also published by Packt [16].

Can you feel it coming in the air? Yes, our quantum journey is about to begin in earnest. Buckle up: the best is yet to come!

Summary

In this chapter, we have studied what makes quantum computers special and how they are different from classical machines, opening up the possibility to develop new, faster algorithms for some important tasks. We have also demystified some of the usual misconceptions about quantum computers and clarified that their advantage over classical computing devices is best explained with asymptotical analysis, since the difference in execution time grows bigger with the size of the problems under consideration.

We have also taken a first qualitative look at some important quantum properties, including superposition, entanglement, and interference, that will be central in the design of useful quantum algorithms. We have also learned that it is possible to simulate quantum operations with classical computers (and that it is useful, for instance, when debugging quantum software) but that all those quantum properties combined make it impossible (or, at least, very, very unlikely) that those simulations will ever be efficient.

Finally, we have had a sneak peek into what is to come in the remainder of this book, and into the kind of protocols, algorithms, and methods that we will be studying in detail throughout the following chapters.

2

Qubits, Gates, and Measurements

If you are receptive and humble,
mathematics will lead you by the hand.

— Paul Dirac

In the previous chapter, we had the chance to informally introduce quantum computing, its advantages and challenges, and some of the intriguing quantum phenomena that quantum algorithms leverage on. Now that we all have a bird's-eye view of where we are heading, it's time to set out to work, and dive into the details. We should warn you that things are soon going to get a little bit mathematical from now on. Thus, if you feel that your linear algebra may be a little bit rusty, this is the perfect moment for you to go over *Appendix A*. The same applies if you need a refresher on complex numbers!

In any computational paradigm, we need three basic ingredients: a way of storing information, a way of retrieving the information that we are storing, and a way of transforming that information. In quantum computing, this is what these ingredients look like: we use **qubits** to store information, we perform **quantum measurements** to extract information from them, and we use **quantum gates** to transform them. The three sections in this chapter discuss these three elements in detail:

- What is a qubit?
- Extracting information from qubits
- How to transform a qubit state

Over the following chapters, we will have plenty of time to build all our way up to the full glory of quantum computing, but Rome wasn't built in a day and we want to take things easy and slow. For this reason, in this chapter, we are going to focus on the simplest systems that can be used for quantum computing: one-qubit systems.

After reading this chapter, you will have a general picture of the role that qubits, measurements, and gates play in quantum computing. You will also understand how one-qubit systems are represented and manipulated from a theoretical point of view. This will give you all the necessary background to understand some exciting applications of one-qubit systems in *Chapter 3* and get your hands dirty programming your first quantum algorithms and protocols in *Chapter 4*.

2.1 What is a qubit?

Qubits are the most fundamental unit of quantum information. That's the headline, and that's what will keep us busy throughout this whole section. But before we get down to that, a small "methodological suggestion" may come in handy. We humans live in a world that, at our scale, is very different from the quantum-mechanical realm. In our daily lives, we don't experience superposition, entanglement, or any of the snazzy phenomena that we discussed in the previous chapter. On the other hand, on a daily basis, we do experience forces, we see the effect of energy and we can somewhat easily corroborate any of the postulates of classical physics. Thus, while it can be healthy and useful to hold on to our intuitive understanding of reality when approaching classical physics or even classical computing, please, we beg you to lose all hope of doing the same here. Things are going to be different. And that's fine.

Thus, before we get started, let's make sure that we've all left our human-sized physical intuition at the door, and let us be ready to trust in the formalism that we are about to introduce. Just follow the mathematics; they will never lead you astray!

2.1.1 Qubits and their states

In classical computing, the most fundamental unit of information is the bit, and at any moment it can be in one of two possible states: 0 or 1. In quantum computing, the most fundamental unit of information is the **qubit** (short for "quantum bit") and it can be in two states denoted as $|0\rangle$ and $|1\rangle$ (more on those weird bars and angles later). However, a qubit may also be in any **superposition** of those two states, that is, in any state of the form

$$\alpha_0 |0\rangle + \alpha_1 |1\rangle$$

where α_0 and α_1 are complex numbers—called **amplitudes**—such that $|\alpha_0|^2 + |\alpha_1|^2 = 1$. We refer to this constraint on the amplitudes as the **normalization condition**, and we say that the state is **normalized**.

A few examples may help us illustrate this. If we take, for example, $\alpha_0 = \alpha_1 = 1/\sqrt{2}$, which clearly satisfy the normalization condition, we can produce the state

$$\frac{1}{\sqrt{2}} |0\rangle + \frac{1}{\sqrt{2}} |1\rangle,$$

which is a perfectly balanced superposition of $|0\rangle$ and $|1\rangle$. In an informal way, we could say that a qubit in this state is both in state $|0\rangle$ and state $|1\rangle$ at the same time, but this vague idea will become more concrete when we introduce measurements in the next section. For now, let us get accustomed to dealing with states.

To learn more...

From a physical point of view, there are many ways in which we could implement a qubit. One of the the conceptually simplest ways would be to encode a qubit state as the **spin** of a **spin-**$1/2$ particle. In essence, these "spin-$1/2$" particles have some "spin" (a quantum-mechanical magnitude, the details of which we will not discuss), which can be in a **spin-up** state, in a **spin-down** state, or in any superposition of these two states. This makes the spin of these particles a good choice for implementing qubit states, perhaps interpreting spin-down as the qubit state $|0\rangle$ and spin-up as the qubit state $|1\rangle$. There are plenty of examples of spin-$1/2$ particles, but the most famous one can be found in the humble electrons that we all know and love. If you would like to learn more about spin (and about quantum mechanics, with a more physical approach), you should read *Quantum Physics of Atoms, Molecules, Solids, Nuclei, and Particles* [17].

Beyond spin-$1/2$ particles, there are other and very promising ways of physically realizing qubits, such as through **superconducting circuits** [18], **ion traps**, or **quantum dots**. If you have a solid background in physics and want to have an in-depth look into these, you might want to read Chapter 6 of *Classical and Quantum Information* [19].

As we mentioned before, the amplitudes α_0 and α_1 can be complex numbers. This means, for example, that this would also be a valid state, as you can easily check:

$$-\frac{i}{\sqrt{3}}|0\rangle + \sqrt{\frac{2}{3}}|1\rangle.$$

We acknowledge that, at a first glance, this state can be strange-looking. However, there is nothing to be feared about complex numbers! We will slowly grow accustomed to them until their presence becomes second nature.

As we discussed before, when approaching this, you need to invest some trust in the formalism and set your intuition aside. However, there are a couple conditions here that may seem rather arbitrary. Firstly, we have the question of why we are imposing this "normalization condition" on α_0 and α_1; this will become apparent in just a few pages, so bear with us. Lastly, you may be confused as to why exactly these coefficients can be complex: why not just take them to be real numbers? For this question, we cannot provide a straightforward answer, but you can accept the use of complex numbers as an axiomatic foundation.

Now that you know how qubit states can be constructed, at least from a theoretical point of view, it might be a good idea for you to practice with an exercise.

Exercise 2.1

Which of the following are (valid) qubit states? Why?

(a) $|0\rangle + |1\rangle$,

(b) $\sqrt{\frac{4}{7}}\,|0\rangle + \sqrt{\frac{3}{7}}\,|1\rangle$,

(c) $2\,|0\rangle$,

(d) $e^{-i}\,|1\rangle$.

Find a suitable value of x that will make this a valid qubit state:

(e) $\frac{1}{3}\,|0\rangle + x\,|1\rangle$.

Find all the possible values of x that will make the following construction a valid qubit state (keep in mind that there is an infinite amount of them):

(f) $\frac{1}{\sqrt{2}}\,|0\rangle + x\,|1\rangle$.

If you find yourself stuck with any of these exercises, you might want to take a look at *Appendix A*. Also, remember that you can find the solutions at the end of book.

By now you should be comfortable enough constructing qubit states. Later in this chapter we will further explore the physical meaning of these states and better understand what role they play in quantum computing. Nevertheless, before heading down that road, we need to take a brief mathematical detour in order to make sense of some of the fancy symbols that we have been using.

2.1.2 The bra-ket notation

There is a crucial fact about qubit states that we have been hiding under the carpet. As it turns out, these mysterious qubit states that we have been discussing are nothing more than good old vectors; vectors in $\mathbb{C}^2$, to be precise. Actually, the states $|0\rangle$ and $|1\rangle$ are the canonical basis vectors

$$|0\rangle := \begin{pmatrix} 1 \\ 0 \end{pmatrix}, \qquad |1\rangle := \begin{pmatrix} 0 \\ 1 \end{pmatrix}.$$

In quantum computing, we don't call this basis the "canonical basis" but the **computational basis**. Much more flashy, isn't it?

Having revealed the true nature of the computational basis vectors, for any complex numbers α_0 and α_1 satisfying the normalization condition, we now have

$$\alpha_0 |0\rangle + \alpha_1 |1\rangle = \begin{pmatrix} \alpha_0 \\ \alpha_1 \end{pmatrix}.$$

This is how any qubit state is, in disguise, a vector of $\mathbb{C}^2$. Since α_0 and α_1 satisfy the normalization condition, we also say that this vector is **normalized**. In the language of linear algebra (which we reviewed in *Appendix A*), we would say that the vector is a unit vector.

If you ever studied physics, you are probably used to denoting vectors with a small arrow on top. For instance, some physicists or engineers like to write $\vec{v}$ to denote that a certain quantity v is a vector. In quantum mechanics (and, by extension, in quantum computing), instead of using these little arrows, we use some funky bars and angles and write $|v\rangle$. Does this sound familiar? This is why we have been writing $|0\rangle$ and $|1\rangle$ all this time. We call these notational artifacts **kets**, and we owe this symbolism, usually called **bra-ket notation**, to the English physicist Paul Dirac [20].

Important note

The state of a qubit is specified through a two-dimensional vector of complex numbers of the form

$$\alpha_0 |0\rangle + \alpha_1 |1\rangle = \begin{pmatrix} \alpha_0 \\ \alpha_1 \end{pmatrix},$$

in which the **amplitudes** α_0 and α_1 satisfy the **normalization condition** $|\alpha_0|^2 + |\alpha_1|^2 = 1$. A vector verifying this condition is said to be **normalized**.

You may now be wondering why we bother to use these "kets". For starters, kets can be very practical. Inside a ket, you can write anything you want without any fear of confusion, which is, for example, convenient in order to avoid mistaking the qubit states $|0\rangle$ and $|1\rangle$ with the actual numbers zero and one! Also, you could write any string of symbols inside a ket, so you could consider a vector

$$|\text{cool}\rangle := \sqrt{\frac{2}{5}}\,|0\rangle + \sqrt{\frac{3}{5}}\,|1\rangle,$$

and use this ket as $|\text{cool}\rangle$ all along. Cool, isn't it? This can be convenient should you want to give memorable names to your kets[1].

OK, so that clarifies the mysteries about kets. But what about the "bra" in "bra-ket" notation? It turns out that, to each column-vector ket you can associate a certain row vector that we call a **bra**. And with a bra and a ket you can (sometimes) form a **bra-ket** or **bracket**, which is nothing more than a **dot**, **scalar** or **inner product**. We love bras and bra-kets, but unfortunately they are not very relevant in the way we have chosen to present quantum computing in this book, so we won't be covering them in the main text. However, in case you are curious, we have written a whole appendix on the topic, so check out *Appendix B* for everything you always wanted to know about bras (but were afraid to ask).

[1]In fact, it is customary to use Greek letters to name kets, such as in $|\psi\rangle$ (pronounced "ket psi"). In this way, you can boast that you are not only learning quantum computing but a foreign language too!

At this point we are more than capable of manipulating one-qubit systems, so it's time for us to give some physical meaning to all this formalism. Let's shift our attention to the surprising ways in which qubits can be measured!

2.2 Extracting information from qubits

If you have a classical bit, its state can be 0 or 1 and, if you want to access that state... the process is rather trivial. In a normal digital circuit, a bit is often physically implemented as a voltage: oversimplifying a bit (no pun intended), if the voltage is above a certain threshold, the bit is in state 1 and, otherwise, it is in state 0. Simple enough! If you are given a bit, you can always find its state.

If you have a qubit, its state is slightly more messy than that of a classical bit and, reasonably, you may now be wondering if we can just determine it. Shockingly, we cannot. When you are given a qubit (with no further information provided, just a qubit), there is no way of knowing its state. All that you can do is perform a **(quantum) measurement** on the qubit, the outcome of which will depend on the original state of the qubit but will not fully reveal it. What is more, performing such a measurement will potentially permanently "corrupt" the state of the qubit. Sounds weird? Let's see how these measurements work in detail.

There are many kinds of measurements that you can perform in a quantum system, but we will only consider the most natural kind of measurement that you can perform in a qubit system, and that's what is known as a **measurement in the computational basis**. If you are given a qubit in a state of the form $\alpha_0 \left|0\right\rangle + \alpha_1 \left|1\right\rangle$ and you perform a measurement in the computational basis, this is exactly what will happen:

1. The outcome of the measurement will be either 0 or 1, and this outcome will be probabilistic. You will get the outcome 0 with a probability $|\alpha_0|^2$ and you will get 1 with a probability $|\alpha_1|^2$.

2. If the outcome of the measurement is 0, the state of the qubit right after the measurement will **collapse** to $\left|0\right\rangle$. If the outcome is 1, the state will collapse to $\left|1\right\rangle$.

This is all depicted in *Figure 2.1*. Now, there is quite a lot to unpack here, so let's go step by step.

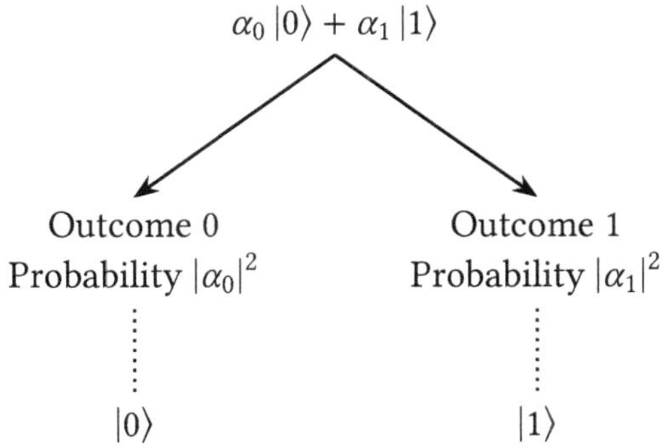

Figure 2.1: Schematic representation of a measurement in the computational basis of a one-qubit system in a general state

The first thing that we need to highlight is that quantum measurements are intrinsically probabilistic. When you have a qubit, even if you fully know its state, there may not always be a way for you to predict what the outcome of a measurement will be. As we have mentioned, the probabilities of getting each of the two possible outcomes 0 and 1 are, respectively, $|\alpha_0|^2$ and $|\alpha_1|^2$. Since qubit states are normalized, these probabilities add up to 1 and everything makes sense—and this is the long-awaited reason for that seemingly arbitrary normalization condition that we imposed back in *Section 2.1*.

Of course, if the state of your qubit is $|0\rangle$ (and therefore $|\alpha_0|^2 = 1$ and $|\alpha_1|^2 = 0$), whenever you perform a measurement in the computational basis, you are always guaranteed to get a 0. Analogously, a measurement will always return 1 in the state $|1\rangle$. Since, for any real number θ, we have $\left|e^{i\theta}\right| = 1$, if we are given qubits in the states $e^{i\theta} |0\rangle$ or $e^{i\theta} |1\rangle$, we will always get, respectively, a 0 or a 1 after a measurement. These are the only situations in which either $\alpha_1 = 0$ or $\alpha_0 = 0$, and they are the only ones in which we can be certain about the outcome of a measurement in the computational basis.

Whenever both α_0 and α_1 are non-zero, there is no way for us to fully predict the outcome of a measurement in the computational basis, and this is connected to the uncertainty principle that we informally described in the previous chapter. In this scenario, we say that the qubit is in a **superposition** of the states $|0\rangle$ and $|1\rangle$, because it behaves as if it were in $|0\rangle$ and $|1\rangle$ at the same time and, upon measurement, it picked one of the two states with a certain probability. Since the amplitudes α_0 and α_1 determine these probabilities for each of these outcomes—and, the bigger their absolute value, the higher the chance of its corresponding outcome—they are also called the **probability amplitudes** of the state of the qubit.

Let us consider a few examples. If we perform a measurement (in the computational basis) of a qubit in state $i\,|1\rangle$, we will always get 1 as outcome, because $|i| = 1$. On the other hand, if the qubit is in state

$$\frac{1}{\sqrt{2}}\,|0\rangle + \frac{1}{\sqrt{2}}\,|1\rangle\,,$$

there will be an equal probability of getting 0 and 1 as outcomes, for $\left|1/\sqrt{2}\right|^2 = 1/2$. In this case, the qubit is in a state of perfectly balanced superposition between $|0\rangle$ and $|1\rangle$.

Exercise 2.2

What is the probability of getting the outcomes 0 and 1 when measuring (with respect to the computational basis) a qubit in each of the following states?

(a) $\frac{1}{\sqrt{2}}(|0\rangle - |1\rangle)$,

(b) $e^{i}\,|0\rangle$,

(c) $\sqrt{\frac{1}{3}}\,|0\rangle + e^{-i}\sqrt{\frac{2}{3}}\,|1\rangle$,

(d) $i\sqrt{p}\,|0\rangle - \sqrt{1-p}\,|1\rangle$ for $0 \leq p \leq 1$.

So far, we have paid attention to what happens at a measurement and how the outcomes are determined. Let's now focus on what happens immediately afterwards. As we mentioned before, if the outcome of the measurement is 0, the state right after the measurement becomes $|0\rangle$ (in quantum-mechanical jargon, we say that the state "collapses" to $|0\rangle$). Analogously, if the outcome is 1, the state collapses to $|1\rangle$. This means that, right after the measurement, all information contained in any original superposition is lost for good: all that you get out of the measurement is a binary output, and you are stuck with it. This means, in particular, that if you measure a qubit and get a 0, there is absolutely no way for you to determine whether the original state was $|0\rangle$ or a superposition between $|0\rangle$ and $|1\rangle$—all that you know is that the original amplitude for $|0\rangle$ was non-zero, and that the state after the measurement is indeed $|0\rangle$.

To learn more...

Being fully rigorous, when measuring a qubit $\alpha_0 |0\rangle + \alpha_1 |1\rangle$ in the computational basis, if the outcome of the measurement is $|0\rangle$, the state of the qubit after the measurement should be $(\alpha_0/|\alpha_0|)\,|0\rangle$. Mathematically, this is called the normalized projection of the original state onto $|0\rangle$. As you can see, this is very similar to $|0\rangle$. In fact, the only difference is the $\alpha_0/|\alpha_0|$ factor. But this what we call a **phase**, because its absolute value is 1. As we will see later, phase factors that multiply a whole state (called **global phases**) can always be ignored, hence why we have simplified this a little bit when talking about measurements and state collapse.

So, isn't there any way in which we could at least approximate the state of a qubit? Well, there kind of is. If you are given a huge amount of qubits, all in the same state, and you measure them all, you can then run some statistics in order to approximate the absolute value of the amplitudes of their state. For example, if you have a sufficient number of copies and, upon measuring them, a third yield 0 while two thirds return 1, you can be reasonably sure that the state of the qubit must be $\alpha_0 |0\rangle + \alpha_1 |1\rangle$ with $|\alpha_0| \approx \sqrt{1/3}$ and $|\alpha_1| \approx \sqrt{2/3}$. That is something! But still, keep in mind that these results would reveal no information about the actual complex values of the amplitudes. For example, measurements in the computational basis would never allow us to distinguish the state $\frac{1}{\sqrt{2}}(|0\rangle + |1\rangle)$ from the state $\frac{1}{\sqrt{2}}(|0\rangle - |1\rangle)$. Nor would they allow us to distinguish any of those states from $\frac{1}{\sqrt{2}}(|0\rangle + i\,|1\rangle)$, although—as we will discuss towards the end of the chapter—they are all very different, both physically and computationally.

Exercise 2.3

Compute the probabilities of obtaining 0 and 1 when measuring the following states in the computational basis:

(a) $\frac{1}{\sqrt{2}}(|0\rangle + |1\rangle)$,

(b) $\frac{1}{\sqrt{2}}(|0\rangle - |1\rangle)$,

(c) $\frac{1}{\sqrt{2}}(|0\rangle + i\,|1\rangle)$

(d) $\frac{1}{\sqrt{2}}(|0\rangle + e^{i\theta}\,|1\rangle)$, with θ any real number.

In the previous exercise, the $+1$, -1, i, and $e^{i\theta}$ factors before $|1\rangle$ are **phases** (complex values whose absolute value is 1), and they are **relative phases** instead of **global** ones, because they are only multiplying part of the state, not the whole state. As we will see later in the book, relative phases will play a rather significant role in quantum algorithms! They are so important that we will devote a whole section at the end of this chapter to them.

Returning to our discussion of measuring a qubit repeatedly, the whole idea of getting many copies of qubits with the same state is problematic. As we mentioned in *Chapter 1*, if someone gives you a qubit (just a physical qubit, providing no information about its state), it is physically impossible to clone it. We will discuss this later in the book, including a simple mathematical proof of this principle.

To learn more…

With what we know so far, we cannot extract all the information from the state of a qubit, even if we have as many copies of the qubit as we may want. However, once we learn how to transform qubit states, we will be able to estimate much more precisely the amplitudes, including their phases. The techniques used for that kind of estimation are known as **quantum tomography** or **quantum state tomography** [21].

In any case, in the context of quantum information theory, **Holevo's theorem** gives an upper bound for the amount of information that can be extracted out of a quantum system using measurements [13]. In particular, when you are given a single copy of a one-qubit state, that bound is exactly one bit.

Thus, if someone gives you a qubit from a suspicious origin and with no further information, there is no way for you to fully determine its state. Nevertheless, if you have produced a certain qubit state through a fixed procedure (such as running a quantum circuit, as we will discuss in the following section), nothing prevents you from running the same procedure as many times as you want and producing as many copies of that state as you want. So not all is lost!

Measurements in the quantum world are very different from measurements in our ordinary world because of two main reasons: they are non-deterministic and they lead to a loss of information. This idea can be uncomfortable and it has led to many theories and conjectures as to what may be hiding behind this apparent counter-intuitive nature of quantum measurements. For practical reasons, in this book, we will strictly adhere to the Copenhagen interpretation of quantum mechanics, which is also known as the "shut up and calculate!" interpretation. Let us succinctly summarize it. For all we know, considering quantum measurements as purely and intrinsically non-deterministic is in perfect accordance with empirical results. Consequently, if it ain't broke, don't fix it. Let's just accept this as true, and let's get things done.

That pretty much sums it up in terms of measurements. Having mastered the art of how to measure qubits, we shall now turn our attention to how to transform them in order to construct quantum algorithms.

To learn more...

What's behind the non-deterministic nature of quantum mechanics? Do we live in a multiverse or is God playing dice with the world? As we've mentioned, in this book, we will not bother pondering these questions. Nevertheless, all these matters—as unfitting as they may be for an introductory quantum computing textbook—are very philosophically relevant and worthy of your interest. We ourselves would be quite happy to talk about them over a cup of tea.

If you want to have some fun philosophizing about the implications of the Copenhagen interpretation and its variants, we invite you to read the science fiction novel *Quarantine*, by Greg Egan [22]. If you would like to dive into a deterministic theory able to explain quantum phenomena, you may wish to learn about Bohmian mechanics [23], [24].

2.3 How to transform a qubit state

In a classical computer, the state of bits is transformed using **logic gates**. For example, the **negation** or **NOT** gate is a one-bit gate that transforms a bit in state 0 to a bit in state 1, and vice-versa. These logic gates are the basic ingredients that are then used to construct classical algorithms. In fact, as you probably know, the chip in your computer is nothing more than a vast mesh of interconnected logic gates!

In quantum computing, we don't use logic gates, but **quantum gates**. These—together with measurement operations—are the building blocks that enable us to construct any quantum algorithm. What are these quantum gates and how can we use them? Let's find out.

2.3.1 Quantum gates

From a mathematical point of view, a quantum gate is any **unitary operator** acting on the vector space of states (in the case of a one-qubit system, $\mathbb{C}^2$). In case you don't remember (we also discuss this in *Appendix A*), a unitary operator is an ordinary linear operator (i.e., "a matrix") whose inverse is its conjugate transpose. Given a matrix U, we can obtain its **conjugate transpose** $U^{\dagger}$ by taking its transpose and then transforming every entry into its conjugate. Thus, an operator U is unitary if and only if $U^{-1} = U^{\dagger}$, which is to say that $U^{\dagger}U = UU^{\dagger} = I$, where I is the identity matrix. Thus, for all practical purposes, we can think of one-qubit gates as just 2×2 unitary matrices.

Exercise 2.4

Find the conjugate transpose of the following matrix:

$$U_1 := \begin{pmatrix} 0 & 1 \\ 1 & 0 \end{pmatrix}.$$

Is U_1 unitary? What about the following matrices?

$$U_2 := \begin{pmatrix} 1 & 0 \\ 0 & -i \end{pmatrix},$$

$$U_3 := \begin{pmatrix} 0 & -i \\ i & 0 \end{pmatrix},$$

$$U_4 := \begin{pmatrix} 1 + 2i & 3 \\ i & 4 \end{pmatrix}.$$

Given any gate U and any qubit in a state $|\psi\rangle$, applying U on the qubit transforms its state to $U|\psi\rangle$, that is, the result of multiplying the matrix U times the vector $|\psi\rangle$. Pretty straightforward, isn't it?

From a physical point of view, unitary operators describe the possible time evolution of a quantum-mechanical system; this is, they capture how a system could change in time. That is the reason why these are the only operations that we are allowed to perform on qubits. There are some interesting properties of unitary operators that serve as justification for why they represent allowed time-evolutions of systems:

- Any unitary operator is invertible (if its conjugate transpose is required to be its inverse, then, by definition, it has an inverse). This means that its action can always be reversed. In particular, this applies to quantum gates and it means that the action of a quantum gate can never lead to a loss of information: quantum computing is reversible! This will be very relevant when we discuss multi-qubit systems and something called **oracles**.

- Unitary matrices take vectors that satisfy the normalization condition to vectors that also satisfy the normalization condition, which means that the result of applying a quantum gate on a valid qubit state will always return a valid qubit state. This will also be true for multi-qubit systems.

Important note

Given a matrix U, its **conjugate transpose** (denoted as $U^\dagger$) is the matrix that is obtained after taking the transpose of U and then taking the complex conjugate of all the entries in the matrix. We say that U is **unitary** if its inverse is $U^\dagger$ or, equivalently, if both $U \cdot U^\dagger$ and $U^\dagger \cdot U$ are equal to the identity matrix.

Quantum gates (the operators that can describe transformations on qubit systems) are represented by unitary operators.

Let us now introduce a few examples of quantum gates. By far, the most simple quantum gate that you can find is the gate that does nothing: the identity gate I. This gate takes a qubit in a certain state and returns it untouched. The matrix of this gate is the identity matrix,

$$\begin{pmatrix} 1 & 0 \\ 0 & 1 \end{pmatrix}.$$

It is trivially unitary and, obviously, for any state $|\psi\rangle$, $I|\psi\rangle = |\psi\rangle$.

We can now discuss some non-trivial examples and, for that, let's begin with a classic. The X, **Pauli-**X, **NOT** or **negation** gate takes a qubit in state $|0\rangle$ to $|1\rangle$, and vice-versa, and its action is extended by linearity to all of $\mathbb{C}^2$. This means that its action on an arbitrary qubit state is the following:

$$X(\alpha_0 |0\rangle + \alpha_1 |1\rangle) := \alpha_0 |1\rangle + \alpha_1 |0\rangle .$$

As you can easily verify, the coordinate matrix of this matrix with respect to the computational basis is

$$X = \begin{pmatrix} 0 & 1 \\ 1 & 0 \end{pmatrix}.$$

Exercise 2.5

In the previous exercise, we already showed X to be unitary. Using the coordinate matrix of X, verify that X, as claimed, $X\left|0\right\rangle = \left|1\right\rangle$ and $X\left|1\right\rangle = \left|0\right\rangle$.

The X gate is a good example of a quantum gate with a clear classical analog. Nevertheless, there are plenty of quantum gates with no classical counterparts! Actually, the set of logic gates on a fixed number of bits is finite, whereas the mere set of one-qubit gates is infinite[2].

In addition to the X gate, we also have some Y (or **Pauli-**Y) and Z (or **Pauli-**Z) gates, whose coordinate matrices are the following:

$$Y := \begin{pmatrix} 0 & -i \\ i & 0 \end{pmatrix}, \qquad Z := \begin{pmatrix} 1 & 0 \\ 0 & -1 \end{pmatrix}.$$

We will use some of these gates throughout the book.

Exercise 2.6

Check that Y and Z are unitary and compute how they transform the states in the computational basis, this is, compute $Y\left|0\right\rangle$, $Y\left|1\right\rangle$, $Z\left|0\right\rangle$ and $Z\left|1\right\rangle$.

Before introducing new gates, just to make sure that you've fully understood what a unitary operator is, let's do a quick exercise.

[2] In case you know about infinite cardinals, let us note that, in fact, the number of unitary gates is uncountably infinite.

Exercise 2.7

Find all the complex numbers x for which the matrix

$$M_x = \begin{pmatrix} \frac{1}{\sqrt{2}} & \frac{1}{\sqrt{2}} \\ \frac{1}{\sqrt{2}} & x \end{pmatrix}$$

defines a unitary operator.

Another very important one-qubit gate is the **Hadamard** gate (denoted as H). This gate has the following coordinate matrix:

$$H := \frac{1}{\sqrt{2}} \begin{pmatrix} 1 & 1 \\ 1 & -1 \end{pmatrix}.$$

You should easily be able to verify that H is indeed unitary[3]. One of the main reasons for its importance is that this gate takes a qubit in state $|0\rangle$ to the superposition

$$H|0\rangle = \frac{1}{\sqrt{2}}(|0\rangle + |1\rangle),$$

hence why it is always (or, at least, almost always!) used at the beginning of quantum algorithms in order to get the qubits to a state of superposition.

The state $H|0\rangle$ is so widely used that it has its own name: it is often written as $|+\rangle$. Analogously, $H|1\rangle$ is usually denoted as $|-\rangle$. They are called, respectively, the **plus** and **minus** states, for obvious reasons. As you can easily check for yourself,

$$|+\rangle := H|0\rangle = \frac{1}{\sqrt{2}}(|0\rangle + |1\rangle),$$
$$|-\rangle := H|1\rangle = \frac{1}{\sqrt{2}}(|0\rangle - |1\rangle).$$

[3] In fact, have you realized that we did exactly that in the previous exercise?

Let us now go through a couple of exercises in order to better understand how quantum gates can be composed.

Exercise 2.8

Prove that, if U and V are unitary matrices, so is UV.

Given two gates U and V, their **composition** is the product UV. This gate acts on any state $|\psi\rangle$ as $UV|\psi\rangle = U \cdot V \cdot |\psi\rangle = U(V|\psi\rangle)$. This means that we can make several gates act on a qubit, one after the other, and that is what we will do in quantum circuits, as you will learn in the next section.

Exercise 2.9

Find the coordinate matrix of the gate HX (the product of a Hadamard gate with an X gate). If a qubit is in state $|-\rangle$, what will its state be after applying the gate HX on it?

Two other one-qubit gates that deserve to be mentioned are the S and T gates. These are their coordinate matrices with respect to the computational basis:

$$S := \begin{pmatrix} 1 & 0 \\ 0 & i \end{pmatrix}, \qquad T := \begin{pmatrix} 1 & 0 \\ 0 & e^{i\pi/4} \end{pmatrix}.$$

The Z, S, and T quantum gates are just particular cases of **phase gates**. For any angle θ, the phase gate $P(\theta)$ parametrized by θ is

$$P(\theta) = \begin{pmatrix} 1 & 0 \\ 0 & e^{i\theta} \end{pmatrix}.$$

Clearly, $Z = P(\pi)$, $S = P(\pi/2)$ and $T = P(\pi/4)$

To learn more...

So there are a lot of quantum gates and some of them even have complex coefficients. But can we get rid of them and use only real numbers? Well, the answer is a superposition of $|\text{yes}\rangle$ and $|\text{no}\rangle$!

In the scientific literature, you can find some reasons why quantum mechanics *needs* to rely on complex numbers [25], [26]. Nevertheless, it could be possible to only use real numbers within quantum computing [27].

These are some of the most widely-used one-qubit gates. Let us now discuss how they can be arranged in order to construct quantum algorithms.

2.3.2 Quantum circuits

In a quantum algorithm, one executes a sequence of gates and measurements on a collection of qubits. Such an arrangements of operations is known as a **quantum circuit**, and it is often described through a diagram. The way these are constructed is best illustrated by example. Consider the following circuit, which is read from left to right:

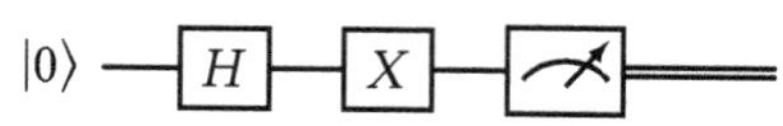

In this circuit, we have a qubit that is initialized to the state $|0\rangle$. Then, the qubit first goes through the gate H and then through the gate X. Finally, the gauge symbol is denoting a measurement in the computational basis. The outcome of such a measurement will be a bit: it will be classical information, not a qubit state, and that's what the double wire after the measurement represents. Single wires are used for quantum information (qubits), whereas double wires are used for classical information (bits).

Usually, given the nature of quantum measurements, quantum circuits are meant to be executed not once, but many times—in order to run some statistics and be able to estimate the state of the qubits, as we discussed earlier. Most software frameworks are actually configured, by default, to run circuits 1024 times. Each of these circuit runs is called a **shot**.

The more shots you have, the better you will be able to approximate what the state of the qubit was when measured.

As an example, consider the following circuit:

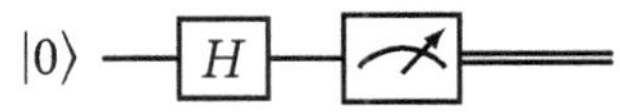

Before the measurement operation in this circuit, the state of the qubit will be $|+\rangle = H|0\rangle = \frac{1}{\sqrt{2)}}(|0\rangle + |1\rangle)$, so the measurement operation will return 0 and 1 with probability $1/2$ each. Thus, as seen from the outside in the physical world, the circuit would be nothing more than a black box producing uniform random bits!

To learn more...

There are many theoretical models for quantum computing, but, by far, the most popular of all is the **quantum circuit model**, which is the only one that we are considering in this book since it is the de facto standard and is used by most quantum software packages.

When tackling optimization problems with quantum computing, in addition to the quantum circuit model, a model known as **adiabatic quantum computation**, which is computationally equivalent to the quantum circuit model, is often used. If you are curious about it, we discuss it in Chapter 4 of our book *A Practical Guide to Quantum Machine Learning and Quantum Optimization: Hands-on Approach to Modern Quantum Algorithms* [16].

To conclude this chapter, let us have a chat about phases.

2.3.3 Global and relative phases

We have seen how to represent qubit states and how to measure them. With that knowledge, we ask you a question. By any odd chance, could it be possible for two mathematically different states to be indistinguishable? Let's find out.

Consider the state $|1\rangle$ and the state $Z|1\rangle = (-1)\cdot|1\rangle = -|1\rangle$. Of course, from a formal point of view, the kets $|1\rangle$ and $-|1\rangle$ are different, but, from a physical point of view—or, maybe, from a computational point of view—are they really distinguishable?

If we apply a quantum gate U on $|1\rangle$, we get $U|1\rangle$. If we apply it on $-|1\rangle$, we just get $-(U|\psi\rangle)$. It appears that this **global phase** -1 is just sticking around and not having much of an effect. Now consider a general state $|\psi\rangle = \alpha_0|0\rangle + \alpha_1|1\rangle$ for some suitable amplitudes. If we measure $|\psi\rangle$, are the probabilities of getting 0 and 1 as outcomes different from the ones we would have if measuring $-|\psi\rangle$? Not at all! Indeed,

$$|-\alpha_0|^2 = |-1|^2|\alpha_0|^2 = 1|\alpha_0|^2 = |\alpha_0|^2, \qquad |-\alpha_1|^2 = |\alpha_1|^2.$$

Therefore, a global phase of -1 has no effect on measurements, and it just gets carried along when applying quantum gates. And not just that, but this would be also true for any global phase.

Consider any complex phase ω (a complex number with absolute value one); and keep in mind that, as we mention in *Appendix A*, these phases are always of the form $e^{i\theta}$ for some real θ. Given any state $|\psi\rangle$, if we apply a gate U to $|\psi\rangle$, we get $U|\psi\rangle$, and if we apply U to $\omega|\psi\rangle$, we get $\omega U|\psi\rangle$ (the phase just gets carried). And in terms of measurements, both $|\psi\rangle$ and $\omega|\psi\rangle$ will return 0 or 1 with the same probabilities since

$$|\omega\alpha_0|^2 = |\omega|^2|\alpha_0|^2 = 1|\alpha_0|^2 = |\alpha_0|^2, \qquad |\omega\alpha_1|^2 = |\alpha_1|^2.$$

This means that, for any practical purpose, global phases can be (and actually are) ignored. Any state $|\psi\rangle$ is computationally equivalent to the state $\omega|\psi\rangle$ for any phase ω. Similarly, any gate U is equivalent to the gate ωU. Global phases, for both states and gates, are irrelevant.

Nevertheless, before you get too excited, we need to highlight one thing: while global phases can be ignored, **relative phases** (that is, phases acting on the individual amplitudes of a state) most certainly cannot. To see why, just consider the states

$$|+\rangle = \frac{1}{\sqrt{2}}(|0\rangle + |1\rangle), \qquad |-\rangle = \frac{1}{\sqrt{2}}(|0\rangle - |1\rangle).$$

While they only differ by a relative phase of -1 on $|1\rangle$, on the one hand, we have $H|+\rangle = |0\rangle$, and, on the other, we have $H|-\rangle = |1\rangle$. Relative phases do make a difference, so better be careful with them!

Important note

Global phases can be always be ignored, whereas relative phases are very significant.

That's it for one-qubit gates, circuits, and phases. With this, we can finish our discussion about one-qubit systems, so let's wrap things up before moving on.

Summary

In this chapter, we have taken our first steps in the theoretical formalism behind quantum computing. We have explored what a qubit is and what its similarities and differences are with respect to (classical) bits.

We first focused on how the state of a qubit can be represented as a normalized linear combination of the two computational basis states $|0\rangle$ and $|1\rangle$, and we took that opportunity to discuss how, in fact, qubit states are vectors of complex numbers.

Building on this foundation, we gave some physical meaning to the states of qubits by introducing quantum measurements and discussing some of their odd properties. We saw how, unlike measurements in the classical world, quantum measurements are probabilistic, may lead to a loss of information, and cannot allow us—at least when measuring in the computational basis—to fully determine the state of a system.

Finally, we explored how quantum circuits can be constructed using quantum gates in combination with measurement operations. We discussed how quantum gates are represented by unitary operators and how, while some have classical analogues, most quantum gates don't really replicate the behaviour of any classical logic gate.

This has been an intense chapter, loaded with mathematical content. Congratulations on making it up until this point! We now have some exciting content ahead. In the next few pages, we will put all our knowledge into practice by discussing a few exciting applications of one-qubit systems. We can't wait to get started!

3

Applications and Protocols with One Qubit

Those who think that the small doesn't exist
can't see the greatness of that which is good.

— José Ortega y Gasset

In the last chapter, we learned how to work with single qubits: we discussed how to represent the state of a qubit, how to extract information from a qubit using quantum measurements, and how to manipulate the state of a qubit through quantum gates. With that knowledge, we are ready to introduce our first quantum protocols. They will be simple, because they will be built on one-qubit systems, but it would be very unwise to look down on them, for, as you will soon see, they hold some unexpected surprises!

We will begin with an interesting application of quantum phenomena that will enable us to design currency that cannot be forged. Actually, this idea of *quantum money* is considered by many as the true beginning of quantum information theory. Without going any further, we will notice its influence in the second protocol that we will discuss: a method for exchanging information with perfect security. Finally, we will briefly study other protocols for secure communication, and we will even find ways to determine whether a bomb (well, an imaginary, quantum bomb) is active without risking an explosion.

The topics that we will cover in this chapter are the following:

- Quantum money
- Quantum key distribution with the BB84 protocol
- Other protocols with individual qubits

Once you've finished reading this chapter, you will have mastered several quantum protocols that rely on individual qubits in order to accomplish important information-processing tasks, including transmitting data securely. You will also have a profound understanding of how quantum phenomena such as superposition, interference, the no-cloning theorem, and the uncertainty principle can be used in practice.

All that learning about amplitudes and unitary matrices is about to pay off. Let the show begin!

3.1 Quantum money

Our first application of quantum information processing has to do with money—**quantum money**, to be more precise,—and how it has some amazing properties, such as being impossible to counterfeit.

The idea of quantum money goes back to the 1970's, when Stephen Wiesner first proposed using quantum phenomena to design procedures that make currency secure against forgery attempts. However, this initial proposal was probably too ahead of its time and it took many years for it to be completely understood and appreciated. In fact, it was only in 1983 when Wiesner's protocol was finally published [28].

The security of Wiesner's quantum money relies on several quantum phenomena that we have already described mathematically, including superposition and the impossibility of determining the state of a qubit if we don't have any information about it. To be more specific, in this protocol, some secret information is stored using the the states $|0\rangle$, $|1\rangle$, $|+\rangle$, and $|-\rangle$ that we introduced in the last chapter, in such a way that this information cannot be copied by anyone who does not know it already, but can easily be checked by the bank that issued the money. Sounds neat, doesn't it? In the following section, we will describe in detail the different quantum ingredients that make this possible.

3.1.1 Creating the banknote

Imagine that we have launched our new quantum software company and we've had great commercial success. So much so that we've just made our first million dollars with our amazing quantum algorithms, and, of course, we wouldn't want anyone to steal our hard-earned money. For the examples in this section, we will use an imaginary company that works with quantum money and that we will call Qiggy Bank. We will assume that Qiggy Bank has recently opened an office in our city and, after some waiting in the queue (no pun intended), we get a shiny, secure quantum banknote for our money. It is a small, smart-looking device with 20 qubits on the inside and a mysterious engraving on its metallic outside that reads `EF217CA75`. What does all this mean?

It turns out that, whenever the people at Qiggy Bank create a new banknote, they do a number of things:

1. They select a unique identifier for the banknote. This identifier will be public and can be, for instance, a hexadecimal string like the one we have created in our example above (`EF217CA75`).
2. For each of the 20 qubits in the banknote, they select at random one of the states $|0\rangle$, $|1\rangle$, $|+\rangle$ or $|-\rangle$. As we learned in the previous chapter, starting from $|0\rangle$ (which is always the initial state of any qubit), we can obtain $|1\rangle$ by applying the X gate. We can also obtain $|+\rangle$ by applying H to $|0\rangle$, and we can obtain $|-\rangle$ by first applying X to $|0\rangle$ and then H to the resulting state. For instance, for our example, the first five qubit states could be $|0\rangle$, $|+\rangle$, $|+\rangle$, $|-\rangle$, and $|1\rangle$. This information is private. Notice that there are 4^{20} possible options for selecting these 20 states, because there are 4

different, independent choices for each of them. This is a big number: more than 10^{12} possibilities! It would be very, very difficult to guess it at random.

3. In a secure and private database, they store the banknote identifier together with an annotation of the qubit states that were selected for it.
4. Then, they hand the banknote to the client. The banknote will have its identifier publicly displayed and the qubits set to the states that were selected at random.

Exercise 3.1

Check that, if you have a qubit in state $|0\rangle$, you can change it to $|-\rangle$ by first applying X and then applying H. Can you think of another sequence of two quantum gates that transforms $|0\rangle$ into $|-\rangle$?

And that's it! You now have in your hands a quantum banknote that is practically impossible to counterfeit.

But, what are the identifier and the qubit states used for? Let's find out.

3.1.2 Checking the banknote

Imagine now that you want to cash out your quantum banknote for dollars in order to buy a new and more powerful quantum computer. You only need to go to Qiggy Bank and hand out your banknote. But before they give you dollars in exchange for your quantum banknote, they will perform the following checks:

1. They will see if the public identifier is stored in their private database. If it is not, they will know that the banknote is false... and they will probably call the police on you!
2. If the public identifier is indeed in their database, they will retrieve the sequence of qubit states associated with it. Then, they will check, one by one, if the states of the qubits in your banknote are the ones that appear in their database. If they all are, they will declare the banknote legit, remove it from the database, and give you your

money. Yay! But, again, if any of the qubits fails the test, you will be in big trouble. Crime does not pay!

This all sounds reasonable enough, but there is a small detail we are glossing over. How do they check if a qubit is in a certain state? You surely remember that, in the previous chapter, we insisted that you cannot determine the state of a qubit that you are given, right? And that is correct, but it's only applicable to cases in which you do *not* know what the state of the qubit should be. If you do know, you can indeed check it. Let's see how to do it in this case. There are two possible situations:

- If the qubit should be in state $|0\rangle$ or $|1\rangle$, you can just measure it in the computational basis and check the result. Easy peasy!
- If, on the contrary, the qubit is expected to be in either $|+\rangle$ or $|-\rangle$, then a simple computational basis measurement will not do. In fact, as you showed in *Exercise 2.2*, in both cases you would get 0 half of the time, and 1 the other half... and you would be unable to distinguish between the two situations. However, if you first apply an H gate, then you will get either $H|+\rangle = |0\rangle$ or $H|-\rangle = |1\rangle$, and now you can perform a measurement (in the computational basis) and get 0 (if the initial state was $|+\rangle$) or 1 (if it was $|-\rangle$) to solve your problem with total certainty.

Behind this process, there is something very significant going on, so let's stop for a minute to contemplate it. In fact, let's unpack what happens when you apply the H gate to a qubit in the $|+\rangle$ state. By linearity, we have that

$$\begin{aligned}
H|+\rangle &= H\left(\frac{1}{\sqrt{2}}(|0\rangle + |1\rangle)\right) = \frac{1}{\sqrt{2}}(H|0\rangle + H|1\rangle) \\
&= \frac{1}{\sqrt{2}}\left(\frac{1}{\sqrt{2}}(|0\rangle + |1\rangle) + \frac{1}{\sqrt{2}}(|0\rangle - |1\rangle)\right) \\
&= \frac{1}{2}(|0\rangle + |1\rangle + |0\rangle - |1\rangle) = \frac{1}{2}(2|0\rangle) = |0\rangle .
\end{aligned}$$

Did you notice how that $|1\rangle$ canceled with that $-|1\rangle$? That is negative interference in action! In the same way, those two $|0\rangle$ states that got added together just positively reinforced each other. And, as you can easily check, something similar happens when you apply an H gate to $|-\rangle$ to obtain 1.

To learn more...

The process of applying an H gate and then measuring in the computational basis can be seen as a different type of measurement in itself. In fact, it is usually referred to as a **measurement in the** $\{|+\rangle, |-\rangle\}$ **basis**.

So legitimate banknotes can always be checked to see if they are valid. But what about detecting fraudulent ones? That is the topic of our next section.

3.1.3 Detecting illegitimate quantum money

The most important (and interesting) property of quantum money is that we can make it practically impossible to counterfeit. Proving it is not completely trivial (you can check the paper by Molina, Vidick and Watrous [29] for all the mathematical details), but in this section we want to discuss why some natural approaches to trying to create fake banknotes are doomed to fail.

The first thing that a crook could try to do is to just create a banknote with random states for the qubits. Okay, this is plain naive, we agree, but explaining why it utterly fails will be very instructive before we can analyse more sophisticated approaches. To get started and create a banknote with random states, the bad guy first needs to use an existing identifier of a valid banknote, because, otherwise, it cannot even pass the first step in the bank verification; this is not hard to do, as this information is public and can be copied from another banknote. Then, the criminal has to select one of the four possible states ($|0\rangle$, $|+\rangle$, $|-\rangle$, and $|1\rangle$) for each of the 20 qubits. Easy. But can it pass for a legit banknote? Well, not likely. Let's see why.

When the fake banknote is taken for verification, the bank employees will find an entry in their database that corresponds to an existing banknote. Then, they will measure the qubit states one by one and check their results against those recorded in their books. And this is very, very likely to fail because the probability of passing all the tests with qubit states selected blindly is fairly low, as we can see by analyzing one of the possible situations.

Imagine that the state selected by the counterfeiter is $|+\rangle$. There are four different cases:

1. If the state expected by the bank is $|+\rangle$, the test will always pass. In this case, the bank employees will apply an H gate to the qubit, transforming its state to $|0\rangle$. They will measure it and obtain 0, as required. Thus, everything will check out and the fake banknote will pass the test for this qubit.

2. If the state expected by the bank is $|-\rangle$, the test will always fail. Indeed, as in the previous case, an H gate will be applied to the fake qubit, which is in state $|+\rangle$, transforming its state to $|0\rangle$ and, when measured, it will give 0 as a result. But 1 was the only acceptable result.

3. If the state expected by the bank is either $|0\rangle$ or $|1\rangle$, they will measure the submitted qubit directly. Since it is in the $|+\rangle$ state, the result will be 0 with probability $1/2$, and 1 also with probability $1/2$. Thus, the probability of passing the test is $1/2$ in both cases.

Since the bank chose the qubit state completely at random, each of the states has probability $1/4$. Thus, we need to multiply the probability of passing the test by $1/4$ for each of the possible values of the real state, and add everything together. This gets us

$$\frac{1}{4} \cdot 1 + \frac{1}{4} \cdot 0 + \frac{1}{4} \cdot \frac{1}{2} + \frac{1}{4} \cdot \frac{1}{2} = \frac{1}{2}.$$

If the crook selects $|-\rangle$, $|0\rangle$, or $|1\rangle$ as the qubit state, the situation is completely analogous and the probability of passing the test is, again, $1/2$.

Exercise 3.2

Prove that the probability of passing the test is $1/2$ when the counterfeiter selects $|-\rangle$, $|0\rangle$ or $|1\rangle$ as the qubit state.

Okay, so the fake banknote will pass an individual qubit test half of the time. But, to be accepted, it has to pass *all* the tests. Since there are 20 qubits in total, the probability is then $(1/20)^{20}$, which is less than one in a million. Not a very good strategy, right? Anyway, Qiggy Bank may still consider this too risky and want to increase the security of their quantum money scheme. It couldn't be easier! If they increase the number of qubits in each note to 30, then the chances of passing all tests go down to less than one in a billion. And if, for example, they use 200 qubits, then the probabilities get so small that it is just practically impossible to forge quantum money with this technique.

Exercise 3.3

Consider a lottery in which there are 100 000 tickets. The winner is selected at random and gets one million dollars. You are given ten tickets, one for each of ten consecutive lottery draws. What is more probable, to win the lottery with each of your tickets or to pass the quantum banknote test with states selected at random and 200 qubits?

Choosing qubit states at random does not seem to be a very intelligent way to forge quantum money. Fair enough, nobody expected it to be. But what about more elaborate strategies? In particular, what if the counterfeiter gains access to a valid quantum note for some time? Maybe they can then extract some information from it to create a fake note, with the legitimate owner none the wiser.

One possible approach for doing this would be to try to copy the qubit states somehow. But, as we mentioned in *Chapter 1*, copying quantum states without having information about their states is impossible because of the no-cloning theorem. We will prove this in detail in *Chapter 5*, but, for now, let's just note that it rules this possibility out.

Finally, the criminal could attempt to measure the qubits of a valid banknote in order to obtain some information and thus try to create a copy. The idea would be the following:

- If the state is $|0\rangle$ (respectively, $|1\rangle$), a measurement in the computational basis will return 0 (respectively, 1) and the forger will have complete knowledge of the state.

What is more, the state after measurement will still be $|0\rangle$ (respectively, $|1\rangle$), not altering the original note. This is important: the counterfeiter wants to create a new valid note without modifying the first one; otherwise it would be much easier to just replace the legit note with a fake one.

- If the state is $|+\rangle$ or $|-\rangle$, the forger will apply H to the qubit to transform it to either $|0\rangle$ or $|1\rangle$. Then, as in the previous case, a computational basis measurement will completely reveal the state without altering it. Finally, the original state can be easily restored by applying, again, an H gate.

This seems promising for the forger, doesn't it? But there's a catch, because they don't know if the actual state is one of $|0\rangle, |1\rangle$ or, rather, one of $|+\rangle, |-\rangle$. Then, when deciding if the H gate should be applied or not, a mistake can be made. Let's illustrate this with an example.

Imagine that, unbeknownst to the crook, the qubit is in state $|+\rangle$. If the forger applies the H gate before the measurement, the correct state will be recovered and returned to its original situation. However, if the qubit is measured directly, then it will collapse to $|0\rangle$ or $|1\rangle$ with a probability of $1/2$ each. In both cases, the qubit state recovered by the criminal will be incorrect. Moreover, the qubit will have also collapsed to an incorrect state in the original banknote, corrupting it, and it will now fail the test for this qubit with a probability of $1/2$. The situation if the original state is $|-\rangle$, $|0\rangle$ or $|1\rangle$ is the same. If the forger wrongly decides what measurement to make, the state will be incorrectly determined and the original banknote will be altered—and there's always a 50% chance of getting the measurement wrong and corrupting the banknote.

Thus, if the counterfeiter decides at random to apply or not apply the H gate on each qubit, the probability of altering the qubit state will be $1/2$, and this will be detected by the bank with a probability of $1/2$. Thus, the total probability of the qubit tampering being detected is $1/4$, and it will go unnoticed with a probability of $3/4$. This is better than selecting the states at random, of course, but still not good at all. With this approach, if there are n

qubits in the banknote, the probability of passing the bank test is $(3/4)^n$, which will be very small if n is moderately big. For instance, it will be less than one in a million if n is 50.

Important note

Here, we can see both superposition and the uncertainty principle in action. Superposition allows us to consider the states $|+\rangle$ and $|-\rangle$ in addition to $|0\rangle$ and $|1\rangle$ as possible values for the secret information in our quantum money. The uncertainty principle makes it impossible for the counterfeiter to determine without error which of the four states the qubit is in.

We want to stress the fact that our discussion here is not, by any means, a complete proof of the security of Wiesner's scheme (for that, please refer to the paper by Molina, Vidick, and Watrous [29]). We only want to give evidence that some of the most straightforward attacks will not work, and to highlight how some of the phenomena that we introduced in *Chapter 1* (including superposition, interference, the no-cloning theorem, and the uncertainty principle) underlie the design of this type of quantum money. It is also good practice to use all those measurements and transformations with the Hadamard gate on the $|0\rangle$, $|1\rangle$, $|+\rangle$ and $|-\rangle$ states before we move to more challenging (and useful!) protocols.

We also want to remark that quantum money, at least as presented here, is not really very practical. First of all, we do not currently possess the technology to create qubits stable nough to keep their states for prolonged periods of time, let alone in a portable format! reover, the scheme that we have considered involves interacting with the bank every we need to check the validity of each note, probably making it not very usable in e situations. However, more advanced and useful quantum money proposals have nsidered and analysed in the literature (you can read about some of them in the Molina, Vidick, and Watrous [29]). In any case, the ideas behind quantum money ied in other domains, such as confidential transmission of information, which our next section.

3.2 Quantum key distribution with the BB84 protocol

As we mentioned in the previous section, the ideas proposed by Wiesner to implement quantum money had a great influence on the development of a very important application of quantum information processing: **quantum key distribution**, abbreviated as **QKD**. QKD is used to create and share secret keys that can be later used to securely transmit information.

In this section, we will discuss the need for key distribution schemes, and we will study the first ever QKD protocol: **BB84**. It is time for us to meet Alice, Bob, and Eve!

3.2.1 Alice, Bob, Eve, and the one-time pad

Cryptography studies protocols meant to enable the secure communication between different parties, usually two, which are customarily called Alice and Bob. One of the main objectives is to prevent a third person, Eve (for "eavesdropper") from getting any information from the communication between Alice and Bob. Cryptography is a vast and extremely interesting subject (for a thorough but accessible introduction, we recommend the book by Katz and Lindell [30]), but we will focus on just one protocol: the **one-time pad** or **Vernam's cipher**. This cipher does not use quantum information at all—it actually runs just on classical communication channels and computers,—but bear with us, because it will help us discuss the need for quantum key distribution.

Suppose that Alice wants to send a secret message to Bob using the one-time pad cipher. How does it work? First, Alice will represent the message using a binary string (this is not at all problematic because, as you know, in a computer, everything is encoded in bits). Let's call m the message binary string and let n be its length in bits. In order to use the one-time pad, Alice and Bob must use a previously agreed secret key k consisting of n bits chosen at random. Then, Alice computes the encrypted message c by adding, modulo 2, each bit of k to each bit of m. Remember that addition modulo 2 (also called the XOR logical operation) is usually represented by the $\oplus$ symbol and that $0 \oplus 0 = 0$, $0 \oplus 1 = 1$,

$1 \oplus 0 = 1$, and $1 \oplus 1 = 0$. Thus, if $m = 11010$ and $k = 01011$, then the encrypted message is

$$c = m \oplus k = 11010 \oplus 01011 = 10001.$$

Now, Alice sends c to Bob, who, in order to recover the original message, only needs to, again, apply the XOR operation to each bit of k and c. In our example, Bob will obtain

$$c \oplus k = 10001 \oplus 01011 = 11010,$$

which is, indeed, m. This is not a happy coincidence. Notice that

$$c \oplus k = m \oplus k \oplus k = m,$$

because $k \oplus k = 0$ for any k. Thus, Bob will always recover the original message m, as intended.

Exercise 3.4

Compute the encrypted message c when m is 100110 and the key is 001101. What is the original message if the encrypted message is 1110011 and the key is 0101101?

The beauty of this method lies in the fact that Alice can send c over a not necessarily secure channel. This means that Eve may have access to it and capture all its traffic, including all the bits of c. But if every bit of k was selected independently and uniformly at random, was kept secret by Alice and Bob, and is never used again for any encryption (hence the name "one-time" pad), it can be proved that Eve can't learn any information about m from observing c. In cryptographic jargon, this is called **perfect secrecy**... and it certainly lives up to its name!

To learn more...

The one-time pad encryption was patented by Gilbert Vernam in 1919 and, for this reason, it is sometimes called Vernam's cipher. However, it was later discovered that Frank Miller had invented basically the same method back in 1882 (see the paper by Bellovin [31] for a detailed account). The fact that this encryption achieves perfect secrecy was mathematically proved by Claude Shannon in 1945 in a classified report, made public four years later [32]. Independently, Kotelnikov had also proved the same result in 1941 in the Soviet Union, but it was kept secret for security reasons [33].

So we have a simple method that achieves perfect communication secrecy. That's wonderful, isn't it? Well, as you may have noticed, there is a big drawback in the practical application of the one-time pad. In order for it to be secure, the key must be as long as the message, it must be known by both Alice and Bob and nobody else, it must be perfectly random, and it can never be reused. That seems like a lot to ask for, but, at least in principle, there is a way in which we can achieve this.

If Alice and Bob physically meet at some point, they may spend some time flipping a lot of coins (or using another good source of randomness) and writing down the results in their notepads (hence the name one-time "pad") the results for later use as binary keys. In fact, this method was used during the 20th century by governments of different countries, including the USA and the USSR during the Cold War.

Nevertheless, as safe as the one-time pad may be, if you are making an online purchase and you want to securely send your credit card details, using the Vernam's cipher might not be the most practical of ideas. It would be unfeasible for every possible customer and every possible Internet shop to meet beforehand in order to create one-time pads, should they wish to make transactions in the future. As fun as those coin-flipping sessions at the shop could be, that wouldn't keep the world running.

The question is... can Alice and Bob systematically and repeatedly create, at a distance, long, random keys known only to them? One option would be to use a secure communication channel. But since the key needs to be the same size as the message, if Alice and Bob had a secure channel in the first place, they could use it to directly send the message. This seems like an unsolvable problem, but you should take nothing for granted when quantum physics is involved! In the next section, we will study a way of securely sharing secret keys with the help of our beloved qubits.

3.2.2 BB84: the protocol

Influenced by the ideas introduced by Wiesner in his quantum money scheme, Charles Bennett and Gilles Brassard proposed the first protocol for quantum key distribution [34]. It later became known as the BB84 protocol because, well, "Bennet" and "Brassard" both start with "B", and it was proposed in 1984. Not especially original, but the name stuck.

The goal of any QKD scheme is to create, with the help of quantum information, a secret key shared by two parties (Alice and Bob are the usual suspects here). This key can later be used in the one-time pad cipher, for instance, or for other purposes. The main objective is to be able to do this securely, even if Alice and Bob cannot meet physically and even if they do not have a secure communication channel.

It turns out that, even if this may seem impossible, it can be achieved if we assume that we have at our disposal both a quantum channel (i.e., some way of sending qubit states between Alice and Bob) and a classical communication channel. Neither of the two channels is required to be secure, but the classical communication channel is assumed to be authenticated, meaning that Bob can be sure that the information he receives from Alice really comes from Alice and nobody else, and vice versa.

We will assume that Alice initiates the protocol. This is what happens:

1. She creates, at random, a long sequence of uniform, independent bits $s_1, s_2, \ldots, s_m$ from which the key will be extracted. As we will see later, about half of the bits will be selected for the final key k, so if the desired key length is n, then m should be at

least $2n$. However, some of the bits may be used for verification and other purposes, so, the longer m, the better.

2. For each of the random bits s_i, Alice creates a qubit q_i. If s_i is 0, she sets $q_i = |0\rangle$; if s_i is 1, she sets $q_i = |1\rangle$. After that, with probability $1/2$, she applies a Hadamard gate to q_i, or else she does nothing. In the end, if $s_i = 0$, she will have $q_i = |0\rangle$ or $q_i = |+\rangle$, both with a probability of $1/2$. And if $s_i = 1$, she will have $q_i = |1\rangle$ or $q_i = |-\rangle$, both with a probability of $1/2$. She writes down whether she applied H to q_i or not.
3. Alice sends, in order, all the q_i qubits to Bob over the quantum channel.

Now, it's Bob's turn in the protocol. He needs to do the following:

1. After he receives a qubit q_i from Alice, he decides uniformly at random whether he applies an H transformation to it or not. He writes down his decision.
2. He then measures the qubit in the computational basis. He makes a note of the result.

The final phase of the protocol requires that Alice and Bob communicate over the classical channel. They do the following:

1. For each qubit q_i, Bob tells Alice whether he applied the Hadamard transformation or not. Alice confirms if she took the same decision as Bob. They do *not* talk about the value of s_i or the result of Bob's measurement.
2. If both Alice and Bob's decisions were equal for qubit q_i, Alice keeps the bit s_i for the final key and Bob keeps his measurement result of q_i for the key. If their decisions were different, they discard s_i and Bob's result.

The possible outcomes of this protocol are summarized in *Figure 3.1*.

Alice picks the bit	0				1			
Alice sends the qubit	$\|0\rangle$		$\|+\rangle$		$\|1\rangle$		$\|-\rangle$	
Bob applies the gate		H		H		H		H
Alice and Bob both have the bit	0	$\varnothing$	$\varnothing$	0	1	$\varnothing$	$\varnothing$	1

Figure 3.1: Possible outcomes in the BB84 protocol, depending on the sequential (random) choices made by Alice and Bob. Here, $\varnothing$ denotes that they do not share a bit (they discarded whatever they had)

Notice that, if Alice did not apply the H gate, then she sent either $|0\rangle$ (if $s_i = 0$) or $|1\rangle$ (if $s_i = 1$). If Bob measured q_i directly, then he surely obtained the correct value of s_i. If Alice used the Hadarmard gate on q_i, then she sent either $|+\rangle$ or $|-\rangle$. In that case, if Bob also used the H gate before measuring, he reversed the state back to either $|0\rangle$ (if $s_i = 0$) or $|1\rangle$ (if $s_i = 1$), and he recovered the correct value of s_i. This proves that, after the protocol, both Alice and Bob have the same key. The probability that they both make the same decision for a qubit is $1/2$, so they will keep half of the s_i bits on average, hence the reason for choosing m to be at least $2n$, where n is the desired number of qubits in the key.

To learn more...

In our description of the BB84 protocol, we are assuming that the creation, transmission, transformation, and measurement of qubits are perfect at every step. However, in practice, there will be noise and errors. This may cause Bob's results not to coincide with Alice's bits. Also, Eve may capture some information (we will talk more about this in the next section). If the rate of these errors and information leakages are below a certain threshold, they can be overcome with post-processing techniques called **information reconciliation** and **privacy amplification**. If you want to know more about them, you can read the 1988 paper by Bennett and Brassard [35].

Let's consider an example to illustrate how the protocol works. Imagine that Alice generates five random bits and obtains the sequence 11010. She decides to apply H to the second and fifth positions. Thus, she will send $|1\rangle$, $|-\rangle$, $|0\rangle$, $|1\rangle$, and $|+\rangle$ to Bob, in that order. Suppose that, upon receiving the qubits, Bob decides to apply the Hadamard transformation to the qubits in positions 1, 2, and 4. Thus, he will have the following sequence of qubits: $|-\rangle$, $|1\rangle$, $|0\rangle$, $|+\rangle$, $|+\rangle$. When he measures them in the computational basis, the results for the second and third qubits will be 1 and 0, with certainty. For the other positions, the result can be either 0 or 1, because the qubits are in superposition states. Let's then imagine that his measurements are 11001.

Then, Alice and Bob talk over the classical communication channel (a simple phone would do in this case), and they realize that they only made the same decisions for qubits 2 and 3. Thus, they keep those values for the key, which in this case would be 10. Notice that they did not communicate the actual value of those positions, but (assuming the implementation is perfect) they can be sure that they both have the same bits. Notice also that they cannot be sure if they agree on the rest of the qubits, so they discard them all, even if, in this case, Bob's first bit does coincide with Alice's. Finally, note that they have obtained a key of size 2 from an initial sequence of 5 bits. Starting with a longer bit sequence would probably have gotten them a longer final key.

Exercise 3.5

Alice starts with the binary sequence 1100110. She uses the Hadamard transformation on qubits 2, 3 and 6. Bob uses the H gate on qubits 1, 3, 4, and 6. What positions should they keep for the key? What will the final key be?

We have, thus, a communication protocol that allows Alice and Bob to agree on a binary key. But is it secure? Can Eve get information about the key if she has access to the quantum and classical channels used by Alice and Bob? This is exactly what we will discuss in the following section.

3.2.3 Security of BB84

One of the most appealing properties of the BB84 is that it can be proved that it is secure with just a few weak and reasonable assumptions, namely that Eve has no access to

Alice and Bob's encoding and decoding devices, that the sources of randomness are truly random, and that the classical channel is authenticated (otherwise, Eve could perform a person-in-the-middle attack, impersonating Alice in front of Bob, and Alice in front of Bob). Giving a complete mathematical proof is out of the scope of this book (you can check the paper by Shor and Preskill [36] for more details), but as we did in the case of Wiesner's quantum money scheme, we want to at least give some intuition for the reasons why some straightforward attacks cannot succeed.

To learn more...

In addition to QKD schemes, there exist classical methods for secure key exchange or establishment such as the famous Diffie-Hellman protocol (initially proposed in the works of Ralph Merkle [37], and Whitfield Diffie and Martin Hellman [38]). However, these methods usually rely on assumptions about Eve's computational power. In the case of the Diffie-Hellman protocol, the security of the method is based on the difficulty of solving something called the discrete logarithm problem. This problem is hypothesized to be unfeasible to solve for adversaries with just classical computers. But, incidentally, it can be solved efficiently with quantum computers, as we will discuss in *Chapter 11*.

BB84, on the other hand, is unconditionally secure in the sense that it does not assume any limitation on the computational power available to Eve, and it remains secure even if she has access to a fault-tolerant quantum computer (and, in fact, any kind of computer allowed by the laws of quantum physics).

Since, in the last phase of the BB84 protocol, Alice and Bob will make public their decisions about their use or not of the H gate, Eve could try to capture the qubits sent by Alice, wait until Alice and Bob's decisions are disclosed, and repeat the measurements conducted by Bob to obtain the bits of the key. The problem with this strategy is that, as in the case of quantum money, the no-cloning theorem prevents Eve from copying the qubit states without further information. Close, but no cigar!

Another option could be to capture the qubits sent by Alice, measure them, determine their states, and send new qubits with the same states to Bob. This, however, leaves us in the very same situation that we analyzed when we considered the possibility of a counterfeiter trying to measure the qubits of a quantum banknote. As in that case, Eve will determine the incorrect state half of the time without her realizing it. Moreover, in that case, she will send an incorrect state to Bob. And this can be used to Alice and Bob's advantage. Indeed, our previous analysis for quantum money exactly applies here, and, for each qubit that Eve measures, there is a $1/4$ probability that Bob gets an incorrect result when he measures it in the same basis that Alice prepared it (that is, if Bob gets the decision of applying H or not right). Thus, for each bit that Alice and Bob keep, if Eve tampered with it, there is a probability of $1/4$ that there will be a discrepancy in Alice and Bob's results. Then, they can select at random a certain number r of bits from the ones that they have kept, publicly announce their results on them (and discard them later, of course) and see if they match. Assuming that Eve measured all the qubits, the probability that Eve's manipulation goes unnoticed is $(3/4)^r$, which can be made arbitrarily small by choosing r to be sufficiently large.

To learn more...

This kind of check can also be used to test the procedure for noise and measurement errors. In fact, Eve's tampering is indistinguishable from hardware errors and unwanted interactions with the environment. In any case, Alice and Bob should always set aside some qubits to estimate the amount of deviation of the results from the expected values, and to inform their application of information reconciliation and privacy amplification methods.

Let us insist once again that our discussion in this section is not a full proof of security of BB84. For that, please refer to the paper by Shor and Preskill [36]. Also, we would like to mention that, unlike quantum money, QKD is very much of practical interest. In fact, experimental (and even commercial) implementations of QKD have existed for the last two decades, using BB84 and other protocols.

This wraps up our study of BB84. In the next and last section of this chapter, we will discuss some alternative QKD methods as well as other quantum protocols built on individual qubits.

3.3 Other protocols with individual qubits

In the final section of this chapter, we will briefly discuss some additional protocols that involve individual qubits. We will start by talking a little bit about alternatives to BB84 when trying to agree on a secret key with quantum methods, and then we will cover a colorful protocol that has to do with quantum bombs.

3.3.1 Alternative QKD protocols

In addition to BB84, there exist a number of quantum key distribution protocols that rely on the use of individual qubits, such as the **six-state protocol** and **SARG04**. In general, they try to improve some figure of merit, like the eavesdropping detection rate or the fraction of qubits that are retained for the final key. To that end, they modify some parts of the protocol, including the number or type of states that are used in the transmission.

For instance, the six-state protocol uses the states $|i\rangle := \frac{1}{\sqrt{2}}(|0\rangle + i\,|1\rangle)$ and $|-i\rangle := \frac{1}{\sqrt{2}}(|0\rangle - i\,|1\rangle)$ in addition to $|0\rangle, |1\rangle, |+\rangle$, and $|-\rangle$. This makes it more difficult for Eve to go unnoticed and, thus, provides increased security. For details, please check the papers by Bruß [39], and Bechmann-Pasquinucci and Gisin [40].

The SARG04 scheme (proposed by Scarani, Acín, Ribordy, and Gisin in 2004, hence the protocol's name), uses the same states as BB84, but encodes information in a different way. This leads to a more robust protocol that also shows improved resistance to certain types of attacks. For more information, refer to the original paper by Scarani et al. [41] and to the detailed analysis provided by Branciard et al. [42].

To learn more...

There exist QKD protocols that exploit entanglement, a quantum phenomenon produced by certain interactions of quantum systems. For instance, in 1991, Ekert proposed a method that uses entangled pairs of photons and is known as the E91 protocol [43]. We will study entanglement in detail when we introduce two-qubit systems in *Chapter 5*.

And now, to close this chapter, it would be nice to briefly cover a method that is especially peculiar. And that is a lot to say when talking about quantum physics!

3.3.2 The Elitzur–Vaidman bomb tester

In 1993, Avshalom Elitzur and Lev Vaidman proposed a way of testing whether a bomb is active or not without any risk of exploding it [44]. Specifically, the setting of this weird thought-experiment is the following: you are given a bomb and you do not know if it works or not, but this is a special "quantum bomb", that you can interact with by sending qubits to it. If the bomb is a dud, nothing will happen, and the qubit will be returned with no changes to its state. But if the bomb is active, the qubit will be measured. If the result is 1, the bomb will explode. If the result is 0, the bomb will remain active and unexploded. The question is: how can you determine if the bomb works… without making it go off?

The first, naive attempt, could be to send a qubit in either state $|0\rangle$ or $|1\rangle$. But this will get us nowhere. If we send $|0\rangle$, the bomb will not explode, but we won't know if it is working or a dud. If we send $|1\rangle$, we will certainly learn if the bomb works or not… but we will make it explode if it is active!

The solution proposed in **the Elitzur-Vaidman bomb tester** is to start with a qubit in state $|0\rangle$, but then apply to it a quantum gate that is known as a Y-rotation (to learn more about this gate and why it is called a "rotation", please take a look at *Appendix B*). This gate depends on a parameter θ, it is represented as $R_Y(\theta)$, and its action is given by

$$R_Y(\theta) := \begin{pmatrix} \cos\frac{\theta}{2} & -\sin\frac{\theta}{2} \\ \sin\frac{\theta}{2} & \cos\frac{\theta}{2} \end{pmatrix},$$

where θ is a real number.

Exercise 3.6

Check that $R_Y(\theta)$ is unitary for every θ. Prove also that $R_Y(\theta)^n = R_Y(n\theta)$ for any natural number n.

And here comes the trick behind this protocol. Before sending it to test the bomb, you are supposed to apply $R_Y(2\epsilon)$ to $|0\rangle$, thus obtaining a qubit in the state $|\epsilon\rangle := \cos\epsilon\,|0\rangle + \sin\epsilon\,|1\rangle$. In a moment, we will discuss which value of ϵ will be most convenient for our purposes, but, for now, assume it to be some arbitrary real number.

If we send a qubit in the state $|\epsilon\rangle$ through the bomb, what will happen? If the bomb is a dud, the qubit will be returned to us in the exact same state $|\epsilon\rangle$. On the other hand, if the bomb is active, it will be measured: with probability $|\cos\epsilon|^2$, the result will be 0 and the bomb will not explode, and with probability $|\sin\epsilon|^2$, the bomb will explode. Of course, we do not want this to happen, so we will pick an ϵ such that $|\sin\epsilon|^2$ is very small. We know that if ϵ is close to 0, then $\sin\epsilon \approx \epsilon$. Then, if we set ϵ to be, for instance, 10^{-10}, the probability of making the bomb go off will be 10^{-20}, which is very small.

Assuming the bomb did not explode, we can repeat the process. We again apply the $R_Y(2\epsilon)$ to the qubit and send it back to the bomb. If the bomb is a dud, the state will now be

$$R_Y(2\epsilon)\,|\epsilon\rangle = R_Y(2\epsilon)^2\,|0\rangle = |2\epsilon\rangle\,,$$

because $R_Y(2\epsilon)^2 = R_Y(4\epsilon)$, as you proved in *Exercise 3.6*. In this case, since the bomb is not active, the state will be returned to you unchanged.

However, if the bomb is active, and provided it hasn't exploded yet, in the previous interaction the qubit got measured and collapsed to $|0\rangle$. Thus, you are sending the $|\epsilon\rangle$ state again. With probability $|\cos\epsilon|^2$, the bomb will not explode and you will get $|0\rangle$ back. Only with probability $|\sin\epsilon|^2$ will the bomb explode.

You should repeat this process n times (we will determine the value of n in a minute, don't worry). Then, you have two possibilities:

1. If the bomb is a dud, you will have the state $|n\epsilon\rangle$.
2. If the bomb is active, it will have exploded with probability at most $n|\sin\epsilon|^2$ (you tried n times, each with probability $|\sin\epsilon|^2$ of exploding). If it hasn't exploded, your qubit is in the $|0\rangle$ state.

Once you have completed this process, you need to measure the qubit. If the bomb is active (and it hasn't exploded), you will always obtain 0 as a result. If it is a dud, you will obtain 1 with a probability of $|\sin n\epsilon|^2$. Now, we have all the information that we need in order to find an appropriate value for n. We want $n|\sin\epsilon|^2$ to be small (in order to make sure that the bomb doesn't go off) and $|\sin n\epsilon|^2$ to be close to 1 (to be able to distinguish between the dud and the working bomb). If we take n to be the closest integer to $\frac{\pi}{2\epsilon}$, then

$$|\sin n\epsilon|^2 \approx \left|\sin \frac{\pi}{2\epsilon}\epsilon\right|^2 = \left|\sin \frac{\pi}{2}\right|^2 = 1,$$

and

$$n|\sin\epsilon|^2 \approx \frac{\pi}{2\epsilon}|\sin\epsilon|^2 \approx \frac{\pi}{2\epsilon}\epsilon^2 = \frac{\pi}{2}\epsilon,$$

which is very small because we had set $\epsilon = 10^{-10}$. Thus, we carry out this protocol and, at the end, measuring 1 will indicate that the bomb is a dud, while measuring 0 will show that it is working. Moreover, both the probability of making a mistake in the determination and the probability of making the bomb go off are very small. It's da bomb, isn't it?

To learn more...

The $R_y(\theta)$ gate has a nice geometrical interpretation. One-qubit states can be represented as points on the surface of a sphere known as the **Bloch sphere**. Then, $R_Y(\theta)$ transforms quantum states by rotating them θ radians around the Y axis of this sphere. For more details, check out *Appendix B*.

Needless to say, this is just a thought experiment, only proposed to illustrate certain quantum phenomena, and not applicable to real bombs! In any case, we thought that this would be a neat and striking example that would nicely complement our discussion on the possible applications of individual qubits for certain information-processing tasks.

With this, we can wrap things up. We are now ready to use all this knowledge to begin writing some quantum code and run it on simulators and actual quantum computers. That will be the focus of our next chapter.

Summary

In this chapter, we have studied several different applications of individual qubits, including schemes for quantum money that cannot be forged, protocols for distributing secret keys that are practically impossible to break, and methods to distinguish working bombs from duds without making them explode.

With these examples, you now have a deep understanding of how qubit states, gates, and measurements work in practice, and how to exploit certain quantum phenomena, including superposition, interference, the no-cloning theorem, and the uncertainty principle. You are also familiar with some of the applications of quantum systems based on single qubits, including their use in securing communications through the use of Quantum Key Distribution protocols.

In the next chapter, we are going to put all this into practice. We will take our first steps in quantum computer programming using Qiskit, a very comprehensive Python library that will help us manipulate qubits with both quantum gates and measurements. We will also learn how to run simple quantum circuits on simulators and actual quantum computers. Finally, we will show how to implement some of the protocols that we have introduced in this chapter, including the BB84 method.

4

Coding One-Qubit Protocols in Qiskit

Give a man a program, frustrate him for a day.
Teach a man to program, frustrate him for a lifetime.

— Muhammad Waseem

So far, we've managed to get a lot done. After seeing the general landscape of quantum computing from a bird's-eye view, we set out to discuss one-qubit systems at length, seeing how their states can be represented and transformed, and analyzing the crucial yet counter-intuitive role that measurements play in quantum mechanics at large, and in quantum computing in particular. From this point, we were able to explore some interesting applications in one-qubit systems.

All of that theory is about to become practice in this chapter, as we take our first steps in programming quantum computers and getting circuits to run on both simulators and real quantum hardware.

The topics covered in this chapter are the following:

- Quantum software and the case for Qiskit
- How to work with one qubit in Qiskit
- Implementing the BB84 protocol

By the end of this chapter, you will have a general perspective on the software frameworks that are currently available to program and run quantum algorithms. You will also know how to implement one-qubit algorithms using Qiskit and you will be able to simulate them locally or send them to real quantum hardware through the Internet. In addition to this, you will have a better understanding of the BB84 protocol and you will be able to simulate it yourself.

That's a packed agenda, so we better get started. Let's begin by exploring the quantum software landscape and introducing Qiskit.

4.1 Quantum software and the case for Qiskit

In recent years, interest in quantum computing has been growing rapidly, and, as a consequence, so has the number of software frameworks oriented toward programming quantum algorithms. While many share common features, each of these frameworks is, obviously, different: some are better suited than others for different tasks, and all of them have their pros and cons.

From a general perspective, any quantum software framework needs—at the very least—to provide a way for users to represent their quantum algorithms. And, of course, it also must provide a way for those quantum algorithms to be run in some way.

In an ideal world, we would all have access to a quantum processing unit at home, but sadly, we're not there just yet, so the whole business of executing quantum algorithms gets a bit tricky. In this regard, there are two things that one can do:

- While quantum hardware is still not as good as we would like it to be, some quantum computers exist and—despite their limitations—can execute some quantum algorithms. These quantum computers are hosted and made available over the Internet by companies such as IBM, Amazon, or IQM. While most of them charge a fee for the use of their quantum hardware, some companies, such as IBM, provide limited access for free, as we will discover in this chapter.

- Running algorithms on real quantum hardware is, right now, anything but a smooth experience. In addition to the economic cost that it may entail, you may have to not only wait for long periods of time to execute the algorithms but also accept the fact that, nowadays, real quantum computers still make quite a number of errors; this will improve in the future—and in *Chapter 14*, we will discuss how those improvements will come about—but the future isn't now. For this reason, the best way in which the average Jane or Joe can run and test their quantum programs is by running them on **simulators**. Quantum simulators enable the execution of quantum algorithms with a limited number of qubits on any ordinary modern computer, and we will use them extensively throughout this book. While an ordinary classical computer can't handle too many qubits, simulators are powerful enough to illustrate the behavior of most quantum protocols, albeit at a small scale. Moreover, their use has some advantages. Namely, they are free of the errors that may affect real hardware, and they can also easily provide information about the evolution of the state of a system through the execution of an algorithm.

Pretty much all general-purpose quantum software frameworks provide both quantum simulators and interfaces to real quantum hardware. Needless to say, not all simulators are equal: some offer better performance than others, and their features may vary. Likewise, the interfaces for quantum hardware may work better with some devices than with others depending on which framework is used.

To learn more...

Most of the time, software frameworks represent quantum algorithms through the construction of quantum circuits, but some frameworks use different models. Feel free to read our book *A Practical Guide to Quantum Machine Learning and Quantum*

> *Optimization: Hands-on Approach to Modern Quantum Algorithms* [16] to learn more about some of these alternative models.

In this book, we will be using **Qiskit**, which is one of the most popular general-purpose quantum software frameworks out there. Nevertheless, before we start to discuss it in depth, we thought it would be a good idea to give you a brief overview of *some* of the most commonly used quantum software frameworks.

Let us please emphasize that our discussion will be far from exhaustive—we are only covering a handful of the tools that currently exist. With that being said, let's see what the world has to offer for quantum programmers!

PennyLane

We begin our discussion with PennyLane [45], a general-purpose quantum software framework developed by Canadian company Xanadu. Like most other frameworks that we will discuss, PennyLane runs on Python and it can be easily installed with pip, Python's package manager. Within it, quantum circuits can be implemented with a very simple and intuitive API, and they can be run locally with its own simulators. In addition, PennyLane also provides interfaces for circuits to be executed on real hardware.

In terms of simulators, it includes a basic simulator implemented in Python as well as a high-performing "Lightning" simulator that runs on a C++ backend. For real hardware, it provides interfaces to Amazon Braket, Rigetti, AQT, Honeywell, Quantum Inspire, and IonQ, among other quantum hardware platforms.

One of the many virtues of PennyLane is its interoperability with other quantum software frameworks. The team behind PennyLane supports interfaces to Qiskit and Cirq among many others. Another great feature of PennyLane is the stability of its API.

While PennyLane could be used to implement any kind of quantum algorithm or protocol, it is most commonly used within the domain of quantum machine learning, and we actually use it extensively in our book *A Practical Guide to Quantum Machine Learning and Quantum*

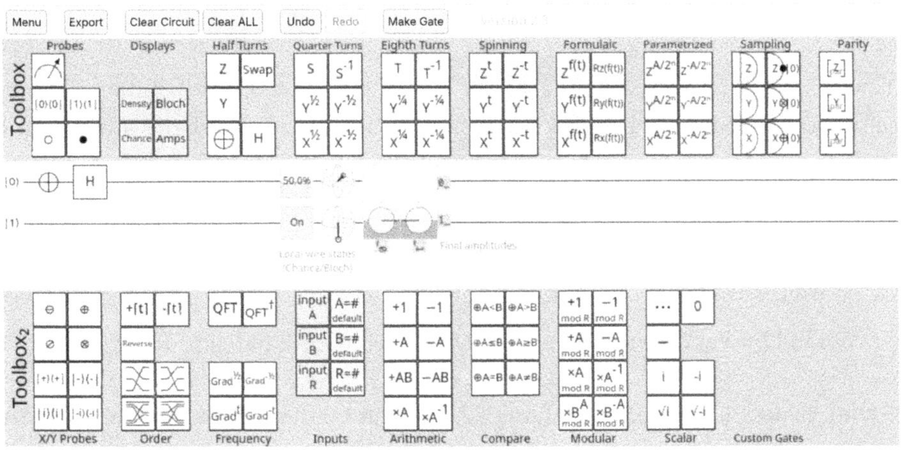

Figure 4.1: A simple circuit constructed with Quirk, a web-based simulator with a graphical user interface that allows operations to be dragged and dropped into a quantum circuit

Optimization: Hands-on Approach to Modern Quantum Algorithms [16]. In quantum machine learning, the interoperability of PennyLane truly shines, as it provides interfaces to some of the most commonly used machine learning frameworks—thus making the training of quantum machine learning models a very smooth experience.

Quirk

While not a full-fledged quantum software framework, Quirk [46] is as simple as it is useful—which is why we have chosen to include it here. Quirk is a web-based platform on which you can construct and simulate quantum circuits using a drag-and-drop interface. It can be accessed freely on `https://algassert.com/quirk`, and it was developed by Craig Gidney. In spite of its modesty, it is a very valuable tool for quick prototyping; in *Figure 4.1*, you can see a screenshot of its interface, in which we have constructed a simple circuit with a NOT and a Hadamard gate.

Quirk can graphically display the amplitudes of the state at any point through the execution of the circuit, as well as showing the probabilities of a measurement at different points.

Cirq

Moving on to a more conventional framework, Cirq [47] is Google's general-purpose quantum software toolkit and it shares some similarities with PennyLane. Like most other frameworks, it runs on Python and can be installed through pip.

It can implement any quantum algorithm through the construction of quantum circuits. It includes its own simulator and it has interfaces to run circuits on the real quantum hardware provided by vendors such as AQT, Azure, IonQ, Pasqal, and Rigetti.

A feature that makes Cirq special among other frameworks is its support for **qudits**: quantum systems that can be in a number of states other than two.

QuEST

The Quantum Exact Simulation Toolkit [48] (abbreviated as QuEST) is a high-performance C library for simulating quantum circuits. It is designed to take full advantage of computers' hardware—including support for GPUs through CUDA—in order to simulate quantum circuits efficiently and smoothly. It is developed and maintained by a team at the University of Oxford.

Qiskit

Last—but certainly not least—we have Qiskit [49], one of the best-known quantum software frameworks out there, which is actively developed and maintained by IBM and by a strong ommunity of volunteers. In some ways, Qiskit works in a similar way to PennyLane. runs on Python and provides an API for constructing quantum circuits, and it comes dled with some high-performance simulators that are written in C++ and (increasingly)

s a very strong community all over the world and, most importantly, it is the best pare circuits in order to send them to IBM's own quantum hardware, for which ted access free of charge.

It also comes loaded with features and makes it easy to implement even the most complex quantum algorithms and protocols. For anything related to quantum computing, there most likely already is some Qiskit module implementing it.

In this book, we will exclusively use Qiskit, and we will be working with version 2.1. If you haven't already, please have a look at *Appendix D* and make sure that you have all the required packages—with the appropriate versions—installed on your machine.

Now that is enough of an introduction. Without further ado, let's get coding! Our qubits are about to have their "hello world!" moment.

4.2 How to work with one qubit in Qiskit

We are ready to begin working with Qiskit. For that, please launch your Python interpreter or, preferably, create a Jupyter notebook. Alternatively, you can run this code on the cloud with a service such as Google Colab.

Important note

We will be using Qiskit 2.1. Please refer to *Appendix D* for instructions on how to install the tools and frameworks that we use in this book.

Remember, by the way, that the code that we will be using can be downloaded as a Jupyter notebook from the book's GitHub repository:

`https://github.com/PacktPublishing/A-Practical-Guide-to-Quantum-Computing`

To get started, we will first import the `QuantumCircuit` class from the Qiskit package as follows:

```
from qiskit import QuantumCircuit
```

All quantum circuits will be represented as objects of the `QuantumCircuit` class.

To learn more...

If you are not familiar with object-oriented programming and the words "class", "object", and "method" don't sound too familiar, that's alright. For the most part, you can follow along and try to infer the meaning from how we use our code. If, however, you would like to learn a little bit more about this, feel free to check out the free online course *CS50's Introduction to Programming with Python*, by Harvard University.

Thus, to create a new quantum circuit, we just have to create a new `QuantumCircuit` object, and its initializer will take the number of qubits that we want to use as an argument. Since we only know how to work with one-qubit circuits, we will create a circuit as follows:

```
circuit = QuantumCircuit(1) # We specify "1" for one qubit.
```

And voilà! Now the `circuit` object represents an empty one-qubit circuit in Qiskit, whose state will be initialized to $|0\rangle$ as that is the default.

Having an empty circuit is exciting... but not too exciting, so let's add a few gates and see what we get. In order to add gates to a quantum circuit in Qiskit, you have to call some specific methods. For instance, in order to add a NOT gate, you must call the `x` method, providing as an argument the qubit on which the gate must be applied. In our case, we are only working with a single qubit, so this argument will always be `0`; keep in mind that qubits are 0-indexed, so the first qubit is referred to as 0, and the second one (when we get to that in *Chapter 7*) would be identified by 1. Similarly, in order to include a Hadmard gate, you can use the `h` method.

With this knowledge, let us try to turn our Qiskit `circuit` into the following circuit:

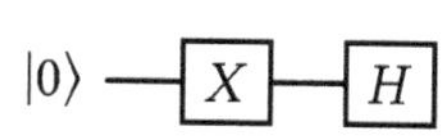

For that, we only have to execute the following lines of code:

```
circuit.x(0) # Apply an X gate on the fist qubit (0).
circuit.h(0) # Apply an H gate.
```

That's all that it would take! After doing this, in order to check whether our construction is correct, we can print our circuit using the `print(circuit)` command, which will show us a text-based representation of our circuit. That would be okay for most purposes (and maybe our only way of printing the circuit in some scenarios), but Qiskit also offers an alternative way of representing circuits that produces better-looking results.

To represent our circuit in all its glory, we can use the `draw` method with the optional `"mpl"` argument (short for `matplotlib`):

```
circuit.draw("mpl")
```

Upon running this, we will get a beautiful representation of our circuit, which confirms that it has been implemented correctly, with a NOT gate followed by a Hadamard gate. You can find this representation below:

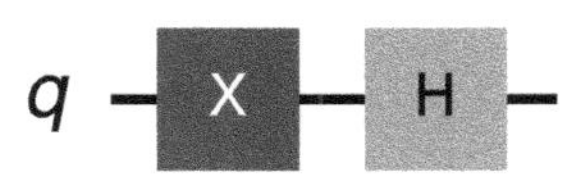

Using the `draw` method, we can get a visual representation of our circuit, but maybe you'd like to save that representation in order to use it in a document or presentation. To do that, you can run the following instructions:

```
drawing = circuit.draw("mpl")
drawing.savefig("circuit.pdf")
```

This will save the circuit representation in PDF format in the `circuit.pdf` file, and it will be stored as a vector image. If you would instead like to have a rasterized image, you can save the representation in PNG format—all it would take would be changing the extension in the filename, and the method will take care of everything else for you.

We have just seen how to apply the X and H gates using the `x` and `h` methods, respectively, but there are plenty of other one-qubit gates that we don't yet know how to use in Qiskit.

Well, these don't hold too much of a mystery. In order to use the gates Z, Y, and S and T, we can use, respectively, the `z`, `y`, `s`, and `t` methods, which work in full analogy to `x` and `h`. Recall that we introduced all of these gates back in *Chapter 2*.

Exercise 4.1

Create a Qiskit circuit named `circuit_yt` implementing the following quantum circuit:

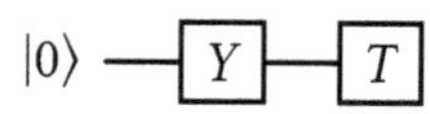

Represent the circuit using the `draw` method in order to verify that you have constructed it correctly.

There is another one-qubit gate that we introduced in *Chapter 2* and that works in a slightly different way to the ones we've discussed so far, and that is the phase gate. Remember that, for any real number θ, we defined the phase gate parametrized by θ as

$$P(\theta) = \begin{pmatrix} 1 & 0 \\ 0 & e^{i\theta} \end{pmatrix}.$$

Since this gate actually depends on a parameter, we will have to specify said parameter when using the gate. In this way, if we want to apply the phase gate parametrized by `theta` on the first qubit of `circuit`, we will have to run `circuit.p(theta, 0)`. In the following example, we apply $P(2)$ on a one-qubit circuit, and we print the result:

```
# Apply the gate P(2) on a one-qubit circuit.
circuit = QuantumCircuit(1)
circuit.p(2,0)
circuit.draw("mpl")
```

Upon running this, we get the following output:

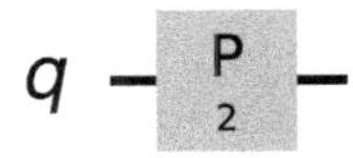

This shows that, indeed, we have constructed our circuit as we intended. Notice, by the way, that we have reinitialized the `circuit` object, deleting the one we had before, and thus starting on a blank canvas.

Important note

Quantum circuits in Qiskit are represented as objects of the `QuantumCircuit` class, which can be initialized by specifying the number of qubits of the circuit. Quantum gates can then be sequentially applied by using different methods, on which it must be specified the qubit on which the gate is going to be applied, and the parameters of the gate, should it accept any.

It is worth pointing out that quantum circuits can be manipulated like any other object in Python. They can be used within functions and their construction can rely on loops, conditionals, or any other flow control methods provided by the language. To illustrate this, we will create a simple function with a single argument. If the argument is smaller than 1, the function will apply a one-qubit circuit with a phase gate parametrized by it; otherwise, it will return a one-qubit circuit with a Hadamard gate:

```
def phased_circuit(phase):
    circuit = QuantumCircuit(1)
    if phase < 1:
        circuit.p(phase, 0)
    else:
        circuit.h(0)
    return circuit
```

We can draw the output of this function when `theta = 0.4`, as follows:

```
phased_circuit(0.4).draw("mpl")
```

This shows a circuit with a $P(0.4)$ gate. On the other hand, if `phase >= 1`, the output will be a circuit with an H gate, as you can easily check by executing this instruction:

```
phased_circuit(1.1).draw("mpl")
```

Exercise 4.2

Implement a function `apply_gate` that will take two arguments: a circuit `circ` and a string `gate`. If the string is `"X"`, `"Y"`, or `"Z"`, the function must apply, respectively, the X, Y, or Z gate on the first qubit of the circuit `circ`. If the `gate` string takes any other values, the function must do nothing.

That is how one-qubit circuits can be constructed in Qiskit, so let's now analyze how they can be run. We will begin by discussing how to run circuits on simulators, and we will then take care of real quantum hardware.

4.2.1 Simulating the evolution of a state

One of the main advantages of simulating the execution of quantum circuits is that—as opposed to running them on real hardware—it can give us full access to the state of the qubits, which is exceptionally useful for debugging and for understanding the behavior of circuits. We will now see how Qiskit allows us to compute the final state of a quantum circuit.

To get started, let us create a circuit with a NOT gate followed by a Hadamard gate, as we did at the beginning of this section:

```
circuit = QuantumCircuit(1)
circuit.x(0)
circuit.h(0)
```

Once we have any circuit, such as the one we have just defined, using Qiskit to find the state of the system at the end of the execution of the circuit couldn't be easier. All we have to do is import the `Statevector` class and initialize it by passing the circuit as an argument:

```
from qiskit.quantum_info import Statevector

state = Statevector(circuit)
print(state)
```

Upon running the preceding code, we get the amplitudes of the prepared state, which is $|-\rangle$, as we expected:

```
Statevector([ 0.70710678+0.j, -0.70710678+0.j],
            dims=(2,))
```

In the preceding representation, the first entry on the list is the amplitude of $|0\rangle$, while the second one is that of $|1\rangle$. Keep in mind that $1/\sqrt{2}$ is approximately equal to `0.70710678`, and remember that $|-\rangle = (1/\sqrt{2})(|0\rangle - |1\rangle)$. Notice, by the way, that the returned object has an attribute `dims` specifying the dimensions of the state vector. In this case, as we are dealing with a one-qubit system, the dimension of this vector is 2, as shown in the output.

Exercise 4.3

Use Qiskit to obtain the state of a one-qubit system at the end of the circuit that you constructed in *Exercise 4.1*.

Getting the final state of a circuit is useful and convenient, but not very well aligned with what we can expect from quantum hardware. On a real quantum computer, all we can hope to get are results from measurements, and Qiskit also allows us to simulate that, as we will discuss next.

4.2.2 Getting samples from a simulator

If we want to get samples from a measurement, the first thing that we must do is include a measurement operation in our circuit. In order to do that, we can use the `measure_all` method, which—as the name suggests—measures all the qubits in our circuit and stores the result on some classical bits. Thus, we can run this piece of code:

```
circuit.measure_all()
circuit.draw("mpl")
```

With this, we've added a measurement operation and represented our updated circuit. The result that we get is the following:

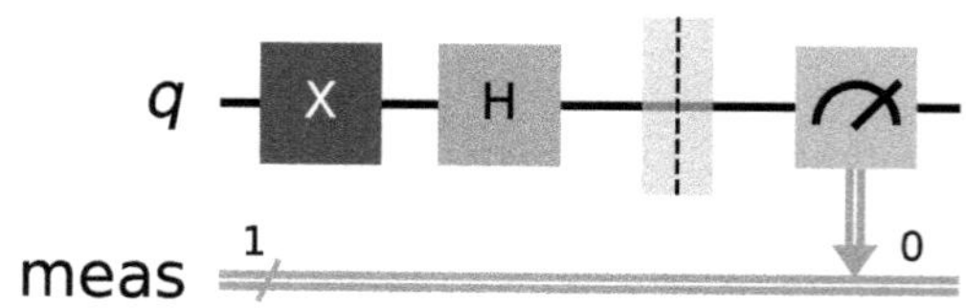

Here, we have the same circuit that we used before, but with a new measurement operation that feeds its output to a classical bit that has been added below; this bit has the default name `meas`. Notice, by the way, that there is a gray rectangle with a dashed line separating the measurement operation from the rest of the circuit; this is called a **barrier**. This has been added automatically to our circuit by Qiskit upon calling the `measure_all` method. Barriers are used not only to visually separate the different components of a circuit, but also to prevent them from being merged in a process known as **transpilation**, which consists in decomposing the circuit into gates that real quantum computers can execute. We will come back to this later.

In order to be able to sample our circuit, we first need to instantiate an `AerSimulator` object, which will be the **backend** on which the circuit will run; we will create this object specifying a seed in order for our results to be reproducible as follows:

```
from qiskit_aer import AerSimulator
backend = AerSimulator(seed_simulator = 18620123)
```

Once we have our simulator ready, we need to create a `Sampler` object, having it as a backend:

```
from qiskit_ibm_runtime import SamplerV2 as Sampler
sampler = Sampler(backend)
```

When using real hardware, we will see that the workflow is analogous but for the use of a different backend object, representing a real quantum computer.

To learn more...

The `Sampler` class is an instance of a Qiskit **primitive**. Informally speaking, these primitives are the mechanisms that enable Qiskit to return outputs from a circuit, and they are used to interact with quantum hardware as well as with simulators. In addition to sampler primitives, which return samples from measurements, there are others such as estimator primitives, which return expectation values. To learn more about these, you can take a look at the Qiskit documentation.

With our sampler ready, we can run it on our circuit, asking it to get 8 shots for us. If we don't provide a value for the shots, it will default to 1024.

```
job = sampler.run([circuit], shots = 8)
result = job.result()[0].data.meas # Get the results!
```

Notice that we have passed `circuit` within a list: that's because we could've passed a list of several circuits and run them all at once in a single job. In the preceding code, you can see how, after executing the circuit with `run`, we have called the `result` method to retrieve the results; this returns a list with the results for every circuit that is passed when running the job. Since we only sent one circuit, we only have to get the first element, hence why we have the `[0]` after the call to the `result` method. From there, we simply extracted the data corresponding to the classical bit, which, as we mentioned before, goes by the name of `meas`.

Once our results are ready, we can extract them in several ways. The first and most convenient one is getting a dictionary with the counts of all the measurement outcomes, as follows:

```
print(result.get_counts())
```

The preceding instruction returns a dictionary telling us how many times each outcome was observed:

```
{'0': 5, '1': 3}
```

In this case, the measurement in our circuit turned out to be 0 a total of five times, and it was 1 a total of three times.

Alternatively, we may wish to obtain the raw list of measurements outcomes. This can be achieved with this piece of code:

```
print(result.get_bitstrings())
```

The output is the following list of outcomes:

```
['0', '1', '0', '1', '0', '1', '0', '0']
```

Keep in mind that this list preserves the order in which the results were measured.

Exercise 4.4

Use Qiskit to obtain 4 measurement samples from the final state of the circuit that you constructed in *Exercise 4.1*.

There are still other ways in which the results can be retrieved. For example, the `array` parameter contains an array with the results stored as integers. We can run this piece of code:

```
print(result.array)
```

This gives the following output:

```
[[0]
 [1]
 [0]
 [1]
 [0]
 [1]
 [0]
 [0]]
```

Lastly, it is worth mentioning that it's possible to get the number of shots that were used by accessing `results.num_shots`.

That's how Qiskit can simulate getting samples from a measurement operation. Now, let's go wild. Let's see how we can actually use real hardware.

4.2.3 Getting results from real quantum hardware

IBM provides free and open access to some of their quantum computers through the IBM Quantum Platform. In order to use it, you first need to create an IBM account, access the IBM Quantum Platform, create an instance, and retrieve your unique API key (we describe this process in *Appendix D*). This token is the equivalent of a password and allows anyone to connect to IBM's quantum computers on your behalf, so it's very important to keep it safe and not share it with anyone.

To get started, we need to authenticate ourselves and launch an instance of the Qiskit Runtime service. This can be done as follows (setting `token` to a string with your own API key):

```
from qiskit_ibm_runtime import QiskitRuntimeService
token = "YOUR API KEY GOES HERE"
service = QiskitRuntimeService(token=token, channel="ibm_quantum_platform")
```

In the call to the `QiskitRuntimeService` constructor, you can also specify an `instance` as an optional argument (for example, passing `instance = "PGQC_instance"`, if you've followed the instructions in *Appendix D*). Nevertheless, if you don't specify one, Qiskit will eventually pick one for you automatically, provided that you have created at least one.

If you are running this on your own personal computer and want to have your token saved on your machine (to avoid having to paste it in every time), you can run this command:

```
QiskitRuntimeService.save_account(token = token,
                                  channel = "ibm_quantum_platform")
```

With it, you can from then on launch `QiskitRuntimeService` without the need to specify a token and a channel.

Now that we have our Qiskit Runtime service up and running, we can use it to get a backend object pointing to one of IBM's real quantum computers. We will use the `least_busy` method to get the least busy quantum computer, and we will ask for it to be operational and an actual quantum computer (not a simulator).

```
backend = service.least_busy(simulator = False, operational = True)
print(backend)
```

In our case, we got access to IBM's quantum computer named Kyiv, but you may have gotten a different one.

Now, we're almost set to send our circuit to the backend, but before we do that, we need to transpile the circuit. Real quantum computers can only run a limited set of instructions, and transpiling a circuit simply means decomposing it into instructions from that limited set. With Qiskit, we can use `generate_preset_pass_manager` to transpile a circuit for its execution in a specific backend. We can do it as follows:

```
from qiskit.transpiler.preset_passmanagers import (
    generate_preset_pass_manager)
pm = generate_preset_pass_manager(backend=backend, optimization_level=1)
transpiled = pm.run(circuit)
transpiled.draw("mpl", idle_wires = False)
```

Notice that `generate_preset_pass_manager` requires an `optimization_level` argument. This is an integer from 0 to 3, and, the greater it is, the more optimized the circuit will be. Upon executing the preceding code, we can see a representation of our transpiled circuit:

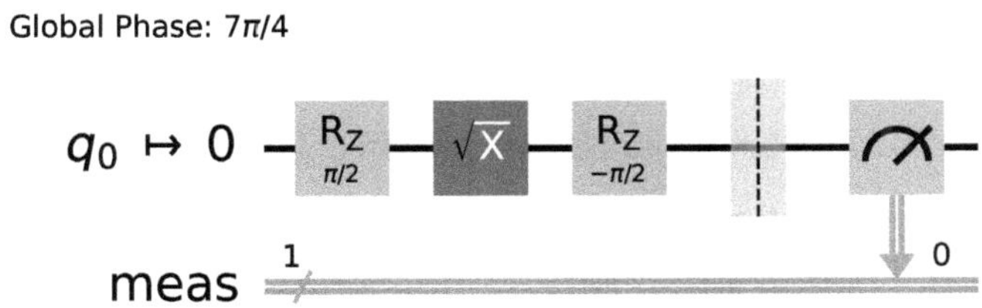

Don't worry if you get a different circuit. If you are using a different backend, your circuit may be decomposed using a different set of instructions.

To learn more...

In order to illustrate how barriers affect transpilation, let us recreate our circuit, adding a barrier between the H and X gates:

```
circuit_barrier = QuantumCircuit(1)
circuit_barrier.x(0)
circuit_barrier.barrier()
circuit_barrier.h(0)
circuit_barrier.measure_all()
circuit_barrier.draw("mpl")
```

After running this, we get the following representation:

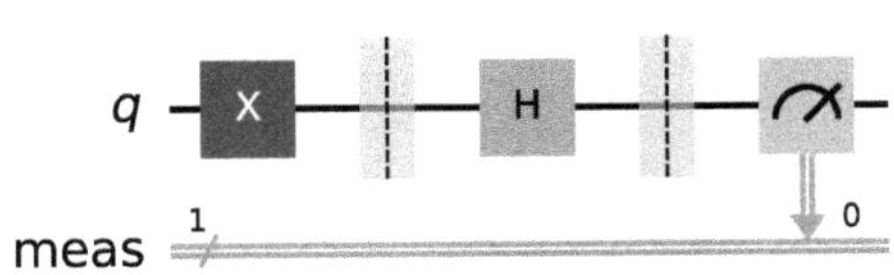

We can now transpile the circuit as follows:

```
transpiled_barrier = pm.run(circuit_barrier)
transpiled_barrier.draw("mpl", idle_wires = False)
```

After running this, we can see in the following output how, instead of three gates, we now get four, and they are separated by a barrier. This is, of course, less optimal than the transpilation that we got without the barriers, but it ensures that the X and H gates are not merged together in the transpilation process.

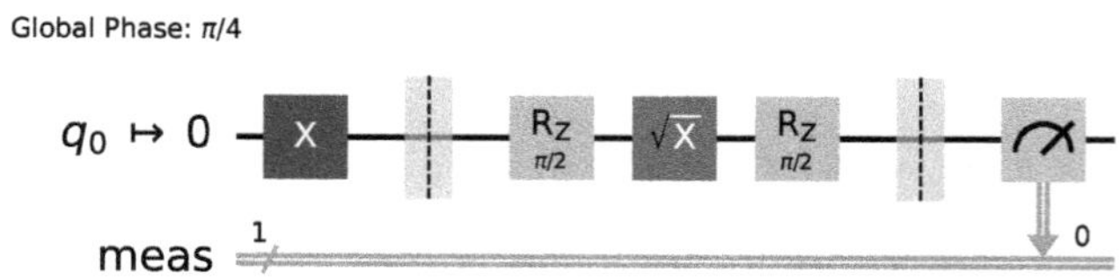

Now we have all the pieces gathered to run our circuit on our quantum hardware backend with IBM Runtime. All we need to do is initialize the `Sampler` class with our backend and run our job just as we did with our simulator before:

```
from qiskit_ibm_runtime import SamplerV2 as Sampler
sampler = Sampler(mode=backend)
job = sampler.run([transpiled], shots = 1024)
result = job.result()[0].data.meas
print(result.get_counts())
```

After a few moments, we get the following counts:

```
{'0': 505, '1': 519}
```

These align very well with the results that we obtained in the simulation, and with the even distribution we would expect from a theoretical point of view. Notice, by the way, how we have asked for 1024 shots in this case. If the number of shots is left unspecified, in this case, it will default to 4096.

Of course, if you got a different result from ours, there's nothing for you to worry about! Since we are running on real hardware and measurements are probabilistic, it is more than expected to get slightly different results on each execution.

There you have it! That is how you can run a circuit on a real quantum computer using Qiskit. It wasn't that difficult, was it? With your free IBM account, you get a total of 10 minutes of computing time every month—now it's up to you if and how you want to use them!

Having discussed how to construct and run circuits in Qiskit, we will now implement one of the applications that we considered in *Chapter 3*: the BB84 protocol. Let's get to it!

4.3 Implementing the BB84 protocol

We are about to prepare a simple implementation of the BB84 protocol using Qiskit, following our discussion from *Chapter 3*. Since the protocol involves Alice and Bob choosing to apply a Hadamard gate at random, we will import the `random` module to take care of that, and we will give it a seed for our results to be reproducible. In addition, we will also initialize a simulator backend with a fixed seed and a sampler, as follows:

```
import random
random.seed(18620123)
backend = AerSimulator(seed_simulator = 18620123)
sampler = Sampler(backend)
```

For the protocol implementation, we will prepare a short list of bits meant to represent Alice's key. We will then send them one by one applying the protocol, and we will store the results that Bob receives in another list. In addition, we will create two lists to store whether, for the transmission of each bit, Alice and Bob applied or did not apply a Hadamard gate:

```
# The key we want to send:
alice_bits = [0,1,0,1,1,1,0,0,1,1,0]
# A list for the bits that bob will receive:
bob_bits = []
```

```
# We will create lists to store wheteher Alice/Bob used H.
# If they do, we add "True", otherwise we add "False".
alice_used_h = []
bob_used_h = []
```

Now that we are all set, let's simulate the transmission of each of the bits using the BB84 protocol! For each of them, we will create a one-qubit circuit. Since these are initialized to $|0\rangle$, we will apply an X gate if the bit that Alice wants to send is 1 (and otherwise we will do nothing). Then, we will let Alice and Bob decide whether or not to apply a Hadamard gate, and we will have Bob measure the qubit and retrieve his result. We will then store all the relevant information in the lists that we have just created, as follows:

```
for bit in alice_bits:
    circuit = QuantumCircuit(1)
    # If we are going to send 1, we apply an X gate.
    # Remember that the state is initialized to |0>.
    if bit:
        circuit.x(0)

    # Choose at random if Alice applies H, and do it.
    alice_h = random.choice([True, False])
    if alice_h:
        circuit.h(0)

    # Apply a barrier and choose whether Bob does H, and do it.
    circuit.barrier()
    bob_h = random.choice([True, False])
    if bob_h:
        circuit.h(0)
```

```
    # Measure the qubit (on Bob's end).
    circuit.measure_all()
    job = sampler.run([circuit], shots = 1)
    bob_bit = int(job.result()[0].data.meas.get_bitstrings()[0])

    # Add the measured bit to bob_bits and record who used H.
    bob_bits.append(bob_bit)
    alice_used_h.append(alice_h)
    bob_used_h.append(bob_h)
```

Once the transmission of the bits is completed, we find the occasions on which Alice and Bob both applied (or did not apply) a Hadamard gate:

```
# List of indices where Alice and Bob agree on whether to apply H.
agree = [i for i in range(len(alice_bits)) if alice_bits[i] == bob_bits[i]]
```

Lastly, we keep the bits corresponding to the occasions in which their choices agreed, as follows:

```
# Get the sent/received bits when they agree on applying H.
alice = [alice_bits[i] for i in agree]
bob = [bob_bits[i] for i in agree]

print(alice)
print(bob)
```

With this, we can see how Alice and Bob now have an identical sequence of bits that they can use to communicate securely! Upon running the preceding code, this is the resulting output:

```
[0, 0, 1, 1, 0, 0, 1, 0]
[0, 0, 1, 1, 0, 0, 1, 0]
```

That shows how the BB84 protocol can be simulated using Qiskit. There you have it, our first implementation of a quantum protocol! With this, we can now wrap up and bring this chapter to an end.

Summary

In this chapter, we have had our first contact with quantum software frameworks. We began by introducing what these frameworks are, and what they are meant to offer, and we also got an overview of some of the most commonly used quantum software toolkits out there. In this context, we had a chance to introduce Qiskit and discuss what advantages can be found in using it.

Leaving that general introduction behind, we set out to write our first lines of code. First we discussed how to construct one-qubit circuits in Qiskit and we saw how Qiskit circuits can be treated and manipulated like any other Python object. Once familiar with this, we explored how these circuits could be run on both simulators and quantum hardware.

Finally, we implemented a simulation of the BB84 protocol on Qiskit, leveraging on all the skills that we acquired throughout the chapter.

That was Qiskit and that's all we have to say about one-qubit systems. Now it's time to go bigger and explore what two qubits enable us to do. One-qubit systems are indeed interesting, but let us tell you: you have seen nothing yet!

Part 2

Qubit Meets Qubit: Two Qubits and Entanglement

In this part, we introduce two-qubit systems. We begin by laying down the mathematical foundations of working with two qubits, including how to describe the system state, how to measure it, and how to transform it with quantum gates. We also show how to use new quantum phenomena, such as entanglement, to design surprising protocols, including quantum teleportation. We finally put all this in practice by learning how to work with two-qubit systems in Qiskit.

This part includes the following chapters:

- *Chapter 5*, How to Work with Two Qubits
- *Chapter 6*, Applications and Protocols with Two Qubits
- *Chapter 7*, Coding Two-Qubit Algorithms in Qiskit

5

How to Work with Two Qubits

Great things are done by a series of small things brought together.

— Vincent van Gogh

In this chapter, we are going to introduce two-qubit systems. At first, this may seem like a modest incremental improvement after having learned how to process information with one qubit, but beware! Adding a second qubit into our systems will open up a world of possibilities and, in particular, will allow us to exploit one of the strangest and most useful phenomena in all of quantum physics: entanglement. As you will see, from a mathematical point of view, entanglement is a rather simple concept, but it has profound physical implications and it will become one of our key tools for designing quantum algorithms.

To be able to understand two-qubit systems, we will first discuss how to describe their states. Then, we will learn what happens when we measure the qubits of a two-qubit system and we will introduce new quantum gates that are able to operate on two qubits at once. This will also lead us to describe entanglement and to (finally!) prove the famous no-cloning theorem that we have invoked so many times in previous pages. Believe us, this is going to be exciting!

In this chapter, we cover the following topics:

- One plus one is more than two: two-qubit states
- Measuring two-qubit systems
- Two-qubit gates

After reading this chapter, you will master the theoretical concepts needed to work with two-qubit systems: you will understand how to describe their states, how to transform them with quantum gates and how to extract information from them through measurements. You will also be familiar with entangled and product states, and you will be able to explain why it is impossible to create independent copies of qubits with unknown states.

Are you ready to get entangled in the magic of two-qubit systems? Let's go!

5.1 One plus one is more than two: two-qubit states

Thanks to all that you learned about one-qubit systems in *Chapter 2*, you already have most of the mathematical tools needed to understand two-qubit systems. As you surely remember, a one-qubit system can be in one of the two states from the computational basis $\{|0\rangle, |1\rangle\}$ (or in a superposition of them, we will come back to that soon). Unsurprisingly, when we have two qubits, each of them can also be in one of those two states, so we have four different possibilities in total: both qubits are $|0\rangle$, the first one is $|0\rangle$ and the second is $|1\rangle$, the first is $|1\rangle$ and the second is $|0\rangle$, or both are $|1\rangle$.

Mathematically, these four options are described by something called **tensor products** (more on this later in this same section), which we represent as follows:

$$|0\rangle \otimes |0\rangle, \qquad |0\rangle \otimes |1\rangle, \qquad |1\rangle \otimes |0\rangle, \qquad |1\rangle \otimes |1\rangle.$$

Most of the time, we will omit the tensor product symbol ⊗ and simply write

$$|0\rangle\,|0\rangle\,, \qquad |0\rangle\,|1\rangle\,, \qquad |1\rangle\,|0\rangle\,, \qquad |1\rangle\,|1\rangle\,,$$

or even just

$$|00\rangle\,, \qquad |01\rangle\,, \qquad |10\rangle\,, \qquad |11\rangle\,.$$

These tensor products are just fancy ways of writing the four-dimensional column vectors that form the computational basis of two-qubit systems. In fact, it holds that

$$|00\rangle = \begin{pmatrix} 1 \\ 0 \\ 0 \\ 0 \end{pmatrix}, \qquad |01\rangle = \begin{pmatrix} 0 \\ 1 \\ 0 \\ 0 \end{pmatrix}, \qquad |10\rangle = \begin{pmatrix} 0 \\ 0 \\ 1 \\ 0 \end{pmatrix}, \qquad |11\rangle = \begin{pmatrix} 0 \\ 0 \\ 0 \\ 1 \end{pmatrix}.$$

Important note

As a quick reference, the following table shows the different ways in which the computational basis states of two-qubit systems can be represented. In every cell, you will find several expressions for the four computational basis states:

$\|0\rangle \otimes \|0\rangle = \|0\rangle\,\|0\rangle = \|00\rangle = \begin{pmatrix} 1 \\ 0 \\ 0 \\ 0 \end{pmatrix}$	$\|0\rangle \otimes \|1\rangle = \|0\rangle\,\|1\rangle = \|01\rangle = \begin{pmatrix} 0 \\ 1 \\ 0 \\ 0 \end{pmatrix}$
$\|1\rangle \otimes \|0\rangle = \|1\rangle\,\|0\rangle = \|10\rangle = \begin{pmatrix} 0 \\ 0 \\ 1 \\ 0 \end{pmatrix}$	$\|1\rangle \otimes \|1\rangle = \|1\rangle\,\|1\rangle = \|11\rangle = \begin{pmatrix} 0 \\ 0 \\ 0 \\ 1 \end{pmatrix}$

But if these were the only possible states of two-qubit systems, they would be rather dull, wouldn't they? We would be confined to a fixed set of states, missing out on many rich and complex behaviors enabled by quantum systems. Fortunately, superposition comes to the rescue. As you may be suspecting, a two-qubit system can also be in a superposition state, which is nothing more than a linear combination of the form

$$\alpha_{00}\left|00\right\rangle + \alpha_{01}\left|01\right\rangle + \alpha_{10}\left|10\right\rangle + \alpha_{11}\left|11\right\rangle,$$

where α_{00}, α_{01}, α_{10} and α_{11} are complex numbers (called amplitudes, as in the one-qubit case) satisfying the normalization condition

$$\left|\alpha_{00}\right|^2 + \left|\alpha_{01}\right|^2 + \left|\alpha_{10}\right|^2 + \left|\alpha_{11}\right|^2 = 1.$$

Of course, this constraint has to do with the probabilities of obtaining different results when measuring the system, but we will discuss that in more detail in the next section.

Valid two-qubit states include

$$\frac{1}{2}\left(\left|00\right\rangle + \left|01\right\rangle + \left|10\right\rangle + \left|11\right\rangle\right),$$

because $|1/2|^2 + |1/2|^2 + |1/2|^2 + |1/2|^2 = 1/4 + 1/4 + 1/4 + 1/4 = 1$, and

$$\frac{1}{\sqrt{2}}\left(\left|00\right\rangle + \left|11\right\rangle\right),$$

because $\left|1/\sqrt{2}\right|^2 + \left|1/\sqrt{2}\right|^2 = 1/2 + 1/2 = 1$.

Exercise 5.1

Which of the following are valid two-qubit states?

(a) $\frac{1}{\sqrt{3}}\left(\left|00\right\rangle + \left|10\right\rangle + \left|11\right\rangle\right)$

(b) $\frac{1}{2}\left(\left|00\right\rangle + \left|01\right\rangle + \left|10\right\rangle - \left|11\right\rangle\right)$

(c) $\frac{1}{\sqrt{2}}\left(\left|00\right\rangle - \left|11\right\rangle\right)$

(d) $\frac{1}{\sqrt{2}}(|00\rangle + i\,|11\rangle)$
(e) $\frac{2}{\sqrt{3}}|00\rangle + \frac{2}{\sqrt{3}}|10\rangle - \frac{1}{\sqrt{3}}|11\rangle$
(f) $i\,|10\rangle$
(g) $2\,|00\rangle - i\,|11\rangle$
(h) $\sqrt{\frac{2}{3}}\,|00\rangle - \sqrt{\frac{1}{3}}\,|10\rangle$

Find all the values of x that make $\frac{1}{2}|01\rangle + x\,|10\rangle$ a valid two-qubit state.

Okay, so this wasn't that different from one-qubit states, was it? The only thing that may be unsettling you at this point is that mysterious tensor product that we have introduced, so let's clarify what it actually is. In general, the tensor product of two column vectors

$$a = \begin{pmatrix} a_1 \\ a_2 \end{pmatrix}, \qquad b = \begin{pmatrix} b_1 \\ b_2 \end{pmatrix},$$

is another column vector defined by

$$a \otimes b := \begin{pmatrix} a_1 b_1 \\ a_1 b_2 \\ a_2 b_1 \\ a_2 b_2 \end{pmatrix}.$$

This may seem somewhat arbitrary, but the logic here is that we are taking two copies of b, putting one on top of the other, and multiplying the first copy by a_1 and the second one by a_2. More graphically, this can be seen as follows:

$$a \otimes b = \begin{pmatrix} a_1 \begin{pmatrix} b_1 \\ b_2 \end{pmatrix} \\ a_2 \begin{pmatrix} b_1 \\ b_2 \end{pmatrix} \end{pmatrix} = \begin{pmatrix} a_1 b_1 \\ a_1 b_2 \\ a_2 b_1 \\ a_2 b_2 \end{pmatrix}.$$

To learn more...

The tensor product can be defined not only for column vectors of size 2 but for any two vectors. In fact, it can even be defined for matrices. More on this later in this chapter and in *Chapter 8*.

With this definition, our choice for the column-vector representation of $|00\rangle, |01\rangle, |10\rangle$, and $|11\rangle$ makes perfect sense. For instance, we have that

$$|00\rangle = |0\rangle \otimes |0\rangle = \begin{pmatrix} 1 \\ 0 \end{pmatrix} \otimes \begin{pmatrix} 1 \\ 0 \end{pmatrix} = \begin{pmatrix} 1 \begin{pmatrix} 1 \\ 0 \end{pmatrix} \\ 0 \begin{pmatrix} 1 \\ 0 \end{pmatrix} \end{pmatrix} = \begin{pmatrix} 1 \\ 0 \\ 0 \\ 0 \end{pmatrix}.$$

Exercise 5.2

Prove that

$$|01\rangle = \begin{pmatrix} 0 \\ 1 \\ 0 \\ 0 \end{pmatrix}, \qquad |10\rangle = \begin{pmatrix} 0 \\ 0 \\ 1 \\ 0 \end{pmatrix}, \qquad |11\rangle = \begin{pmatrix} 0 \\ 0 \\ 0 \\ 1 \end{pmatrix}.$$

Of course, the tensor product can be computed for any vector, not just for those of the computational basis. For example, it holds that

$$|0\rangle \otimes |+\rangle = \begin{pmatrix} 1 \\ 0 \end{pmatrix} \otimes \begin{pmatrix} \frac{1}{\sqrt{2}} \\ \frac{1}{\sqrt{2}} \end{pmatrix} = \begin{pmatrix} 1 \begin{pmatrix} \frac{1}{\sqrt{2}} \\ \frac{1}{\sqrt{2}} \end{pmatrix} \\ 0 \begin{pmatrix} \frac{1}{\sqrt{2}} \\ \frac{1}{\sqrt{2}} \end{pmatrix} \end{pmatrix} = \begin{pmatrix} \frac{1}{\sqrt{2}} \\ \frac{1}{\sqrt{2}} \\ 0 \\ 0 \end{pmatrix}.$$

Working with column vectors to compute tensor products is, in general, a tiny bit awkward, but there's an easier way, a better way! As it turns out, the tensor product is linear in both of its components. That is, it holds that

$$(\alpha_1 |\psi_1\rangle + \alpha_2 |\psi_2\rangle) \otimes |\varphi\rangle = \alpha_1 |\psi_1\rangle \otimes |\varphi\rangle + \alpha_2 |\psi_2\rangle \otimes |\varphi\rangle,$$

and that

$$|\varphi\rangle \otimes (\alpha_1 |\psi_1\rangle + \alpha_2 |\psi_2\rangle) = \alpha_1 |\varphi\rangle \otimes |\psi_1\rangle + \alpha_2 |\varphi\rangle \otimes |\psi_2\rangle.$$

Exercise 5.3

Prove the two previous identities in the case where α_1 and α_2 are complex numbers, and $|\psi_1\rangle, |\psi_2\rangle$, and $|\varphi\rangle$ are column vectors of size 2.

With this useful property, we can leave column vectors behind and easily compute tensor products through the convenience of Dirac's notation. Let's repeat our previous computation of $|0\rangle \otimes |+\rangle$ to illustrate this. It holds that

$$|0\rangle \otimes |+\rangle = |0\rangle \otimes \left(\frac{1}{\sqrt{2}} |0\rangle + \frac{1}{\sqrt{2}} |1\rangle \right) = \frac{1}{\sqrt{2}} |0\rangle |0\rangle + \frac{1}{\sqrt{2}} |0\rangle |1\rangle = \frac{1}{\sqrt{2}} (|00\rangle + |01\rangle),$$

which is the same result that we previously obtained when working with column vectors. Neat, isn't it? Incidentally, notice that, as it is customary, we have omitted the $\otimes$ symbol and written, for example, $|0\rangle |0\rangle$ instead of $|0\rangle \otimes |0\rangle$.

Exercise 5.4

Compute, using Dirac's notation, the following tensor products:

(a) $|-\rangle |1\rangle$

(b) $|+\rangle |+\rangle$

(c) $|+\rangle |-\rangle$

This kind of computation will become very handy when we talk about entanglement later in this chapter. But, before that, let's focus on what happens when we measure two-qubit systems.

5.2 Measuring two-qubit systems

As in the case of one-qubit systems, when we are working with two qubits and we want to extract information from them, we need to perform a (quantum) measurement. In this case, though, we have a decision to make: shall we measure both qubits or just one of them? Let's study both scenarios, starting with the one in which the two qubits are measured, since it puts us in a situation analogous to that of *Chapter 2*.

If we have a two-qubit system in a general state

$$\alpha_{00}\left|00\right\rangle + \alpha_{01}\left|01\right\rangle + \alpha_{10}\left|10\right\rangle + \alpha_{11}\left|11\right\rangle$$

and we measure both qubits at the same time, then

- we will obtain 00 with probability $\left|\alpha_{00}\right|^2$, and then the state will collapse to $\left|00\right\rangle$;
- we will obtain 01 with probability $\left|\alpha_{01}\right|^2$, and then the state will collapse to $\left|01\right\rangle$;
- we will obtain 10 with probability $\left|\alpha_{10}\right|^2$, and then the state will collapse to $\left|10\right\rangle$;
- or we will obtain 11 with probability $\left|\alpha_{11}\right|^2$, and then the state will collapse to $\left|11\right\rangle$.

Of course, the normalization condition $\left|\alpha_{00}\right|^2 + \left|\alpha_{01}\right|^2 + \left|\alpha_{10}\right|^2 + \left|\alpha_{11}\right|^2 = 1$ accounts for the fact that the total probability must add up to 1.

Let's see some examples of two-qubit measurements in action. If we consider the state

$$\frac{1}{2}\left(\left|00\right\rangle + \left|01\right\rangle + \left|10\right\rangle + \left|11\right\rangle\right),$$

then all possible results (00, 01, 10 and 11) have exactly the same probability: $1/4$. If we instead consider the state

$$\frac{1}{\sqrt{2}}\left(\left|00\right\rangle + \left|11\right\rangle\right),$$

then only two results are possible: 00 and 11, each with probability $1/2$. Lastly, in the state

$$\frac{3i}{5}\left|00\right\rangle - \frac{4}{5}\left|01\right\rangle,$$

we will obtain 00 with probability 9/25 and 01 with probability 16/25.

Just as in a one-qubit system, after the measurement, the states collapse to the computational basis state that corresponds to the result that we have obtained. Hence, if we measure $\frac{3i}{5}|00\rangle - \frac{4}{5}|01\rangle$ and obtain 00, the state right after the measurement will be $|00\rangle$. Consequently, if we measured the state again, we would obtain 00 with probability 1.

Exercise 5.5

Compute the probability of obtaining 00, 01, 10, or 11 as a result when measuring both qubits of a system in the following states:

(a) $\frac{1}{2}(|00\rangle + |01\rangle + |10\rangle - |11\rangle)$
(b) $\frac{1}{\sqrt{3}}(|00\rangle - |01\rangle + |10\rangle)$
(c) $\frac{1+i}{\sqrt{2}}|11\rangle$
(d) $\frac{1}{2}|01\rangle - \frac{1}{\sqrt{2}}|10\rangle + \frac{i}{2}|11\rangle$
(e) $|0\rangle|+\rangle$

Okay, this was very similar to the one-qubit case. But, what if we decide to measure just one of the qubits? In this case, we will only obtain a value for the qubit that we measure, and the system will collapse partially. Let's see exactly how.

If we have a two-qubit system in the state

$$\alpha_{00}|00\rangle + \alpha_{01}|01\rangle + \alpha_{10}|10\rangle + \alpha_{11}|11\rangle$$

and we measure its first qubit, then

- we will obtain 0 as a result with probability $|\alpha_{00}|^2 + |\alpha_{01}|^2$ and the state will collapse to

$$\frac{\alpha_{00}}{\sqrt{|\alpha_{00}|^2 + |\alpha_{01}|^2}}|00\rangle + \frac{\alpha_{01}}{\sqrt{|\alpha_{00}|^2 + |\alpha_{01}|^2}}|01\rangle;$$

- or we will obtain 1 as a result with probability $|\alpha_{10}|^2 + |\alpha_{11}|^2$ and the state will collapse to

$$\frac{\alpha_{10}}{\sqrt{|\alpha_{10}|^2 + |\alpha_{11}|^2}} |10\rangle + \frac{\alpha_{11}}{\sqrt{|\alpha_{10}|^2 + |\alpha_{11}|^2}} |11\rangle .$$

This may seem complicated, but it is actually quite reasonable. In order for a measurement on the first qubit to give 0 on the first qubit, a measurement of the whole system must yield $|00\rangle$ or $|01\rangle$. We know that the first outcome has probability $|\alpha_{00}|^2$, and the second, $|\alpha_{01}|^2$. If we add both probabilities, we obtain the total probability of $|\alpha_{00}|^2 + |\alpha_{01}|^2$ for a 0 in the first qubit. If this is indeed the outcome of the measurement, the first qubit then collapses to $|0\rangle$, but the second one can still be in a superposition of the $|0\rangle$ and $|1\rangle$ states, with amplitudes α_{00} and α_{01}, respectively. The $\sqrt{|\alpha_{00}|^2 + |\alpha_{01}|^2}$ value in the denominator guarantees that the normalization condition is still satisfied. The case when the result is 1 is completely analogous.

Of course, if we instead measure the second qubit, the situation is symmetrical. In fact, If we have a two-qubit system in the state

$$\alpha_{00} |00\rangle + \alpha_{01} |01\rangle + \alpha_{10} |10\rangle + \alpha_{11} |11\rangle$$

and we measure its second qubit, then

- we will obtain 0 as a result with probability $|\alpha_{00}|^2 + |\alpha_{10}|^2$ and the state will collapse to

$$\frac{\alpha_{00}}{\sqrt{|\alpha_{00}|^2 + |\alpha_{10}|^2}} |00\rangle + \frac{\alpha_{10}}{\sqrt{|\alpha_{00}|^2 + |\alpha_{10}|^2}} |10\rangle ;$$

- or we will obtain 1 as a result with probability $|\alpha_{01}|^2 + |\alpha_{11}|^2$ and the state will collapse to

$$\frac{\alpha_{01}}{\sqrt{|\alpha_{01}|^2 + |\alpha_{11}|^2}} |01\rangle + \frac{\alpha_{11}}{\sqrt{|\alpha_{01}|^2 + |\alpha_{11}|^2}} |11\rangle .$$

Let's see some examples of how to use this in order to determine the probability of different results when we measure a qubit in a two-qubit system.

Imagine that we are given the state

$$\frac{1}{2}\left(\left|00\right\rangle + \left|01\right\rangle + \left|10\right\rangle + \left|11\right\rangle\right),$$

and we measure its first qubit. We will obtain 0 with probability $1/4 + 1/4 = 1/2$ and, in that case, the state will become

$$\frac{1}{\sqrt{2}}\left(\left|00\right\rangle + \left|01\right\rangle\right).$$

Likewise, we will obtain 1 with probability $1/4 + 1/4 = 1/2$ and the state will collapse to

$$\frac{1}{\sqrt{2}}\left(\left|10\right\rangle + \left|11\right\rangle\right).$$

If we instead have the state

$$\frac{1}{\sqrt{2}}\left(\left|00\right\rangle + \left|11\right\rangle\right),$$

and we we measure its second qubit, we will obtain 0 with probability $1/2$ and the state will collapse to $\left|00\right\rangle$; and we will obtain 1 with probability $1/2$ and the state will then become $\left|11\right\rangle$.

Exercise 5.6

Compute the probability of obtaining 0 or 1 as a result, and the collapsed states after each outcome, when measuring the following:

(a) The first qubit of $\frac{1}{2}\left(\left|00\right\rangle + \left|01\right\rangle + \left|10\right\rangle - \left|11\right\rangle\right)$

(b) The second qubit of $\frac{1}{\sqrt{3}}\left(\left|00\right\rangle - \left|01\right\rangle + \left|10\right\rangle\right)$

(c) The first qubit of $\frac{1+i}{\sqrt{2}}\left|11\right\rangle$

(d) The second qubit of $\frac{1}{2}\left|01\right\rangle - \frac{1}{\sqrt{2}}\left|10\right\rangle + \frac{i}{2}\left|11\right\rangle$

(e) The first and second qubit of $\left|0\right\rangle\left|+\right\rangle$

You may now be wondering what would happen if you had a two-qubit system and you measured one of the qubits and then the other. Would that be any different from measuring both qubits at the same time? That would be quite problematic, wouldn't it? Fortunately, the final results are exactly the same in either situation. Let's prove it.

Imagine, again, that we have a two-qubit system in the general state

$$\alpha_{00}\left|00\right\rangle + \alpha_{01}\left|01\right\rangle + \alpha_{10}\left|10\right\rangle + \alpha_{11}\left|11\right\rangle.$$

Suppose that we measure its first qubit and that we obtain 1 as a result. We know that this happens with probability $\left|\alpha_{10}\right|^2 + \left|\alpha_{11}\right|^2$ and that the new state will be

$$\frac{\alpha_{10}}{\sqrt{\left|\alpha_{10}\right|^2 + \left|\alpha_{11}\right|^2}}\left|10\right\rangle + \frac{\alpha_{11}}{\sqrt{\left|\alpha_{10}\right|^2 + \left|\alpha_{11}\right|^2}}\left|11\right\rangle.$$

Assume that we now measure the second qubit and obtain 0 as a result. This will happen with probability $\frac{\left|\alpha_{10}\right|^2}{\left|\alpha_{10}\right|^2+\left|\alpha_{11}\right|^2}$. Taking into account that the probability of measuring 1 in the first qubit was $\left|\alpha_{10}\right|^2 + \left|\alpha_{11}\right|^2$, this means that the probability of obtaining 1 in the first qubit and 0 in the second, when measuring them in that order, is

$$\left(\left|\alpha_{10}\right|^2 + \left|\alpha_{11}\right|^2\right) \cdot \frac{\left|\alpha_{10}\right|^2}{\left|\alpha_{10}\right|^2 + \left|\alpha_{11}\right|^2} = \left|\alpha_{10}\right|^2,$$

which is exactly the probability of obtaining 10 as a result when we measure both qubits at the same time. Moreover, the collapsed state will be $\frac{\alpha_{10}}{\left|\alpha_{10}\right|}\left|10\right\rangle$, which is equivalent to $\left|10\right\rangle$ up to an unimportant global phase (go back to *Section 2.3.3* if this global phase makes you uncomfortable!).

As you can easily check yourself, in all other cases, there is no difference between measuring both qubits sequentially or simultaneously, so the order in which the measurements are performed—and whether they are performed at the same time or one right after the other—is completely irrelevant.

Okay, that was more than enough about measuring two-qubit systems. Let's now move on to studying two-qubit gates.

5.3 Two-qubit gates

As in the case on one-qubit systems, the allowed transformations on two-qubit states are given by unitary matrices. In this case, however, the matrices will be of size 4, because we will be multiplying them by column vectors of dimension 4. We will begin by studying the simplest such transformations, which will be given by two one-qubit gates acting independently on each of the qubits of the system. Then, we will learn about the all-important CNOT gate. This will lead us to discuss entanglement and, finally, to prove the no-cloning theorem. Ready to start?

5.3.1 Tensor products of gates

As we have just mentioned, to act on a two-qubit state, we need a unitary matrix of size 4. The easiest way of constructing such a matrix is by combining the action of two one-qubit gates, each acting on one of the qubits of the system. For instance, *Figure 5.1* shows an H gate applied to the top qubit and an X gate acting on the bottom one.

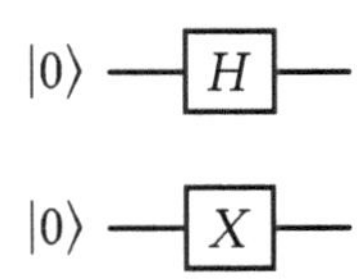

Figure 5.1: A two-qubit circuit

Incidentally, this is our first example of a two-qubit circuit. Notice how we now have two lines representing each of the two qubits, each of them starting on state $|0\rangle$.

The state of the qubits after applying the two individual one-qubit gates will be

$$H|0\rangle \otimes X|0\rangle = |+\rangle \otimes |1\rangle = |+\rangle |1\rangle .$$

Important note

When representing the states given by a circuit with two or more qubits, we have adopted the convention of identifying the topmost qubit of the circuit with the leftmost qubit of the expression. However, some other authors (and even some software packages!) take the opposite convention and read the qubits from bottom to top. Always be sure to know what convention you are working with!

We can consider the combined action of the two one-qubit gates in the circuit of *Figure 5.1* as a single two-qubit gate acting on two-qubit states. We denote it by $H \otimes X$ and call it the **tensor product** of the one-qubit gates H and X. In order to obtain its explicit matrix, we need to use an extension of the tensor product of column vectors that we introduced in *Section 5.1*. Namely, if we have two square matrices

$$A = \begin{pmatrix} a_{11} & a_{12} \\ a_{21} & a_{22} \end{pmatrix}, \qquad B = \begin{pmatrix} b_{11} & b_{12} \\ b_{21} & b_{22} \end{pmatrix}$$

of size 2, their tensor product is given by

$$A \otimes B := \begin{pmatrix} a_{11}\begin{pmatrix} b_{11} & b_{12} \\ b_{21} & b_{22} \end{pmatrix} & a_{12}\begin{pmatrix} b_{11} & b_{12} \\ b_{21} & b_{22} \end{pmatrix} \\ a_{21}\begin{pmatrix} b_{11} & b_{12} \\ b_{21} & b_{22} \end{pmatrix} & a_{22}\begin{pmatrix} b_{11} & b_{12} \\ b_{21} & b_{22} \end{pmatrix} \end{pmatrix} = \begin{pmatrix} a_{11}b_{11} & a_{11}b_{12} & a_{12}b_{11} & a_{12}b_{12} \\ a_{11}b_{21} & a_{11}b_{22} & a_{12}b_{21} & a_{12}b_{22} \\ a_{21}b_{11} & a_{21}b_{12} & a_{22}b_{11} & a_{22}b_{12} \\ a_{21}b_{21} & a_{21}b_{22} & a_{22}b_{21} & a_{22}b_{22} \end{pmatrix}.$$

It turns out that if U_1 and U_2 are unitary gates, then $U_1 \otimes U_2$ is also unitary and its action on tensor product states is given by $(U_1 \otimes U_2)(|\psi_1\rangle \otimes |\psi_2\rangle) = (U_1 |\psi_1\rangle) \otimes (U_2 |\psi_2\rangle)$, just as we needed.

Exercise 5.7

Let U_1, U_2, U_3 and U_4 be square matrices of size 2, and $|\psi_1\rangle$ and $|\psi_2\rangle$ be two one-qubit states. Prove that

$$(U_1 \otimes U_2)(|\psi_1\rangle \otimes |\psi_2\rangle) = (U_1 |\psi_1\rangle) \otimes (U_2 |\psi_2\rangle).$$

Use that to prove that the matrix product $(U_1 \otimes U_2)(U_3 \otimes U_4)$ is equal to $(U_1U_3) \otimes (U_2U_4)$. Deduce that, if U_1 and U_2 are unitary, then so is $U_1 \otimes U_2$ and, in fact, $(U_1 \otimes U_2)^\dagger = U_1^\dagger \otimes U_2^\dagger$.

For instance, the matrix for $|H\rangle \otimes |X\rangle$ is

$$\begin{pmatrix} \frac{1}{\sqrt{2}} & \frac{1}{\sqrt{2}} \\ \frac{1}{\sqrt{2}} & -\frac{1}{\sqrt{2}} \end{pmatrix} \otimes \begin{pmatrix} 0 & 1 \\ 1 & 0 \end{pmatrix} = \begin{pmatrix} 0 & \frac{1}{\sqrt{2}} & 0 & \frac{1}{\sqrt{2}} \\ \frac{1}{\sqrt{2}} & 0 & \frac{1}{\sqrt{2}} & 0 \\ 0 & \frac{1}{\sqrt{2}} & 0 & -\frac{1}{\sqrt{2}} \\ \frac{1}{\sqrt{2}} & 0 & -\frac{1}{\sqrt{2}} & 0 \end{pmatrix}.$$

Notice that, if we have a two-qubit system and we apply a gate to only one of its qubits, we can still consider this gate to be a two-qubit gate: the tensor product of the identity I (the gate that does not change the state of a qubit) with the actual one-qubit gate we have applied. For instance, if we apply Z to the second qubit of a two-qubit system, we are, in fact, applying $I \otimes Z$ to the whole system.

Exercise 5.8

Explicitly compute the matrix expressions of the following tensor products of gates:

(a) $X \otimes H$

(b) $X \otimes X$

(c) $I \otimes Z$

Not all two-qubit gates, though, can be obtained as the tensor product of one-qubit gates; that would have been too boring, wouldn't it? In the next section, we focus on one of the most important and interesting two-qubit gates that cannot be represented as the combined action of one-qubit gates: the CNOT gate.

5.3.2 The CNOT gate

The **CNOT**, **controlled-NOT**, or **controlled-X** gate is a two-qubit gate that cannot be obtained as the tensor product of two one-qubit gates. Its name is derived from the way it acts on the computational basis, which is the following:

$$\begin{aligned} \mathrm{CNOT}\left|00\right\rangle &= \left|00\right\rangle, & \mathrm{CNOT}\left|01\right\rangle &= \left|01\right\rangle, \\ \mathrm{CNOT}\left|10\right\rangle &= \left|11\right\rangle, & \mathrm{CNOT}\left|11\right\rangle &= \left|10\right\rangle. \end{aligned}$$

As you can observe, the value of the second qubit is reversed (or negated) if and only if the value of the first qubit is 1. In this way, we can say that the first qubit controls whether an X (also called NOT) gate is applied to the second qubit or not, hence the name of the operation. Usually, the first qubit is called the **control qubit** and the second one is called the **target qubit**. The action on non-basis states is obtained by linearity. Indeed, we have that

$$\begin{aligned} &\mathrm{CNOT}(\alpha_{00}\left|00\right\rangle + \alpha_{01}\left|01\right\rangle + \alpha_{10}\left|10\right\rangle + \alpha_{11}\left|11\right\rangle) \\ &\qquad = \alpha_{00}\mathrm{CNOT}\left|00\right\rangle + \alpha_{01}\mathrm{CNOT}\left|01\right\rangle + \alpha_{10}\mathrm{CNOT}\left|10\right\rangle + \alpha_{11}\mathrm{CNOT}\left|11\right\rangle \\ &\qquad = \alpha_{00}\left|00\right\rangle + \alpha_{01}\left|01\right\rangle + \alpha_{11}\left|10\right\rangle + \alpha_{10}\left|11\right\rangle. \end{aligned}$$

The CNOT gate is usually represented in a circuit as in the following figure, where the control qubit is the top one and the target qubit is the bottom one:

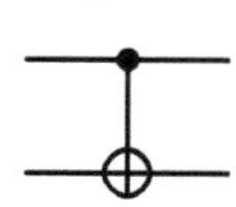

The solid black dot in the figure represents a control, and we will see in later chapters that it can be used to indicate that a certain qubit controls a specific operation, not necessarily always a NOT gate. Since the NOT gate is also called the X gate, an alternative way of representing the CNOT gate is as follows:

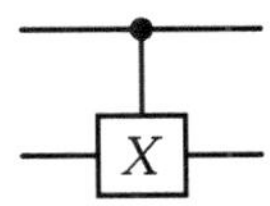

The matrix for the CNOT gate with the control on the first (top) qubit and the target on the second (bottom) is

$$\begin{pmatrix} 1 & 0 & 0 & 0 \\ 0 & 1 & 0 & 0 \\ 0 & 0 & 0 & 1 \\ 0 & 0 & 1 & 0 \end{pmatrix}.$$

Exercise 5.9

Check that the matrix for the the CNOT gate with control on the first (top) qubit and target on the second (bottom) is indeed

$$\begin{pmatrix} 1 & 0 & 0 & 0 \\ 0 & 1 & 0 & 0 \\ 0 & 0 & 0 & 1 \\ 0 & 0 & 1 & 0 \end{pmatrix}.$$

Exercise 5.10

Prove that the CNOT gate cannot be constructed as the tensor product of two one-qubit quantum gates.

Of course, you can also consider a CNOT gate in which the control qubit is the second (bottom) one and the target qubit is the first (top) one. In a quantum circuit, it would be represented as follows:

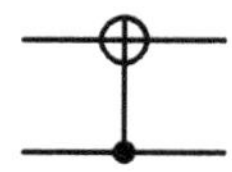

The matrix of the CNOT whose control is the second (bottom) qubit is

$$\begin{pmatrix} 1 & 0 & 0 & 0 \\ 0 & 0 & 0 & 1 \\ 0 & 0 & 1 & 0 \\ 0 & 1 & 0 & 0 \end{pmatrix}.$$

Exercise 5.11

Check that the matrix for the the CNOT gate with control on the second qubit and target on the first is indeed

$$\begin{pmatrix} 1 & 0 & 0 & 0 \\ 0 & 0 & 0 & 1 \\ 0 & 0 & 1 & 0 \\ 0 & 1 & 0 & 0 \end{pmatrix}.$$

Now that we have defined the CNOT gate, we can begin to unravel its mysteries. One of its most important properties is that it allows us to create entanglement, a phenomenon so relevant that it merits its own discussion. Let's get to it!

5.3.3 Entanglement

As we have seen earlier in this chapter, the tensor product of two one-qubit states always yields a two-qubit state. Such states, which can be decomposed as the tensor product of other states, are said to be **product states**. For example, $|00\rangle$ is a product state because its just a shortened way of writing $|0\rangle \otimes |0\rangle$. Likewise, and even if it is not so apparent, the state

$$\frac{1}{2}(|00\rangle + |01\rangle + |10\rangle + |11\rangle)$$

is also a product state, because it is equal to $|+\rangle|+\rangle$. Something similar happens with

$$\frac{1}{\sqrt{2}}(|00\rangle + |01\rangle),$$

which is the same state as $|0\rangle|+\rangle$.

Nevertheless, not all two-qubit states are product states. For instance, the seemingly innocent state

$$\frac{1}{\sqrt{2}}(|00\rangle + |11\rangle)$$

cannot actually be written as the product of two one-qubit states, even if it looks quite similar to $\frac{1}{\sqrt{2}}(|00\rangle + |01\rangle) = |0\rangle|+\rangle$. Let's see why.

Assume that there exist two states $|\psi_1\rangle = a|0\rangle + b|1\rangle$ and $|\psi_2\rangle = c|0\rangle + d|1\rangle$ such that $|\psi_1\rangle \otimes |\psi_2\rangle = \frac{1}{\sqrt{2}}(|00\rangle + |11\rangle)$. Then, we would have

$$\frac{1}{\sqrt{2}}(|00\rangle + |11\rangle) = (a|0\rangle + b|1\rangle) \otimes (c|0\rangle + d|1\rangle) = ac|00\rangle + ad|01\rangle + bc|10\rangle + bd|11\rangle.$$

Since $\frac{1}{\sqrt{2}}(|00\rangle + |11\rangle)$ has no $|01\rangle$ component, this implies that $ad = 0$. But we know that $ac = bd = \frac{1}{\sqrt{2}}$, so neither a nor d can be 0, reaching a contradiction.

Any state that can't be written as the tensor product of two smaller states is said to be **entangled**. Hence, we have just proved that $\frac{1}{\sqrt{2}}(|00\rangle + |11\rangle)$ is an entangled state.

When measured, entangled states can show correlations that go beyond what is explainable with just classical physics. This can be extraordinarily valuable for information processing, as we will see throughout the rest of the this book. As an example, if we measure the first qubit of a system in the state $\frac{1}{\sqrt{2}}(|00\rangle + |11\rangle)$, we will obtain 0 with probability $1/2$ and 1 also with probability $1/2$. If we obtain 0, the state will collapse to $|00\rangle$; thus, if we then measure the second qubit, we will unequivocally obtain 0. On the contrary, if the first qubit is measured to be 1, any measurement of the second qubit thereafter will always yield a

1 (unless, of course, we manipulate the state!). And the same happens if we measure the second qubit first: the result will be 0 or 1 with equal probability, and the outcome will fully determine the state of the first qubit right after the measurement.

According to the postulates of quantum mechanics, this collapse will occur instantaneously even if the entangled qubits are very far apart. This sounded fishy to Albert Einstein, who called it "spooky action at a distance". However, the effects of entanglement have been verified in the lab on countless occasions, and entanglement will be one of the key ingredients in many of our quantum protocols and algorithms. For example, in *Chapter 6*, we will study something called **quantum teleportation**, which relies on the creation of entangled states such as $\frac{1}{\sqrt{2}}(|00\rangle + |11\rangle)$.

Exercise 5.12

For each of the following states, determine whether they are entangled or not:

(a) $|10\rangle$

(b) $\frac{1}{\sqrt{2}}(|00\rangle - |11\rangle)$

(c) $\frac{1}{2}(|00\rangle - |01\rangle + |10\rangle - |11\rangle)$

(d) $\frac{1}{2}(|00\rangle + |01\rangle + |10\rangle - |11\rangle)$

As we mentioned previously, we can use the CNOT gate to prepare entangled states. For instance, the circuit in *Figure 5.2* prepares the state $\frac{1}{\sqrt{2}}(|00\rangle + |11\rangle)$. Indeed, we start with $|00\rangle$. Then, we apply H to the top qubit to obtain $|+\rangle|0\rangle = \frac{1}{\sqrt{2}}(|00\rangle + |10\rangle)$. If we now apply the CNOT gate to this state, by linearity, we get

$$\begin{aligned}\text{CNOT}\left(\frac{1}{\sqrt{2}}(|00\rangle + |10\rangle\right) &= \frac{1}{\sqrt{2}}(\text{CNOT}\,|00\rangle + \text{CNOT}\,|10\rangle)\\ &= \frac{1}{\sqrt{2}}(|00\rangle + |11\rangle),\end{aligned}$$

as desired.

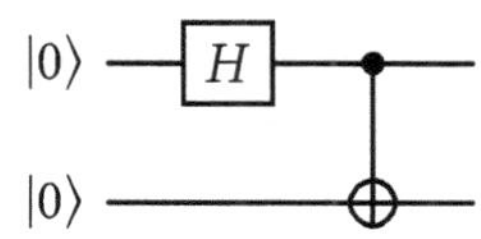

Figure 5.2: Preparing an entangled state

The $\frac{1}{\sqrt{2}}(|00\rangle + |11\rangle)$ state is so important that it has its own a name and notation. It is represented by $|\Phi^+\rangle$ and it is one of the **Bell states**, the other ones being

$$|\Phi^-\rangle := \frac{1}{\sqrt{2}}(|00\rangle - |11\rangle), \qquad |\Psi^+\rangle := \frac{1}{\sqrt{2}}(|10\rangle + |01\rangle), \qquad |\Psi^-\rangle := \frac{1}{\sqrt{2}}(|10\rangle - |01\rangle).$$

We can easily create the rest of the Bell states from $\frac{1}{\sqrt{2}}(|00\rangle + |11\rangle)$. In fact, as you can easily check, it holds that:

- $(Z \otimes I)(\frac{1}{\sqrt{2}}(|00\rangle + |11\rangle)) = (I \otimes Z)(\frac{1}{\sqrt{2}}(|00\rangle + |11\rangle)) = \frac{1}{\sqrt{2}}(|00\rangle - |11\rangle)$
- $(X \otimes I)(\frac{1}{\sqrt{2}}(|00\rangle + |11\rangle)) = (I \otimes X)(\frac{1}{\sqrt{2}}(|00\rangle + |11\rangle)) = \frac{1}{\sqrt{2}}(|10\rangle + |01\rangle)$
- $(X \otimes Z)(\frac{1}{\sqrt{2}}(|00\rangle + |11\rangle) = (Z \otimes X)(\frac{1}{\sqrt{2}}(|00\rangle + |11\rangle) = \frac{1}{\sqrt{2}}(|10\rangle - |01\rangle))$

Exercise 5.13

Show that the Bell states are entangled.

Now that we are familiar with entanglement, this is a great moment to prove a result that we have mentioned quite a number of times already: the no-cloning theorem.

5.3.4 The no-cloning theorem

Surprising as it may sound, it is not possible in general to create an independent copy of an unknown quantum state, even if we are allowed to manipulate the state however we please. But, what do we really mean by "creating an independent copy of" or "cloning" a quantum state? Ideally, given a qubit in a unknown state $|\psi\rangle$ and another, auxiliary qubit in state $|0\rangle$, we would like to end having the initial qubit still in state $|\psi\rangle$ and the auxiliary one also in state $|\psi\rangle$. That is, we would like to go from $|\psi\rangle \otimes |0\rangle$ to $|\psi\rangle \otimes |\psi\rangle$.

We could accomplish this if we had a two-qubit quantum gate U such that, for any quantum state $|\psi\rangle$, $U(|\psi\rangle \otimes |0\rangle) = |\psi\rangle \otimes |\psi\rangle$. But this is, in fact, impossible! Let's see why.

Imagine that we have such a cloning quantum gate U. Then, in particular, it should work for states $|0\rangle$, $|1\rangle$, and $|+\rangle$. That is, we should have

- $U(|00\rangle) = |00\rangle$,
- $U(|10\rangle) = |11\rangle$,
- $U(|+\rangle|0\rangle) = |+\rangle|+\rangle$.

But notice that $|+\rangle|0\rangle = \frac{1}{\sqrt{2}}(|00\rangle + |10\rangle)$. Then, by the linearity of U, it should hold that

$$U(|+\rangle|0\rangle) = U(\frac{1}{\sqrt{2}}(|00\rangle + |10\rangle)) = \frac{1}{\sqrt{2}}(U|00\rangle + U|10\rangle) = \frac{1}{\sqrt{2}}(|00\rangle + |11\rangle),$$

which is not equal to $|+\rangle|+\rangle$, as we needed, because $\frac{1}{\sqrt{2}}(|00\rangle + |11\rangle)$ is entangled and $|+\rangle|+\rangle$ is a product state. Hence, we have a contradiction that comes solely from assuming the existence of the gate U.

This means that we cannot clone general unknown states. But this does not imply that we cannot clone *particular* states. In some cases, when given a qubit, we can operate on a another qubit so that it ends in the same state as the first one. For instance, we know that $\text{CNOT}|00\rangle = |00\rangle$ and $\text{CNOT}|10\rangle = |11\rangle$, so the CNOT gate *does* copy the states $|0\rangle$ and $|1\rangle$ or, as it's usually said, it *copies classical information.* Let's see another example.

The circuit in *Figure 5.3* implements a gate that copies the states $|+\rangle$ and $|-\rangle$. Indeed, if we start from $|+\rangle|0\rangle$ and we apply the H gate to the first qubit, we obtain $|0\rangle|0\rangle = |00\rangle$. Then, with the CNOT gate, we obtain $|00\rangle$. Now, when we apply an H gate to each of the qubits, we obtain $|+\rangle|+\rangle$. If we start with $|-\rangle|0\rangle$, we obtain the sequence of states $|10\rangle$, $|11\rangle$ and, finally, after the H gates, $|-\rangle|-\rangle$ as required.

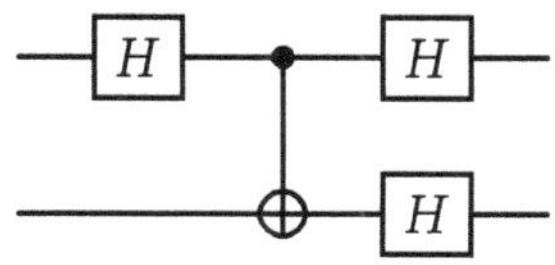

Figure 5.3: Copying the $|+\rangle$ and $|-\rangle$ states

To learn more...

Copying gates can be implemented not only for $\{|0\rangle, |1\rangle\}$ and $\{|+\rangle, |-\rangle\}$, but for any pair of one-qubit orthogonal states. If you want to learn what orthogonal states are, please refer to *Appendix A* and *Appendix B*.

This wraps up this chapter on two-qubit systems. In the next one, we will apply all we have learned here to implement surprising, fun and useful quantum protocols and algorithms. We are going to get really entangled!

Summary

In this chapter, we have studied two-qubit systems. You now know how to describe the state of a pair of qubits, how to extract information from them with simultaneous and sequential measurements, and how to transform their states with two-qubit quantum gates.

In particular, you now have a deep understanding of the CNOT gate, one of the most important operations in all of quantum computing. You know that its action cannot be obtained as the combination of one-qubit gates and that it can be used to create entanglement. In that regard, you also know how to distinguish entangled from product states. Finally, you understand the reasons why it is impossible to clone general unknown quantum states, but you know that certain pairs of the states can always be copied with appropriate quantum gates.

All this will be extremely helpful to us in the rest of the book, starting with next chapter, in which we will put our new knowledge into practice by implementing some key quantum protocols and algorithms.

6

Applications and Protocols with Two Qubits

Everything we call real is made of things that cannot be regarded as real.

— Niels Bohr

In *Chapter 3*, we covered some interesting applications of one-qubit systems, which gave us a first glimpse into the kind of things that quantum computers can do. And, certainly, those applications proved one-qubit systems to be interesting and useful. Nevertheless, if you believe that what we could do with them was impressive... you haven't seen anything yet. This ride is about to get wild.

Having a qubit in a state of superposition is exciting and mysterious enough, but adding more qubits to the recipe and mixing it all with entanglement is a game changer. In this chapter, you will see how quantum algorithms with multiple qubits can be used to efficiently condense classical information into qubits, and you will also get to explore a simple game that shows the power of entanglement. Were this not enough, we will also introduce a landmark quantum algorithm: Deutsch's algorithm. If you have ever heard the infamous saying that "quantum computers are more efficient because they evaluate all possible solutions at once", well, with Deutsch's algorithm, you will begin to understand where that is coming from.

The contents of this chapter are the following:

- Superdense coding
- The CHSH game
- Deutsch's algorithm

Before we begin, please make sure that you have a solid understanding of all the material covered in *Chapter 5*. Trust us, there's no way of getting around Bell states and CNOT gates! With that being said, let's get started in here.

6.1 Superdense coding

Let's put ourselves in the following scenario: two people—who, in accordance with our old customs, we shall name Alice and Bob—want to communicate, and it would suffice for them to do so by sending two bits of information. We know that this is not particularly realistic, as acts of communication often require more information, but we're doing a thought experiment here. To make things more specific, assume it's Alice that wants to send two bits of information to Bob, so she might want to send him the message 00 (two bits set to zero), the message 01, 10 or 11—one of those four.

If Alice and Bob chose to keep things classical, how many bits would Alice have to send to Bob? Well, obviously and unsurprisingly, two. But now imagine that they instead choose to go quantum. If Alice wanted to communicate two bits of information to Bob… how many *qubits* would she have to send him? This is the question that will keep us busy throughout this whole section.

Obviously, Alice could send two bits of information using two qubits. If she wanted to send 00, 01, 10 or 11, she could just send two qubits in the states $|00\rangle$, $|01\rangle$, $|10\rangle$ or $|11\rangle$ respectively. All that Bob would then have to do is performing a measurement in the computational basis et voilà. Problem solved! But why on Earth would we want to use qubits to do something for which bits already work beautifully? That wouldn't make a lot of sense, but, at least, with this we know that at most we need two qubits to get the job done, and obviously we need more than zero. Therefore, if we want to do something interesting, this narrows our options down to sending just one qubit in order to communicate those two bits that Alice so eagerly wants Bob to receive. Luckily for us, there is a way to achieve this, and it is through **superdense coding**. All that it will require is for Alice and Bob to initially share a pair of qubits in a Bell state.

Important note

Superdense coding enables a party (Alice) to send two bits of information to another party (Bob) by just sending them one qubit. The protocol assumes that they initially have one qubit each, and that the pair of qubits is in a Bell state.

So that's what superdense coding is. Let's now see how this protocol actually works.

For superdense coding to enable Alice to send two bits to Bob, we first need Alice and Bob to share a pair of qubits (each with a qubit from the pair) in a Bell state. Thus, we would need an initial two-qubit system in the state

$$\frac{1}{\sqrt{2}}(|00\rangle + |11\rangle),$$

although any of the other Bell states would work equally well (more on that later in this section). Here and in what follows, when writing down quantum states and circuits, we will always assume that the first qubit is Alice's and the second is Bob's. Thus, in this initial setup, we are assuming that Alice has one qubit, that Bob has another qubit, and that the state of these two qubits (considered as a single two-qubit system) is the one above. To further clarify this, if Alice's qubit were in state $|1\rangle$ and Bob's qubit were in state $|0\rangle$, we would say that the total state of the system is $|10\rangle$. Naturally, the state of our system can't be written that simply as it is not a product state, but an entangled state.

Exercise 6.1

You want to get the superdense coding protocol started, but you only have two qubits which are both in state $|0\rangle$ (so the total state is $|00\rangle$). How could you generate the pair of qubits that Alice and Bob need? Which quantum gates could you apply and how? Try to come up with two different quantum circuits that would get the job done.

This pair of entangled qubits in a Bell state that Alice and Bob are assumed to share could have been created by a third-party who then handed them over to them. Or maybe Alice and Bob met in advance and prepared and shared the Bell state. In any case, this is our initial setup and this is all we need to get started. Now how could Alice manage to send to two bits of information to Bob by sending just one qubit? Well, it depends on the message she wants to send. Each particular situation will require a slightly different manipulation of her part of the Bell state and, afterwards, she will be ready to send it to Bob. This is what she would need to do, case by case:

- If Alice wants to send the bits 00, she does not have to do anything at all. She just leaves her qubit as it is.
- If Alice wants to send the bits 01, she has to apply the gate Z to her qubit. The state of the whole system will then be

$$\frac{1}{\sqrt{2}}(|00\rangle - |11\rangle).$$

- If Alice wants to send the bits 10, she has to apply the gate X to her qubit. The state of the whole system will then be

$$\frac{1}{\sqrt{2}}(|10\rangle + |01\rangle).$$

- If Alice wants to send the bits 11, she has to apply the gate Z and then the gate X (or, equivalently the gate[1] XZ) to her qubit. The state of the whole system will then be

$$\frac{1}{\sqrt{2}}(|10\rangle - |01\rangle).$$

Once these transformations are performed, Alice must send her qubit to Bob, so that now Bob has the whole system in his hands.

Exercise 6.2

Verify that the states of the system after Alice applies the gates on her qubit are the ones we have claimed.

At this stage, Bob can easily retrieve the two bits that Alice meant to send him. All he has to do is apply a CNOT gate controlled by his original qubit and with a target on Alice's original qubit, and then apply a Hadamard gate on his original qubit; this is all depicted in *Figure 6.1*.

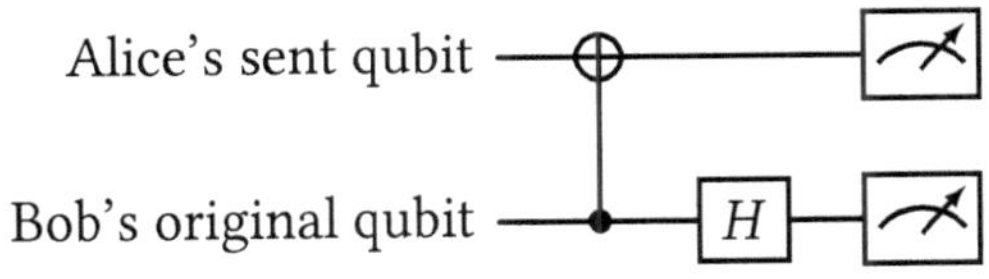

Figure 6.1: This is the circuit that Bob must run upon receiving Alice's qubit and being in possession of the whole system

Actually, you may have noticed that these gates are, in a way, "undoing" the generation of the original Bell state. At this point, all that is left to do is measure the two qubits that Bob has: the results will be the qubits that Alice sent him. The measurement on the first qubit will yield the first bit, and the measurement on the second qubit will return the second bit.

To see why this indeed works, we can analyse each scenario individually.

[1]Notice that when the gate XZ acts on a state, the gate Z acts first and then acts the gate X: $XZ|\psi\rangle = X(Z|\psi\rangle)$.

- If Alice sent the bits 00, the state right before the measurement will be $|00\rangle$, because we are just undoing the creation of the Bell state.
- If Alice sent the bits 01, the action of the CNOT gate transforms the state as

$$\frac{1}{\sqrt{2}}(|00\rangle - |11\rangle) \longrightarrow \frac{1}{\sqrt{2}}(|00\rangle - |01\rangle) = |0\rangle \otimes |-\rangle,$$

  so the application of the H gate will leave the state as $|01\rangle$ and the measurement will return 01.

- If Alice sent the bits 10, the action of the CNOT gate transforms the state as

$$\frac{1}{\sqrt{2}}(|10\rangle + |01\rangle) \longrightarrow \frac{1}{\sqrt{2}}(|10\rangle + |11\rangle) = |1\rangle \otimes |+\rangle,$$

  so the application of the H gate will leave the state as $|10\rangle$ and any measurement will return 10.

- Lastly and analogously, if Alice sent the bits 11, the action of the CNOT and Hadamard gates bring the state of the system to $|11\rangle$, and any measurement will return 11, as desired.

Exercise 6.3

Verify that, indeed, if Alice sent the bits 00 or 11, the action of the CNOT and Hadamard gates, as shown in *Figure 6.1*, leaves the state of the system as $|00\rangle$ or $|11\rangle$, respectively.

And that's pretty much it for the superdense coding protocol. Now when you look at this for the first time it may seem slightly overwhelming, with lots of steps and seemingly arbitrary choices involved. Thus, we shall now invest some time in trying to digest this. Firstly, let us condense the behaviour of the protocol in a compact statement.

Important note

If Alice wants to send Bob two bits a, b through the superdense protocol, she must apply the gate X^aZ^b to her qubit and then send it to Bob. If Bob then executes the circuit depicted in *Figure 6.1*, he will retrieve the two bits (a, b) in the measurement operations. The measurement of the first qubit will yield a while that of the second qubit will yield b.

In summary and looking at the big picture, superdense coding replaces the need for sending two bits with the requirement of getting the two communicating parties to have each a qubit from an entangled pair. From this point, the protocol exploits the fact that, in an entangled pair of qubits, it is enough to act on one of the qubits in order to modify the whole state; and, thus, the sender is able to encode their two-bit message into the whole system by just manipulating their own qubit and then sending it to the receiver.

From a practical point of view, we should highlight that the protocol works regardless of the distance at which the two communicating parties are. In our example, it wouldn't matter if Alice were on the Moon and Bob were on Mars: entanglement works at any distance! This might not be particularly surprising at this point, but this idea will gain relevance in the following section.

Now for those of you who are more comfortable with abstract linear algebra, we would like to share with you a small remark that might prove to be insightful.

To learn more...

Let P be the two-qubit quantum gate that results from combining the CNOT and Hadamard operations in *Figure 6.1*. If I is the identity gate in one qubit, then, clearly,

$P(I \otimes I)P^{\dagger} = I \otimes I$. Moreover, you can easily check for yourself how

$$P(X \otimes I)P^{\dagger} = X \otimes I, \qquad P(Z \otimes I)P^{\dagger} = \begin{pmatrix} 0 & 1 & 0 & 0 \\ 1 & 0 & 0 & 0 \\ 0 & 0 & 0 & -1 \\ 0 & 0 & 1 & 0 \end{pmatrix},$$

where the last operator only differs from $I \otimes X$ by a -1 factor in the image of $|11\rangle$. These are all the ingredients that you would need to trivially deduce why superdense coding works. We leave the rest of the thinking to you!

Earlier in this section, we mentioned that, while we were going to use $\frac{1}{\sqrt{2}}(|00\rangle + |11\rangle)$ as the initial state for our pair of entangled qubits, any other Bell state would do the trick. As a nice way of bringing our discussion of superdense coding to an end, this matter can be worth exploring. But it would be unjust for us to deprive you of the joy of doing it yourself, so we are leaving it as an exercise.

Exercise 6.4

In the superdense coding protocol, assume that—perhaps because of reasons of technical convenience—the initial state of the pair of entangled qubits is

$$\frac{1}{\sqrt{2}}(|10\rangle + |01\rangle\,.$$

If Alice wants to send two bits (a, b), what gates would she have to apply to her qubit (in terms of a and b) for the rest of the protocol to work as intended from Bob's point of view?

And that's it for superdense coding. You may wish to know that this protocol was introduced by Charles H. Bennett and Stephen Wiesner, whom we already met in *Chapter 3* and you can read more about it in their work "Communication via one-and two-particle operators on Einstein-Podolsky-Rosen states" [50].

Now it's time to further dive into the wonders of quantum entanglement. Let's discuss an interesting game!

6.2 The CHSH game

In the previous section, we discussed a simple two-qubit protocol that, leveraging on entanglement, enabled the communication of two classical bits of information through the transmission of a single qubit. When you think about it, entanglement is a rather interesting construction: it implies that two systems, regardless of the physical separation between them, can be perfectly and instantaneously interconnected. In this section, we will dive even deeper into these implications with a simple game.

Imagine that our two favourite characters, Alice and Bob, are very far away from each other. To make things more specific, we will assume that Alice is having lunch in Villalpando while Bob is visiting a park in Poolbeg, and hence they are 1276.8 km away from each other. For some odd reason, they have gotten themselves into an equally odd game. Two referees, Antonio (who is right next to Alice) and Brigid (who is with Bob) will, at the same instant, pick a random bit uniformly: Antonio will pick a bit x and Brigid will pick a bit y. Right after that, Alice and Bob have to pick one bit each: Alice must pick a bit a and Bob must pick a bit b, and they must do so in such a way that

$$a \oplus b = x \cdot y,$$

where $\oplus$ denotes the XOR operation (addition modulo 2). This means that, if $x = 1$ and $y = 1$, then either Alice or Bob must pick their bit to be 1, while the other must set their bit to 0. On the contrary, if $x = 0$ or $y = 0$, either both Alice and Bob must pick 0 or both must pick 1. If they manage to pick a pair of bits that satisfies these constraints, they win! And the good news is that, before the game starts, they can talk over the phone to agree on a strategy to follow and, what is more, they have unlimited access to all technology, quantum and classical. Seems easy, doesn't it? Sadly, there are some bad news as well.

To spice things up, the organizers of the game have decided that, counting from the moment at which the referees choose their bits, Alice and Bob only have 1 microsecond (1×10^{-6} s) to make their choices. Of course they can use a computer to apply whichever strategy they have agreed on in less than a microsecond but there's a catch, which is that this constraint prevents Alice and Bob from communicating at all before making a choice (and after the referees have made theirs). The reason for this is that, travelling at the speed of light, it would take approximately 4.256 microseconds (4.256×10^{-6} s) to go from Villalpando to Poolbeg, and classical information can never travel faster than the speed of light! So this complicates things a little bit.

This is what the **CHSH game** is, and now we need to figure out a way for Alice and Bob to maximise their odds of winning this game. As you may have guessed, entanglement is going to play a role sooner or later, but, for now, let's think classically. Giving up all access to qubits and quantum gates, how could we play this game, and play it right?

To learn more...

In case you were curious, the CHSH game is named after the CHSH inequality, which is itself named after the four people who developed it: Clauser, Horne, Shimony and Holt. This inequality was introduced in the 1969 article "Proposed experiment to test local hidden-variable theories" [51].

We were promised that Antonio and Brigid, our dear referees, would be picking their bits perfectly at random. This means that there is a 25% chance that they will both pick a 1 (result $x = y = 1$) and a 75% chance that at least one of them will pick a bit to be 0 (results $(x, y) = (0, 0), (0, 1), (1, 0)$). Consequently, if Alice and Bob agree to both pick 0 regardless of everything, they are guaranteed to win with 75% probability. And this may seem like an innocent strategy, but it can actually be proved that no classical strategy can yield better odds of winning! As always, if you don't believe us, you are more than welcome to try to come up with a better strategy yourself (or perhaps more reasonably, with a proof of our claim).

Exercise 6.5

Prepare a different classical strategy that would also yield a 75% success rate.

And that's all that we can do with access to classical technology, when we are bounded by the speed of light not to communicate with each other. But what if we brought quantum computing into the mix? As it turns out, there is a way in which, using entanglement, we can boost the odds of our strategies within the CHSH game. And that's where our focus is going to be for the remainder of this section.

You may remember that, in *Chapter 3*, we introduced Y-rotation gates. These were parametrised by an angle θ and their coordinate matrices were defined as

$$R_Y(\theta) := \begin{pmatrix} \cos\frac{\theta}{2} & -\sin\frac{\theta}{2} \\ \sin\frac{\theta}{2} & \cos\frac{\theta}{2} \end{pmatrix}.$$

These gates will give us some of the key ingredients that we will need to quantumize and improve our strategy for the CHSH game. But, before we can take full advantage of them, we need to discuss some of their properties.

First, we should notice that the Bell state $\frac{1}{\sqrt{2}}(|00\rangle + |11\rangle)$ is invariant under Y-rotations, which is to say that, for any angle θ,

$$\frac{1}{\sqrt{2}}(R_Y(\theta) \otimes R_Y(\theta))(|00\rangle + |11\rangle) = \frac{1}{\sqrt{2}}(|00\rangle + |11\rangle).$$

Indeed, a direct computation reveals that

$$\frac{1}{\sqrt{2}}(R_Y(\theta) \otimes R_Y(\theta))(|00\rangle + |11\rangle) = \frac{1}{\sqrt{2}}(R_Y(\theta)\,|0\rangle\, R_Y(\theta)\,|0\rangle + R_Y(\theta)\,|1\rangle\, R_Y(\theta)\,|1\rangle)$$

$$= \frac{1}{\sqrt{2}} \left(((\cos\theta/2)^2 + (\sin\theta/2)^2) |00\rangle + ((\sin\theta/2)^2 + (\cos\theta/2)^2) |11\rangle + 0 |01\rangle + 0 |10\rangle \right)$$
$$= \frac{1}{\sqrt{2}} (|00\rangle + |11\rangle).$$

And lastly, we also need to be aware of the following fact that we leave as an exercise.

Exercise 6.6

Show that, for any angles α and β, we have $R_Y(\alpha + \beta) = R_Y(\alpha)R_Y(\beta)$. In a way, this generalizes the result that you proved in *Exercise 3.6.*

And that's pretty much all we need to know about Y-rotation gates. Now, how can these gates lead to a better CHSH strategy? This is the plan: we will now introduce a quantum strategy for the CHSH game, and then we will explore its details and verify whether it provides any advantages over classical approaches (spoiler: it does).

For our quantum CHSH strategy, we first need Alice and Bob to have each (you guessed it) a qubit from a pair in a Bell state, which we will assume to be $\frac{1}{\sqrt{2}}(|00\rangle + |11\rangle$. Those are all the pre-requisites. Now, this is the strategy that Alice and Bob need to follow.

From Alice's side:

- If her referee Antonio picks the bit 0, she does nothing.
- If Antonio picks the bit 1, she applies the gate $R_Y(\pi/2)$ on her qubit.

From Bob's side:

- If his referee Brigid picks the bit 0, he applies the $R_Y(\pi/4)$ gate on his qubit.
- If Brigid picks the bit 1, he applies the $R_Y(-\pi/4)$ gate on his qubit.

For each of them, once they have concluded these operations, they measure their respective qubits and they pick as their bit whichever outcome they get from the measurement. That's it. Plain and simple. At no point is there a need for Alice and Bob to send information to each other about which bits they are picking or what bits their referees have chosen.

For ease of reference, the key figures used in this protocol are summarised in *Table 6.1.*

	Referee picks 0	Referee picks 1
Alice rotates by...	0	$\pi/2$
Bob rotates by...	$\pi/4$	$-\pi/4$

(a) Angles used in the Y-rotations performed by Alice and Bob in terms of the bits picked by their respective referees

Angle differences	Antonio picks 0	Antonio picks 1
Brigid picks 0	$\pi/4$	$\pi/4$
Brigid picks 1	$\pi/4$	$3\pi/4$

(b) Absolute value of the differences between the angles used by Alice and Bob in their Y-rotations, depending on the choices of the referees

Table 6.1: Key figures in the quantum protocol for the CHSH game

So, does this complex set up lead to any gains? There are four possible scenarios: both referees pick 0, both pick 1, Antonio picks 0 and Brigid picks 1, and Antonio picks 1 and Brigid picks 0. According to our assumptions, all of these scenarios are equally likely, so we will compute the probability of success in each of them and then take the average. If the average is bigger than 75%, we can call it a day!

Let's first assume that both Antonio and Brigid choose the bit 0, which means that Alice and Bob will win if they choose the same bits. In this case, the state of the whole system would be

$$\frac{1}{\sqrt{2}}(|0\rangle \otimes R_Y(\pi/4)|0\rangle + |1\rangle \otimes R_Y(\pi/4)|1\rangle),$$

where, as usual, the first qubit is Alice's and the second is Bob's. If we expand this expression as per the definition of the R_Y gates, this yields the state

$$\cos(\pi/8)\left(\frac{1}{\sqrt{2}}(|00\rangle + |11\rangle)\right) + \sin(\pi/8)\left(\frac{1}{\sqrt{2}}(|01\rangle - |10\rangle)\right),$$

which means that the probability that Alice's and Bob's measurements will coincide is $(\cos\pi/8)^2$, whereas the probability that they will get different results is $(\sin\pi/8)^2$. If this isn't obvious, you can further expand this expression and then add up the probabilities of the outcomes 00 and 11, this will yield the probability that the two outcomes coincide.

Let's now assume that Antonio picked a 0 and Brigid picked a 1. In this case, Alice and Bob would win if they pick the same bits, and the state of the system is

$$\frac{1}{\sqrt{2}}(|0\rangle \otimes R_Y(-\pi/4)\,|0\rangle + |1\rangle \otimes R_Y(-\pi/4)\,|1\rangle),$$

which is perfectly analogous to the situation that we had before. You can easily check for yourself that, in this case, the probability of success is, once again $(\cos \pi/8)^2$.

Exercise 6.7

Expand the expression above and prove that, indeed, when Antonio picks 0 and Brigid picks 1, the strategy succeeds with probability $(\cos \pi/8)^2$.

Now let's consider the scenario in which Antonio picks 1 and Brigid picks 0. As before, Alice and Bob would win if they choose the same bits. And in this situation, the state of the system is

$$\frac{1}{\sqrt{2}}(R_Y(\pi/2)\,|0\rangle \otimes R_Y(\pi/4)\,|0\rangle + R_Y(\pi/2)\,|1\rangle \otimes R_Y(\pi/4)\,|1\rangle),$$

which can be rewritten as

$$\frac{1}{\sqrt{2}}(R_Y(\pi/4) \otimes I) \cdot (R_Y(\pi/4) \otimes R_Y(\pi/4)) \cdot (|00\rangle + |11\rangle),$$

where I is the identity gate. Moreover, because of the invariance of Bell states under rotations, the state is equal to

$$\frac{1}{\sqrt{2}}(R_Y(\pi/4\,|0\rangle) \otimes |0\rangle + R_Y(\pi/4)\,|1\rangle \otimes |1\rangle),$$

and by expanding the action of the rotation gates, it can be easily found to be equal to

$$\cos(\pi/8)\left(\frac{1}{\sqrt{2}}(|00\rangle + |11\rangle)\right) + \sin(\pi/8)\left(\frac{1}{\sqrt{2}}(|10\rangle - |01\rangle)\right),$$

hence the probability of success in this scenario is, again, $(\cos \pi/8)^2$

Lastly, when both Antonio and Brigid pick 1, Alice and Bob will win *if they pick different bits.* In this case, the state of the system is

$$\frac{1}{\sqrt{2}}(R_Y(\pi/2)|0\rangle \otimes R_Y(-\pi/4)|0\rangle + R_Y(\pi/2)|1\rangle \otimes R_Y(-\pi/4)|1\rangle),$$

which, through the invariance of Bell states under rotation gates, can be rewritten as

$$\frac{1}{\sqrt{2}}(R_Y(3\pi/4)|0\rangle \otimes |0\rangle + R_Y(3\pi/4)|1\rangle \otimes |1\rangle),$$

and the expansion of the rotation gates yields

$$\cos(3\pi/8)\left(\frac{1}{\sqrt{2}}(|00\rangle + |11\rangle)\right) + \sin(3\pi/8)\left(\frac{1}{\sqrt{2}}(|10\rangle - |01\rangle)\right),$$

hence, in this case, the probability of success will be $(\sin 3\pi/8)^2$, since Alice and Bob will win if they get different measurement outcomes. However,

$$\sin(3\pi/8) = \cos(3\pi/8 - \pi/2) = \cos(-\pi/8) = \cos(\pi/8),$$

so, in this case, the probability of success will be, once again, $(\cos \pi/8)^2$.

Exercise 6.8

Using the invariance of Bell states under Y-rotations, show that, indeed, the state

$$\frac{1}{\sqrt{2}}(R_Y(\pi/2)|0\rangle \otimes R_Y(-\pi/4)|0\rangle + R_Y(\pi/2)|1\rangle \otimes R_Y(-\pi/4)|1\rangle)$$

is equal to

$$\frac{1}{\sqrt{2}}(R_Y(3\pi/4)\left|0\right\rangle \otimes \left|0\right\rangle + R_Y(3\pi/4)\left|1\right\rangle \otimes \left|1\right\rangle).$$

In all the scenarios that we have considered, the probability of winning was $(\cos \pi/8)^2$, so that must be the global success probability of our strategy. And, at the moment of truth, a simple verification reveals that $(\cos \pi/8)^2 \approx 85\%$, which is significantly greater than the 75% success rate that any other classical approach would have been able to provide.

To learn more...

It can be proved that no quantum protocol can yield a success rate bigger than $(\cos \pi/8)^2$; this result is known as Tsirelson's bound [52].

The CHSH game has shown us how quantum entanglement can help us achieve things that are unimaginable from a classical point of view, and it is a nice application of two-qubit systems. However, there's much more to this protocol and behind this protocol, for its physical implications are much deeper than what we can hope to discuss in this book. For our purposes, we will just state that quantum strategies for the CHSH game have been demonstrated experimentally [53], and, vaguely speaking, this shows that entanglement is a very real phenomenon. On the more practical side of things, the CHSH game has been used in cryptographic protocols for generation of certified random bits that can be used, for instance, in cryptographic settings.

To learn more...

The theoretical foundation for the CHSH game was introduced in the paper "Proposed experiment to test local hidden-variable theories" [51], by Clauser, Horne, Shimony and Holt. Most significantly, the empirical verification of the quantum protocol for this game leads to the impossibility of "local realism". If you would like to learn a little bit more about this, Chapter 2 in *Quantum Computation and Quantum Information: 10th Anniversary Edition* [13] can be a good place to get started.

We should highlight that John Clauser was awarded the 2020 Nobel Prize in Physics, together with Alain Aspect and Anton Zeilinger, for conducting some experiments [51], [54]–[56] with entangled photons that build on the principles that we have explored with the CHSH game.

And that's it about the CHSH game. In the following section we will have a chance to explore Deutsch's algorithm: the most simple quantum algorithm that best encapsulates some of the key principles and ideas that make quantum computing the mighty and interesting paradigm that it is. We have some exciting pages ahead!

6.3 Deutsch's algorithm

In this section, we are going to introduce Deutsch's algorithm, which—in spite of its modest applications—will show you new ways in which superposition, entanglement and interference can play together, and it will introduce you to some of the conceptual ideas that we will better develop and exploit in other algorithms throughout the book.

So what does Deutsch's algorithm actually do? It all has to do with a special kind of functions known as **one-bit Boolean** functions. A one-bit Boolean function is, simply, a function that takes as input a bit and returns another bit as output. For example, the function f defined through

$$f(0) := 0, \qquad f(1) := 1$$

is a good example of a one-bit Boolean function. In fact, as you may have guessed already, one-bit functions aren't that complicated and we can even list them all right now (there are only four of them!). These are all the possible one-bit Boolean functions that you can find out there in the wild:

$$f_0(0) := 0, \qquad f_0(1) := 0,$$

$$f_1(0) := 0, \qquad f_1(1) := 1,$$
$$f_2(0) := 1, \qquad f_2(1) := 0,$$
$$f_3(0) := 1, \qquad f_3(1) := 1,$$

Two of these functions are constant, namely f_0 and f_3, because they maps all their inputs to the same output. On the other hand, the remaining two functions, f_2 and f_3, are said to be **balanced**. This name stems from the fact that those functions map half of their inputs to 0 and half of their inputs to 1.

Deutsch's algorithm enables us to determine, in a single shot, whether a function is constant or balanced. And let's think about what this means for a second. If you approached this problem with a good old classical computer, you would necessarily need to evaluate the function twice: once to obtain $f(0)$ and once to evaluate $f(1)$. What Deutsch's algorithm promises to deliver instead is to solve the problem with a single evaluation of the function.

These promises may sound impressive, but all of this begs the question: how on earth is a quantum algorithm going to evaluate a one-bit Boolean function? To understand that, we must resort to an oracle.

6.3.1 Deutsch's oracle

A way in which a quantum computer can evaluate a Boolean function f is through an **oracle**. An oracle for a one-bit Boolean function f is a two-qubit quantum gate O_f that is given to us as a black box: this means that we cannot see how it is implemented, we cannot see what f actually is, but we can use the gate in our circuits. Moreover, O_f promises to transform any state $|x\rangle \otimes |y\rangle$ from the canonical basis (where x, y may take the values $0, 1$) into the state $|x\rangle \otimes |y \oplus f(y)\rangle$, where $\oplus$ denotes the XOR (or addition modulo 2) operation. In a circuit, this is represented as follows:

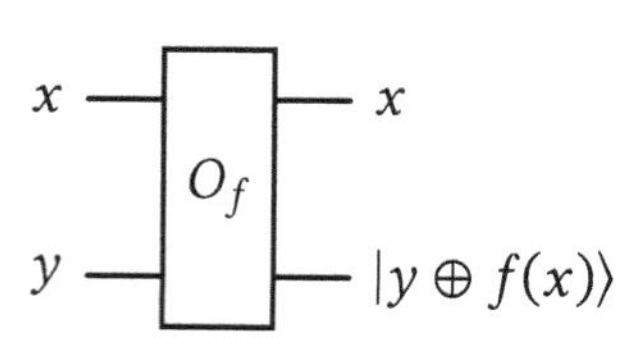

Thus, for example, an oracle for the function f_2 that we introduced before would be a two-qubit gate O_{f_2} that would transform the computational basis states as

$$
\begin{aligned}
O_{f_2}(|00\rangle) &= |0\rangle \otimes |0 \oplus f_2(0)\rangle = |0\rangle \otimes |0 \oplus 1\rangle = |01\rangle\,, \\
O_{f_2}(|01\rangle) &= |0\rangle \otimes |1 \oplus f_2(0)\rangle = |0\rangle \otimes |1 \oplus 1\rangle = |00\rangle\,, \\
O_{f_2}(|10\rangle) &= |1\rangle \otimes |0 \oplus f_2(1)\rangle = |1\rangle \otimes |0 \oplus 0\rangle = |10\rangle\,, \\
O_{f_2}(|11\rangle) &= |1\rangle \otimes |1 \oplus f_2(1)\rangle = |1\rangle \otimes |1 \oplus 0\rangle = |11\rangle\,,
\end{aligned}
$$

and, of course, its action would be extended by linearity to all possible quantum states. Now, when we approach our problem we would be given the oracle O_{f_2} as an obscure quantum gate, and we would not know how it is actually implemented: all that we would be able to do is run circuits using this gate, and Deutsch's algorithm promises us that we will only need to do it once in order to determine whether the function hidden behind the oracle is balanced or constant. However, underneath the hood, if we were tasked with preparing an oracle for f_2, we could easily build it as, for instance, shown in *Figure 6.2.*

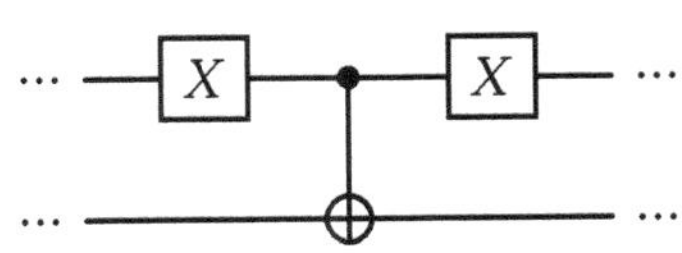

Figure 6.2: This is how an oracle for f_2 could be implemented underneath the hood. In the problem that we are considering, this gate is a device that we are given as a black box, and we are clueless about how it works on the inside

Exercise 6.9

Verify that the circuit shown in *Figure 6.2* actually implements an oracle for f_2.

Exercise 6.10

You have seen how an oracle for f_2 can be implemented. Now it's time for you to get your hands dirty. In this exercise, we invite you to implement an oracle for the one-bit Boolean functions f_0, f_1 and f_3 that we defined previously.

Now that we know what oracles are, we can clearly state what Deutsch's algorithm will be able to do.

> **Important note**
>
> Deutsch's algorithm can determine, in a single evaluation and with full certainty, whether a one-bit function f, provided as an oracle, is balanced or constant.

At this stage, oracles may seem like a mere funny device, but in *Chapter 9* you will see and understand why they are a fundamental abstraction in the design and analysis of quantum algorithms—far beyond what they can achieve in Deutsch's algorithm. In any case, for now, we are more than covered in terms of oracle knowledge, so let's discuss what the deal is with this fancy algorithm we've been talking so much about.

6.3.2 The inner workings of Deutsch's algorithm

In order to run Deutsch's algorithm, we need a pair of qubits initialized to the state $|01\rangle$ (if we are given a system initialized to $|00\rangle$, we would only need to apply an X gate on the second qubit in order to reach the state $|01\rangle$). Our first operation on them will be to apply a Hadamard gate to each qubit, as shown in the following figure:

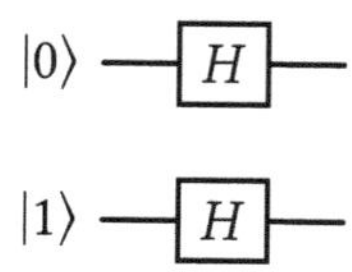

After this step, the state of our system will be

$$\frac{1}{\sqrt{2}}(|0\rangle + |1\rangle) \otimes \frac{1}{\sqrt{2}}(|0\rangle - |1\rangle) = \frac{|0\rangle \otimes (|0\rangle - |1\rangle)}{2} + \frac{|1\rangle (|0\rangle - |1\rangle)}{2}.$$

Following this, we need to apply the oracle of f on our two qubits, so our circuit so far will be as follows:

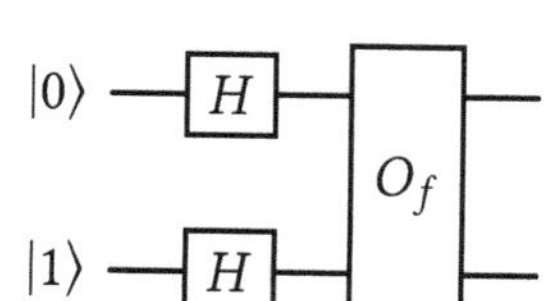

This will transform the state of our system into

$$\frac{|0\rangle \otimes (|0 \oplus f(0)\rangle - |1 \oplus f(0)\rangle)}{2} + \frac{|1\rangle \otimes (|0 \oplus f(1)\rangle - |1 \oplus f(1)\rangle)}{2},$$

which is... a somewhat complicated expression to deal with, isn't it? Luckily, it can be greatly simplified if we examine it closely. To this end, consider any bit b. If $b = 0$, then

$$|0 \oplus b\rangle - |1 \oplus b\rangle = (|0\rangle - |1\rangle),$$

whereas, if $b = 1$, we have

$$|0 \oplus b\rangle - |1 \oplus b\rangle = |1\rangle - |0\rangle = -(|0\rangle - |1\rangle).$$

Since $(-1)^0 = 1$ and $(-1)^1 = -1$, we can write, for any bit b,

$$|0 \oplus b\rangle - |1 \oplus b\rangle = (-1)^b(|0\rangle - |1\rangle).$$

In particular, this must be true when b is $f(0)$ or $f(1)$, so we can simplify our state above as

$$\frac{|0\rangle \otimes (-1)^{f(0)}(|0\rangle - |1\rangle)}{2} + \frac{|1\rangle \otimes (-1)^{f(1)}(|0\rangle - |1\rangle)}{2}.$$

We have mentioned in the past that states that differ by a global phase factor are perfectly equivalent from a computational point of view (head back to the end of *Chapter 2* if you need to meditate on this matter). Thus, it would be completely harmless to multiply the whole state of our system by a factor $(-1)^{f(0)}$; we can think of it as a mere change of notation. If we do it, we have the following representation of the state of our system

$$\frac{|0\rangle \otimes \cancel{(-1)^{2f(0)}}(|0\rangle - |1\rangle)}{2} + \frac{|1\rangle \otimes (-1)^{f(0)+f(1)}(|0\rangle - |1\rangle)}{2}.$$

where we have cancelled the $(-1)^{2f(0)}$ factor since it is equal to $((-1)^2)^{f(0)} = 1^{f(0)} = 1$.

With this, we already have all the ingredients to neatly finish the algorithm and solve our problem. If f is constant, $f(0) = f(1)$ and hence $(-1)^{f(0)+f(1)} = 1$, so the state of our system would be

$$\frac{|0\rangle \otimes (|0\rangle - |1\rangle)}{2} + \frac{|1\rangle \otimes (|0\rangle - |1\rangle)}{2} = \frac{1}{\sqrt{2}}(|0\rangle + |1\rangle) \otimes \frac{1}{\sqrt{2}}(|0\rangle - |1\rangle) = |+\rangle \otimes |-\rangle.$$

On the other hand, if f is balanced, $f(0) \neq f(1)$, which means that one of them must be 0 while the other must be 1, so $(-1)^{f(0)+f(1)} = -1$. Thus, our system would be in the state

$$\frac{|0\rangle \otimes (|0\rangle - |1\rangle)}{2} - \frac{|1\rangle \otimes (|0\rangle - |1\rangle)}{2} = \frac{1}{\sqrt{2}}(|0\rangle - |1\rangle) \otimes \frac{1}{\sqrt{2}}(|0\rangle - |1\rangle) = |-\rangle \otimes |-\rangle.$$

Incidentally, here we are seeing a manifestation of a phenomenon known as **phase kick-back**: the $(-1)^{f(0)+f(1)}$ phase that seemed to only affect the second qubit, "kickbacks" and ends up affecting the whole system.

Taking all that we've done into account, if we apply a Hadamard gate on the first qubit, the first qubit will be in state $|0\rangle$ if the function is constant and it will be in state $|1\rangle$ if the function is balanced. Voilà! At this stage, we are just a measurement away from having solved our problem.

Important note

Given an oracle O_f for a one-bit Boolean function f, if we run a single shot of the circuit

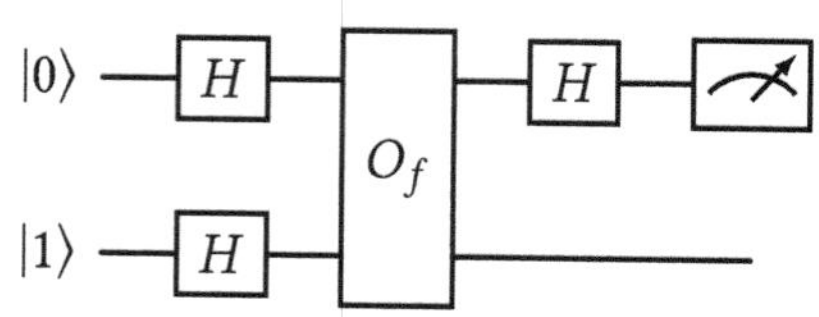

,

the measurement on the first qubit will return 0 if the function is constant and it will return 1 if the function is balanced. This is Deutsch's algorithm.

There are some crucial facts that are worth highlighting about this algorithm that we have just introduced. The first of them is that this algorithm, unlike many other quantum algorithms, **is not probabilistic**. With a single shot, we are guaranteed to get the result that we are expecting. It only takes one execution of the algorithm in order to find out whether the function that we are considering is balanced or constant.

Another detail that is worth mentioning is that this algorithm is the first we have discussed that kind of realises this promise that "quantum computers can evaluate all possible solutions at the same time". We have all heard that statement at some point, and most of the time it is presented as something that is simple to achieve theoretically, and all the credit goes to superposition. However, even with this simple algorithm, you have hopefully seen how, behind this idea, there is much more than superposition at play. In particular, in Deutsch's algorithm, we have used several of the key ingredients of all quantum algorithms:

- The first Hadamard gates created **superposition**.
- The oracle transformed the resulting state.
- The final Hadamard gates used **interference** to lead us to the results that we wanted.

Let this be an appetizer. We will discuss this further in *Chapter 9*.

And that's it about Deutsch's algorithm! Maybe being able to determine whether a one-qubit function is balanced or constant doesn't seem like a game changer, but it is one more step in our journey towards more complex quantum algorithms. We should also mention that this algorithm, which was introduced by David Deutsch in the article "Quantum theory, the Church–Turing principle and the universal quantum computer" [57], is just a particular case of a more general algorithm introduced by Deutsch and Richard Jozsa [58]; we will have a chance to explore it in *Chapter 9*. Anyway, for now, let's wrap up this chapter.

Summary

In this chapter, we have discussed some interesting applications of two-qubit systems, which have allowed us to deepen our understanding of these systems and have helped us illustrate how entanglement (in addition to superposition and interference) is used in quantum algorithms.

We began this chapter with superdense coding, a compact way of transmitting classical information via quantum means; we saw how this protocol enables two parties to send two bits of classical data encoded in a single qubit. Then we introduced the CHSH game, which unraveled some of the hidden power of quantum entanglement. Both of these protocols assumed that the two communicating parties had each a qubit from a pair in a Bell state.

Lastly, we introduced Deutsch's algorithm that allowed us to determine whether a one-bit Boolean function was constant or balanced using a single call to an oracle. This algorithm gave us a first glimpse into some ideas that we will reuse and further develop later in the book, when we get to discuss more sophisticated quantum algorithms.

In the following chapter, we will see all of this in action with Qiskit: everything we've learnt about two-qubit systems and their applications is going to come to life through code. This will be our last stop before taking the big jump into systems with arbitrarily many qubits. Things are moving fast!

7

Coding Two-Qubit Algorithms in Qiskit

Two heads are better than one, not because either is infallible, but because they are unlikely to go wrong in the same direction.

— C.S. Lewis

In the last couple of chapters, we have studied quite a number of algorithms and protocols that work on two-qubit systems. It is thus time for us to learn how to use Qiskit to construct and run circuits that act on two qubits.

We will start with some simple examples that will illustrate how to work with two qubits in Qiskit. As you will soon see, adding a new qubit to the mix won't change things that much. The only difference will be our having to keep track of which qubit we are applying gates to, and the introduction of two-qubit gates such as the CNOT gate.

Then, we will move on to implement some of the protocols that we have already discussed from a theoretical point of view, including superdense coding, the CHSH game, and Deutsch's algorithm.

The topics that we will cover in this chapter are the following:

- Working with two qubits in Qiskit
- Superdense coding
- The CHSH game
- Deutsch's algorithm

Get ready! With Qiskit, entanglement is just a few keystrokes away.

7.1 Working with two qubits in Qiskit

In order to work with two-qubit systems in Qiskit, we need to start by defining a quantum circuit that has a pair of qubits. In quantum computing jargon, these two qubits form what is called a **quantum register**. In fact, a register is nothing more than a bunch of qubits (or bits, if it is a **classical register**) that are grouped together and that we can use for some purpose. In this chapter, we are going to work only with quantum registers of two qubits, but in *Chapter 10* we will extend this to registers with an arbitrary number of qubits.

So, how can we define a quantum circuit with a two-qubit register in Qiskit? It couldn't be easier. We just need to use the following piece of code:

```
from qiskit import QuantumCircuit

circuit = QuantumCircuit(2)
```

As you surely have noticed, this the exact same invocation that we used, back in *Chapter 4*, to define one-qubit circuits, but setting the argument in the call to the `QuantumCircuit` constructor to 2 instead of 1. Makes sense, doesn't it?

We now have a quantum circuit with two qubits, so we can begin applying gates to them. This is done exactly as in the one-qubit case, but we need to be careful with the index of the qubits that we apply the gates to. Let's consider, for instance, the following instructions:

```
circuit.h(0)
circuit.x(1)
```

As you have surely guessed, with this code we have applied a Hadamard gate to the qubit of index 0 (the first or top qubit) and an X gate to the qubit of index 1 (the second or bottom qubit). We can check this by running the `circuit.draw("mpl")` instruction, which will give us the following:

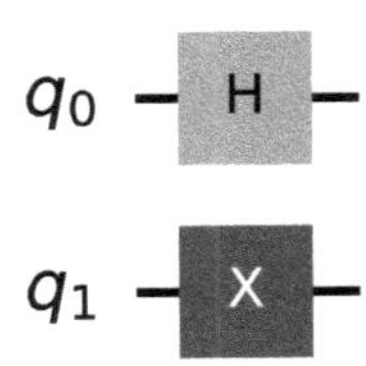

As you can easily check, this circuit prepares the state $\frac{1}{\sqrt{2}}(|01\rangle + |11\rangle)$. To verify that everything is working correctly, we can add measurements to the circuit and run it on a simulator. This is completely analogous to the one-qubit case, so we can use the following instructions, exactly as we did in *Chapter 4*:

```
circuit.measure_all()

from qiskit_aer import AerSimulator
backend = AerSimulator(seed_simulator = 18620123)

from qiskit_ibm_runtime import SamplerV2 as Sampler
sampler = Sampler(backend)
```

```
job = sampler.run([circuit], shots = 8)
result = job.result()[0].data.meas

print(result.get_counts())
```

If you run this code, you will obtain the following output:

```
{'11': 5, '10': 3}
```

This means that we have obtained 11 five times and 10 three times. But... wait a minute! Shouldn't the outcomes be 01 and 11 instead? What is going on here? Well, it turns out that Qiskit uses a very particular convention and reverses the bit strings of measurement outcomes. This is why we obtained 10 instead of 01. Not a big deal, but it's something to always keep in mind.

To learn more...

The rationale behind this idea of reversing measurement outcomes is that, in the minds of the Qiskit development team, a bit string $b_0, b_1, \ldots, b_n$ should represent the number $2^0 b_0 + \cdots + 2^n b_n$, which would be written, in binary, as $b_n \cdots b_0$.

This convention is not universally accepted and, for instance, we do not follow it in this book.

You may remember from *Chapter 4* that there are other ways in which we can access the results of a circuit execution. It turns out that they can also be used with two-qubit circuits and, as we will see in *Chapter 10*, with any circuit in Qiskit. In our particular case, if we execute the `print(result.get_bitstrings())` instruction, we will obtain the following output:

```
['11', '10', '11', '10', '11', '11', '11', '10']
```

Notice that the first character in these strings corresponds, again, to the bottom qubit. We can obtain the same information in numerical form by executing `print(result.array)`. This will display the following data:

```
[[3]
 [2]
 [3]
 [2]
 [3]
 [3]
 [3]
 [2]]
```

Here, 3 is just the integer representation of `'11'` and, similarly, 2 is the integer corresponding to `'10'`. With other circuits, of course, you could also obtain 0 for `'00'` and 1 for `'01'`.

The convention to reverse measurement outcomes also affects the way in which the amplitudes for the system statevector are computed and displayed. To explore this, let's start by removing the measurements from our circuit with the `circuit.remove_final_measurements()` instruction. Then, we can use `Statevector`, as we did back in *Chapter 4*, to compute the amplitudes of our qubits. Altogether, we need to execute the following instructions:

```
from qiskit.quantum_info import Statevector

circuit.remove_final_measurements()

state = Statevector(circuit)
print(state)
```

After running this code, you will obtain the following output:

```
Statevector([0.        +0.j, 0.        +0.j, 0.70710678+0.j,
             0.70710678+0.j],
            dims=(2, 2))
```

Notice how we have an amplitude of $\frac{1}{\sqrt{2}}$ for positions 2 (corresponding to 10) and 3 (corresponding to 11) of the statevector. Also note that the dimensions of the vector are now $(2, 2)$ because we have two qubits.

> **Important note**
>
> Qiskit follows the convention of reversing the bit strings that identify measurement outcomes and computational basis states when giving measurement results and statevector amplitudes. Keep this in mind, or you'll be utterly baffled when the output flips your expectations upside-down!

Now that we know how to define quantum circuits with two qubits and how to use quantum gates on individual qubits, it's time to learn how to apply gates that act on two qubits at once. To illustrate this, we will create a circuit that prepares a pair of qubits in a Bell state. We can achieve this with the following piece of code:

```
circuit = QuantumCircuit(2)
circuit.h(0)
circuit.cx(0,1)
```

Here, we are recycling the name `circuit` to create a new quantum circuit, this time with a Hadamard gate on the top qubit and a CNOT gate controlled by the top qubit and targeting the bottom one. For this, we used the `cx` method, which receives two parameters: the index of the control qubit, and that of the target qubit—in that order. If you draw this circuit with `circuit.draw("mpl")`, you will get the following masterpiece:

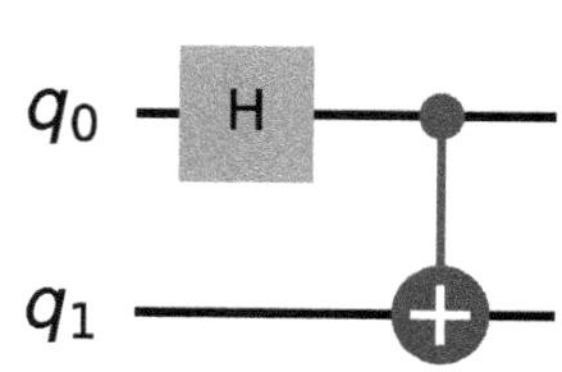

Of course, it is perfectly possible to add measurements to this circuit and to run it on simulators or on real quantum hardware. Also, just as before, we could compute the amplitudes of the state at the end of the circuit. This is done exactly as we did in *Chapter 4* and in this very section, so we invite you to try it on your own in the next easy, yet important exercise.

Exercise 7.1

Write Qiskit code to:

(a) Obtain the amplitudes of the Bell state created by the circuit we have just created.

(b) Run the circuit on a simulator.

(c) Run the circuit on an actual quantum computer.

What results do you obtain? Why?

This covers the fundamentals of working with two-qubit circuits in Qiskit. Let's now put our knowledge to good use by implementing some of our favorite quantum protocols and algorithms!

7.2 Superdense coding

In this section, we are going to use our newly-acquired skills to simulate the superdense coding protocol. This may be a very good moment to go back to *Section 6.1* if you need a refresher on how the protocol works.

To start with, we need Alice and Bob to create and share a Bell pair. This can be accomplished with the following instructions:

```
circuit = QuantumCircuit(2)
# Bell state preparation
circuit.h(0)
circuit.cx(0,1)
```

```
circuit.barrier()
```

Here, the qubit with index 0 will be kept by Alice and the qubit with index 1 will be kept by Bob. We have included a barrier after the creation of the entangled pair to make it clear that this is the first step in the protocol and that it happens before the actual communication takes place.

Later on, Alice wants to send two bits of information, which we will denote by b_1 and b_2, to Bob. Imagine that $b_1 = 1$ and $b_2 = 0$. Then, she needs to perform the following operations:

```
b1 = 1
b2 = 0

# Alice's ops (in reverse because of Qiskit coding)

if b1:
    circuit.x(0)
if b2:
    circuit.z(0)

circuit.barrier() # Qubit sent to Bob
```

In this piece of code, we have included conditional sentences to apply X and Z gates depending on the values of the bits that Alice wants to send. In this way, we only need to change the values of `b1` and `b2` to create the particular circuit that we need for each case. Notice, also, that we have swapped the role of b_1 and b_2 with respect to our explanation of the protocol in *Section 6.1*. This is just to compensate for Qiskit's convention for readout measurements. If we apply the gates exactly as in the theoretical description of the protocol, then we would need to read the measurement result in reserve for it to make sense. Finally, notice that we have added another barrier to mark the point at which Alice sends her qubit (the one with index 0) to Bob.

Now that Alice has performed her operations depending on the bits she wants to send, and that she has sent her qubit to Bob, it is time for Bob to operate on both qubits at once. He will do it with the following instructions:

```
circuit.cx(0,1)
circuit.h(0)
circuit.measure_all()
```

If you draw the circuit with `circuit.draw("mpl")`, you will obtain the following figure:

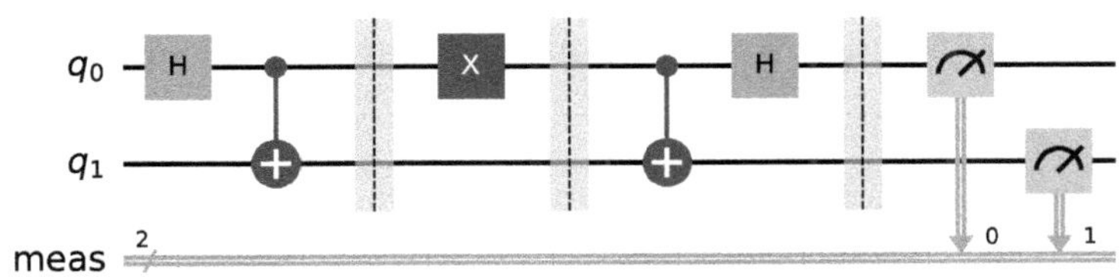

Notice how the three different sections of the protocol are cleanly separated by barriers. In the first section, we have the preparation of the Bell pair, in which both Alice and Bob participate. After this, Alice keeps the top qubit and Bob keeps the bottom one. In the second section, depicted between the two barriers, Alice operates on her qubit depending on the information she wants to send. Lastly, Bob receives Alice's qubit, operates on both qubits and measures them to obtain the result.

Thus, this is exactly the circuit that we need in order to simulate the protocol. If we run it and measure it several times, we will always obtain 10 as the result, as you can easily check. Moreover, if you remove the measurements and compute the amplitudes of the state prepared by the circuit, you will see that they are all zero except for the amplitude corresponding to the state that Qiskit labels as 10.

The cases in which Alice wants to send a different pair of values are left for you to try in the following exercise.

Exercise 7.2

Modify the preceding code to adapt it to the cases in which Alice wants to send 00, 01, and 11.

So, that was our first implementation of a two-qubit protocol in Qiskit. Neat, right? Next, we will use our Qiskit programming skills to play our favourite quantum game.

7.3 The CHSH game

In this section, we are going to simulate the CHSH game that we described and analyzed in *Section 6.2*. As you may remember, it all begins with a referee sending a challenge to Alice and Bob, who are far away from each other so they cannot communicate during the game. The challenge consists of two bits, one for Alice and another one for Bob, and they win the game if they reply with bits of their own that meet the following criteria:

- If Alice and Bob both receive 1, then they need to send back bits that are different from each other
- Otherwise, they need to send back the same bit

To set up the game, we will generate all the challenges in advance. Of course, this is not how the game really works, but it will simplify our computations and will also help us illustrate how to get a bunch of circuits to run at the same time in Qiskit. We can create the bits that the referee will send to Alice and Bob with the following instructions:

```
import numpy as np

seed = 1234
np.random.seed(seed)
reps = 1000

x = np.random.randint(2, size = reps)
y = np.random.randint(2, size = reps)
```

Here, we have just created two arrays, x and y, each with 1000 random bits that will be sent to Alice and Bob respectively. Now, we re going to create all the circuits that Alice and Bob need to use to respond to the challenges. We will need the following code:

```
circuit_list = []

for i in range(reps):

    circuit = QuantumCircuit(2)

    # Bell state preparation
    circuit.h(0)
    circuit.cx(0,1)

    circuit.barrier()

    # Alice and Bob's ops

    if x[i] == 1:
        circuit.ry(np.pi/2,0)
    if y[i] == 0:
        circuit.ry(np.pi/4,1)
    else:
        circuit.ry(-np.pi/4,1)
    circuit.measure_all()

    circuit_list.append(circuit)
```

Notice that we are storing all the circuits in `circuit_list` so we can send them all at once to the simulator or quantum computer. Again, this is not representative of how the game would actually work because Alice and Bob are supposed to respond to each challenge immediately. Nevertheless, we will take this route as the final results will be the same and it will help us to learn a new way of running circuits with Qiskit.

Regarding the circuits themselves, we start by creating a Bell pair that is shared by Alice and Bob. As usual, Alice will keep qubit 0 and Bob will keep qubit 1, and we have separated this part (which is supposed to happen before the game commences) from the rest of the operations with a barrier. Note also that we create the circuits depending on the bits received by Alice and Bob, each of them operating on their own qubit and using Y-rotation gates, which in Qiskit are implemented with the `ry` method: the first parameter to this method is the rotation angle, and the second is the qubit we want to apply the rotation to. Finally, we measure all the qubits to represent the measurements that Alice and Bob perform independently.

We can now run all the circuits and compute how many times Alice and Bob win the game to estimate the probability of success of their quantum strategy. We can achieve this with the following piece of code:

```
sampler = Sampler(AerSimulator(seed_simulator = seed))

job = sampler.run(circuit_list, shots = 1)
results = job.result()

wins = 0

for i in range(reps):
    bits = results[i].data.meas.get_bitstrings()
    a = int(bits[0][1])
    b = int(bits[0][0])

    if x[i]*y[i] == 0:
        if a == b:
            wins+=1
```

```
    else:
        if a != b:
            wins += 1

print("Win percentage:", 100*wins/reps)
```

Notice how, in a single job, we can run a list of circuits (each of them being executed exactly once) with the `job = sampler.run(circuit_list, shots = 1)` instruction. This will impact how we retrieve the results, because `results` will now contain multiple elements. We process them in the `for` loop, first accessing the bits sent by Alice and Bob to the referee (variables `a` and `b`, respectively) and then checking if they satisfy the conditions to win the round of the game. Note that `bits` is a list that contains exactly one string (its values can be, for instance, `['10']` or `['00']`). Thus, we need first to access the string (`bits[0]`) and then the individual bits in that string, which we need to convert to integer values with the `a=int(bits[0][1])` and `b=int(bits[0][0])` instructions.

You may be baffled by our decision to assign `bits[0][1]` to Alice's bits instead of `bits[0][0]`. Wasn't Alice operating on qubit 0 all the time? Yes, that's completely right. But remember that Qiskit reverses the bits when giving measurement results, so in `bits[0]` they appear in the opposite order. In this scenario, the situation is completely symmetrical and it really doesn't matter, but in other cases it can make a huge difference, so you need to be careful.

Upon running this code, you will obtain the following output:

```
Win percentage: 85.9
```

This agrees nicely with the game analysis that we performed in *Section 6.2*.

Exercise 7.3

Modify the preceding code to make it run on an actual quantum computer. If you do it, you will be probing the nature of physical reality with just a mouse click, with experiments along the lines of those that got Aspect, Clauser, and Zeilinger the Nobel Prize in Physics in 2022. How cool is that?

So that's how you can simulate the CHSH game with Qiskit. Next, we move our attention to Deutsch's algoritm. Let's get to it!

7.4 Deutsch's algorithm

For our implementation of Deutsch's algorithm in Qiskit, we will start by creating a sort of "circuit template". If you remember our discussion of this algorithm in *Section 6.3*, every application of Deutsch's algorithm follows the same structure: it starts with two qubits in states $|0\rangle$ and $|1\rangle$; Hadamard gates are applied to both; then, the oracle for the Boolean function that we want to test is applied; finally, a Hadamard gate is applied to the top qubit, which is subsequently measured.

We can implement that shared structure with the following instructions:

```
circuit = QuantumCircuit(2,1)
circuit.h(0)
circuit.x(1)
circuit.h(1)
circuit.barrier()
# PLACE YOUR ORACLE HERE
circuit.barrier()
circuit.h(0)
circuit.measure([0],[0])
```

You may have noticed something different in the way we have defined our quantum circuit here. In addition to the the number of qubits (2) in the call to the `QuantumCircuit` constructor, we have passed a second parameter (1) that indicates the number of classical bits that we will need to receive measurement results. Usually, we rely on the `measure_all()` method and we let Qiskit create the classical bit register used to store the results. However, in this case, we only want to measure one of the qubits, so we only need one classical bit for the measurement.

The rest of the code is straightforward. We start by applying a Hadamard gate to the top qubit, and an X gate (to set its state to $|1\rangle$ from the initial $|0\rangle$) and an H gate to the bottom one. We have then added barriers to leave some empty space as a placeholder where the implementation of your oracle should go. Finally, we apply a Hadamard gate to the top qubit and we measure it.

In this case, since we only want to measure the top qubit, we use `circuit.measure` instead of `circuit.measure_all`. This method takes two parameters: the list of indices of the qubits that we want to measure and the list of indices of the classical bits where we are going to store the measurement results. This gives us a little bit more flexibility than our old `measure_all` method. But be careful! While `measure_all` creates a register of classical bits large enough to store the results, `measure` expects the classical bits to exist beforehand. That's why we included the creation of one classical bit when we declared the quantum circuit. If you draw this circuit template with `circuit.draw("mpl")`, you will obtain the following figure:

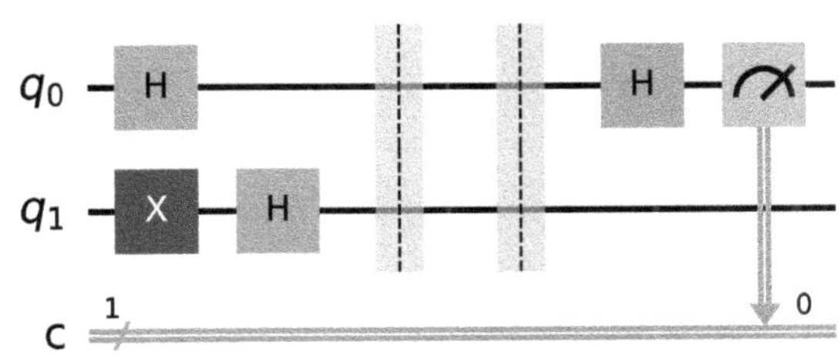

Now, we need to fill in the space for the oracle. We will start by implementing it for the Boolean function that we denoted by f_2 in *Section 6.3.1*. This function is defined by the following action:

$$f_2(0) := 1, \qquad f_2(1) := 0.$$

Obviously, f_2 is balanced, because it returns 0 for one input and 1 for the other. As you checked in *Exercise 6.9*, this function can be implemented with the circuit obtained with the following code:

```
circuit.x(0)
circuit.cx(0,1)
circuit.x(0)
```

Then, our full circuit will look like this:

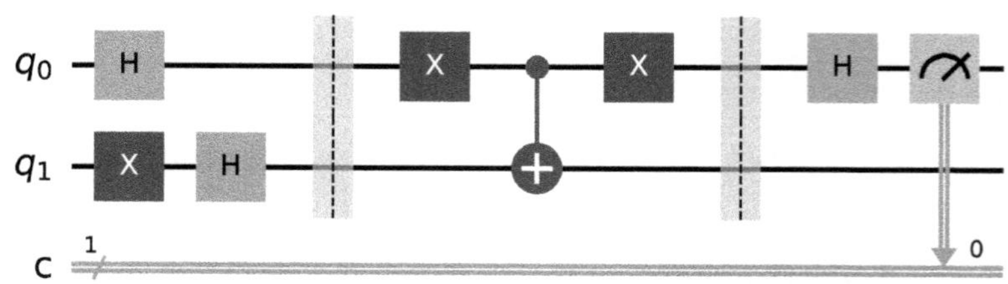

Now, we run the circuit with the following code:

```
backend = AerSimulator(seed_simulator = 18620123)
sampler = Sampler(backend)

job = sampler.run([circuit], shots = 8)
result = job.result()[0].data.c

print(result.get_counts())
```

There is a subtle but extremely important detail here. Notice that we have used the `job.result()[0].data.c` instruction instead of `job.result()[0].data.meas` as we had done up to this point. This is because, when creating a quantum circuit object with some classical bits (as we did with `QuantumCircuit(2,1)`), the name for the classical register is `c`.

But if you use `measure_all` to add the bits in your stead, the name of the classical register is `meas`. What can we say? Appreciating Qiskit's small idiosyncrasies is an acquired taste.

Anyway, the result after running the full circuit is `{'1': 8}`.

This confirms that, indeed, the function is balanced. Yay! Keep in mind that, even though we have used eight shots, one would have sufficed!

The cases of the three other one-bit functions are nice exercises that we leave for you to solve.

Exercise 7.4

Use Qiskit to implement oracles for the following Boolean functions:

$$f_0(0) := 0, \qquad f_0(1) := 0,$$
$$f_1(0) := 0, \qquad f_1(1) := 1,$$
$$f_3(0) := 1, \qquad f_3(1) := 1.$$

Use them with Deutsch's algorithm and check that the measurement results are the expected ones. Check the solution to *Exercise 6.10* if you need a hint.

This concludes our implementation of Deutsch's algorithm in Qiskit and this chapter. Let's wrap it up before climbing to new heights!

Summary

In this chapter, we have studied how to work with two-qubit circuits in Qiskit. We described how to define this type of circuit and how to apply gates to individual qubits. We also discussed how to apply two-qubit operations such as the CNOT gate, and we learned how to retrieve results from our circuits by performing measurements in a number of ways.

We illustrated all of these concepts with three important examples: superdense coding, the CHSH game, and Deutsch's algorithm. We had the opportunity to check the inner workings of these algorithms and protocols on both simulators and actual quantum computers, and we also explored some handy tricks such as sending a bunch of different circuits in the same job.

In the following chapter, we will begin to unlock the full potential of quantum computing by introducing systems with an arbitrary number of qubits. This will allow us to later discuss even more impressive quantum algorithms (such as a generalization of Deutsch's algorithm that works for Boolean functions with any number of qubits). Exciting times ahead!

Part 3

Working with Many Qubits

We have arrived to the part of the book where things are getting seriously big. You will soon learn about multi-qubit systems, that can store an exponential amount of information. You will learn how to work with them, correctly describing their states, transforming them with gates that can be applied to many qubits, and, finally, extracting information from them with different types of measurements.

You will also learn about some quantum algorithm that, albeit not being of much practical interest, will exemplify how superposition, entanglement and interference can be used to solve certain problems with a huge advantage over what is possible with classical computers. And, of course, you will learn all about implementing in Qiskit protocols and algorithms that work with multi-qubit circuits.

This part includes the following chapters:

8

How to Work with Many Qubits

United we stand, divided we fall.

— Aesop

In this chapter, we will study systems with many qubits. This will allow us to fully exploit the potential of superposition, entanglement, and interference in practical quantum algorithms.

In order to learn how to work with multi-qubit systems, we will introduce the mathematical tools needed to describe their states, transform them and measure them. This will all be analogous to what we already know about two-qubit systems, but we will now work in full generality.

Along the way, we will also learn about universal sets of gates, which will help us to implement big quantum circuits with smaller building blocks.

The topics covered in this chapter are the following:

- Multi-qubit states
- Measuring many qubits
- Multi-qubit gates and universality

After reading this chapter, you will be able to describe the states of multi-qubit systems and you will understand how they can implicitly store an exponential amount of information. You will know how to work with quantum gates that can be applied on many qubits at once, and how to extract information from several qubits by either measuring them partially or totally. You will also understand how multi-qubit gates can be assembled from just one- and two-qubit gates. Are you ready to embiggen your quantum systems?

8.1 Multi-qubit states

In previous chapters, we discussed in detail how to work with small quantum systems with up to two qubits, and we saw how they could be used to implement some useful and surprising protocols. But it is time we graduate to the big leagues. We can't unleash the full power of quantum computing unless we start using many qubits at once.

This is the day. Today, we grow big.

But don't worry. You already know most of what you will need in order to understand many-qubit systems. In fact, the maths that we will study in this chapter are just a generalization of what we covered in *Chapter 5*.

For instance, the states in the computational basis of an n-qubit system are simply going to be tensor products of n one-qubit states. This means that if we have n qubits and each of them is in state $|0\rangle$ or $|1\rangle$, then the whole system must be in one of the following states:

$$|0\rangle \otimes |0\rangle \otimes \cdots \otimes |0\rangle \otimes |0\rangle \otimes |0\rangle,$$
$$|0\rangle \otimes |0\rangle \otimes \cdots \otimes |0\rangle \otimes |0\rangle \otimes |1\rangle,$$

$$
\begin{gathered}
|0\rangle \otimes |0\rangle \otimes \cdots \otimes |0\rangle \otimes |1\rangle \otimes |0\rangle\,, \\
|0\rangle \otimes |0\rangle \otimes \cdots \otimes |1\rangle \otimes |0\rangle \otimes |0\rangle\,, \\
\vdots \\
|1\rangle \otimes |1\rangle \otimes \cdots \otimes |1\rangle \otimes |1\rangle \otimes |1\rangle\,.
\end{gathered}
$$

As you probably guessed, we usually omit the $\otimes$ symbol to write

$$
\begin{gathered}
|0\rangle\,|0\rangle \cdots |0\rangle\,|0\rangle\,|0\rangle\,, \\
|0\rangle\,|0\rangle \cdots |0\rangle\,|0\rangle\,|1\rangle\,, \\
|0\rangle\,|0\rangle \cdots |0\rangle\,|1\rangle\,|0\rangle\,, \\
|0\rangle\,|0\rangle \cdots |1\rangle\,|0\rangle\,|0\rangle\,, \\
\vdots \\
|1\rangle\,|1\rangle \cdots |1\rangle\,|1\rangle\,|1\rangle\,,
\end{gathered}
$$

or, even simpler, just

$$
\begin{gathered}
|00 \cdots 000\rangle\,, \\
|00 \cdots 001\rangle\,, \\
|00 \cdots 010\rangle\,, \\
|00 \cdots 100\rangle\,, \\
\vdots \\
|11 \cdots 111\rangle\,.
\end{gathered}
$$

To make full mathematical sense out of this, we need to extend the tensor product of column vectors to situations in which the vector sizes are bigger than 2, which is the only case that we have considered up until this point. It turns out that when you have two column vectors

$$a = \begin{pmatrix} a_1 \\ a_2 \\ \vdots \\ a_j \end{pmatrix}, \qquad b = \begin{pmatrix} b_1 \\ b_2 \\ \vdots \\ b_k \end{pmatrix},$$

their tensor product is

$$a \otimes b := \begin{pmatrix} a_1 \begin{pmatrix} b_1 \\ b_2 \\ \vdots \\ b_k \end{pmatrix} \\ a_2 \begin{pmatrix} b_1 \\ b_2 \\ \vdots \\ b_k \end{pmatrix} \\ \vdots \\ a_j \begin{pmatrix} b_1 \\ b_2 \\ \vdots \\ b_k \end{pmatrix} \end{pmatrix} = \begin{pmatrix} a_1 b_1 \\ a_1 b_2 \\ \vdots \\ a_1 b_k \\ a_2 b_1 \\ a_2 b_2 \\ \vdots \\ a_2 b_k \\ \vdots \\ a_j b_1 \\ a_j b_2 \\ \vdots \\ a_j b_k \end{pmatrix}.$$

For example, it holds that

$$\begin{pmatrix} 0 \\ 1 \\ 0 \\ 0 \end{pmatrix} \otimes \begin{pmatrix} 1 \\ 0 \end{pmatrix} = \begin{pmatrix} 0 \begin{pmatrix} 1 \\ 0 \end{pmatrix} \\ 1 \begin{pmatrix} 1 \\ 0 \end{pmatrix} \\ 0 \begin{pmatrix} 1 \\ 0 \end{pmatrix} \\ 0 \begin{pmatrix} 1 \\ 0 \end{pmatrix} \end{pmatrix} = \begin{pmatrix} 0 \\ 0 \\ 1 \\ 0 \\ 0 \\ 0 \\ 0 \\ 0 \end{pmatrix}.$$

Moreover, the tensor product is associative, so we have, for instance, that

$$|010\rangle = |0\rangle\,|0\rangle\,|0\rangle = (|0\rangle \otimes |1\rangle) \otimes |0\rangle = \left(\begin{pmatrix}1\\0\end{pmatrix} \otimes \begin{pmatrix}0\\1\end{pmatrix}\right) \otimes \begin{pmatrix}1\\0\end{pmatrix} = \begin{pmatrix}0\\1\\0\\0\end{pmatrix} \otimes \begin{pmatrix}1\\0\end{pmatrix} = \begin{pmatrix}0\\0\\1\\0\\0\\0\\0\\0\end{pmatrix},$$

which is the same as

$$|010\rangle = |0\rangle\,|0\rangle\,|0\rangle = |0\rangle \otimes (|1\rangle \otimes |0\rangle) = \begin{pmatrix}1\\0\end{pmatrix} \otimes \left(\begin{pmatrix}0\\1\end{pmatrix} \otimes \begin{pmatrix}1\\0\end{pmatrix}\right) = \begin{pmatrix}1\\0\end{pmatrix} \otimes \begin{pmatrix}0\\0\\1\\0\end{pmatrix} = \begin{pmatrix}0\\0\\1\\0\\0\\0\\0\\0\end{pmatrix}.$$

Since writing a basis state such as $|011010\rangle$, with all those zeros and ones flying around, is tedious and prone to error, it is a common practice to consider the decimal number represented by the binary string in the ket and work with that instead. For instance, we usually write $|13\rangle$ rather than $|011010\rangle$, because the decimal number 13 is 011010 when represented in binary with six bits. Thus, the computational basis states for a system with n qubits can be also written much shorter and neater as

$$|0\rangle\,, |1\rangle\,, \ldots, |2^n - 1\rangle\,.$$

For example, the computational basis states of a two-qubit system would be $|0\rangle\,, |1\rangle\,, |2\rangle$, and $|3\rangle$, and when working with 3 qubits, they would be $|0\rangle\,, |1\rangle\,, |2\rangle\,, |3\rangle\,, |4\rangle\,, |5\rangle\,, |6\rangle$, and $|7\rangle$.

It is very important to notice that something such as, for instance, $|3\rangle$, only has a concrete meaning when we know how many qubits there are in the system we are working with. If we have 2 qubits, then $|3\rangle = |11\rangle$; if we have 3 qubits, then $|3\rangle = |011\rangle$; and if we have, say, 7 qubits, then $|3\rangle = |0000011\rangle$. For this reason, we will only use this notation when the system size is clear from context.

This way of representing computational basis states is very convenient, not only because it is much more succinct than using the binary expansion but also because in a system with n qubits, the state $|j\rangle$ corresponds to a column vector of size 2^n whose components are all zeros but a single 1 in the $(j+1)$-th position (or in the j-th position if you, like any self-respected computer scientist, start counting from 0). That's too big a vector for such a small amount of information.

As an example, using column vectors, when we have three qubits,

$$|0\rangle = \begin{pmatrix}1\\0\\0\\0\\0\\0\\0\\0\end{pmatrix}, \quad |1\rangle = \begin{pmatrix}0\\1\\0\\0\\0\\0\\0\\0\end{pmatrix}, \quad |2\rangle = \begin{pmatrix}0\\0\\1\\0\\0\\0\\0\\0\end{pmatrix}, \quad |3\rangle = \begin{pmatrix}0\\0\\0\\1\\0\\0\\0\\0\end{pmatrix},$$

$$|4\rangle = \begin{pmatrix}0\\0\\0\\0\\1\\0\\0\\0\end{pmatrix}, \quad |5\rangle = \begin{pmatrix}0\\0\\0\\0\\0\\1\\0\\0\end{pmatrix}, \quad |6\rangle = \begin{pmatrix}0\\0\\0\\0\\0\\0\\1\\0\end{pmatrix}, \quad |7\rangle = \begin{pmatrix}0\\0\\0\\0\\0\\0\\0\\1\end{pmatrix}.$$

Of course, this is all consistent with what we know about systems with one or two qubits. In fact, as you surely remember, if we have one qubit,

$$|0\rangle = \begin{pmatrix} 1 \\ 0 \end{pmatrix}, \qquad |1\rangle = \begin{pmatrix} 0 \\ 1 \end{pmatrix}.$$

Likewise, if we have two qubits, then

$$|0\rangle = \begin{pmatrix} 1 \\ 0 \\ 0 \\ 0 \end{pmatrix}, \qquad |1\rangle = \begin{pmatrix} 0 \\ 1 \\ 0 \\ 0 \end{pmatrix}, \qquad |2\rangle = \begin{pmatrix} 0 \\ 0 \\ 1 \\ 0 \end{pmatrix}, \qquad |3\rangle = \begin{pmatrix} 0 \\ 0 \\ 0 \\ 1 \end{pmatrix}.$$

Exercise 8.1

Prove that the tensor product of column vectors is associative. Use that to prove that, in a system with n qubits, $|j\rangle$ is a column vector of size 2^n in which all the elements are 0 except the one in position $j + 1$, which is 1.

So we now know how to construct computational basis states for n-qubit systems. But, to nobody's surprise, basis states are not the end of the story. As you were surely expecting, general states of a system with n qubits are linear combinations of the basis states where the coefficients (again called amplitudes) satisfy a normalization condition. More explicitly, a general state of an n-qubit system is of the form

$$\alpha_0 |0\rangle + \alpha_1 |1\rangle + \cdots + \alpha_{2^n-1} |2^n - 1\rangle,$$

where each α_i is a complex number and

$$|\alpha_0|^2 + |\alpha_1|^2 + \cdots + |\alpha_{2^n-1}|^2 = 1.$$

Obviously, the situations with one and two qubits are just particular cases of this general formulation. Also, let us note that, to save our tired fingers some keystrokes, we will usually use summation symbols so that a general state with n qubits will be

$$\sum_{j=0}^{2^n-1} \alpha_j |j\rangle,$$

under the normalization condition $\sum_{j=0}^{2^n-1} |\alpha_j|^2 = 1$.

Important note

Let's stop here for just a minute and contemplate what all of this implies. When we are working with n qubits, the amount of implicit information that we have is 2^n numbers, because there is one amplitude for each basis state.

That is an incredibly, spectacularly, amazingly, breathtakingly, astonishingly awful lot. And we are being conservative with our adjectives, believe us. For instance, to represent the state of a quantum computer with 1000 qubits (some of which have already been constructed), we would need so many amplitudes that, if we could store a complex number in every atom, we wouldn't have enough storage space even if we could use all the atoms in a trillion universes like ours.

That is BIG. And that is, of course, one of the reasons why quantum computing can be so powerful.

Examples of valid quantum states include

$$\frac{1}{\sqrt{2}}(|000\rangle + |111\rangle) = \frac{1}{\sqrt{2}}(|0\rangle + |7\rangle)$$

when we have 3 qubits,

$$\frac{1}{\sqrt{2}}(|0000\rangle + |1111\rangle) = \frac{1}{\sqrt{2}}(|0\rangle + |15\rangle)$$

when we have 4, and

$$\frac{1}{\sqrt{2}}(|0\rangle + |2^n - 1\rangle)$$

when we have n qubits.

An n-qubit state that we will profusely use is

$$\frac{1}{\sqrt{2^n}} \sum_{j=0}^{2^n-1} |j\rangle,$$

which is a balanced superposition of all computational basis states of n qubits, each with the same amplitude.

Exercise 8.2

Prove that $\frac{1}{\sqrt{2^n}} \sum_{j=0}^{2^n-1} |j\rangle$ is the tensor product of n qubits, each of them in the $|+\rangle$ state.

Exercise 8.3

Determine which of the following are valid quantum states:

(a) $\frac{1}{\sqrt{2}}(|000\rangle - |111\rangle)$
(b) $\frac{1}{\sqrt{2}}(|000\rangle + |11\rangle)$
(c) $\frac{1}{\sqrt{3}}(|001\rangle + |010\rangle + |100\rangle)$
(d) $\frac{1}{\sqrt{2}}|00101\rangle + \frac{i}{2}|01101\rangle - \frac{1}{2}|10101\rangle$
(e) $\frac{1}{2}(|00000\rangle + |00001\rangle + |00010\rangle)$

States of n qubits can also be entangled or product states. The definition is exactly the same as in the case of two-qubit states (check out *Section 5.3.3* if you need to revise this). Namely, an n-qubit state $|\psi\rangle$ is a product state if there exist two states $|\psi_1\rangle$ and $|\psi_2\rangle$ of n_1 and n_2 qubits, respectively, such that $n_1 + n_2 = n$ and $|\psi\rangle = |\psi_1\rangle \otimes |\psi_2\rangle$.

For example, $\frac{1}{\sqrt{2}}(|000\rangle+|010\rangle) = \left(\frac{1}{\sqrt{2}}(|00\rangle + |01\rangle)\right)|0\rangle$ is a product state, and so is $\frac{1}{\sqrt{2^n}} \sum_{j=0}^{2^n-1} |j\rangle$ when $n \geq 2$, because it is the product of n copies of $|+\rangle$, as you proved in *Exercise 8.2*. However, if $n \geq 2$, the state of n qubits $\frac{1}{\sqrt{2}}(|0\rangle + |2^n - 1\rangle)$ is always entangled. The proof is very similar to that for the Bell state that we discussed in *Section 5.3.3*, so we won't be repeating it here.

Exercise 8.4

Prove that the GHZ state (named after Greenberger, Horne, and Zeilinger) $\frac{1}{\sqrt{3}}(|001\rangle + |010\rangle + |100\rangle)$ is entangled.

That is enough for now about multi-qubit states. Next, we are going to learn how to perform quantum measurements when we have n qubits at our disposal.

8.2 Measuring many qubits

Just as when we considered two-qubit systems, in an arbitrary multi-qubit system, we can choose to measure all qubits at once or to measure just one of them. We will start our study with the simplest case, which is when we choose to measure all the qubits in the system.

Assume that we have n qubits in some state $\sum_{j=0}^{2^n-1} \alpha_j |j\rangle$ and we measure all of them. Then, we will obtain a result k with probability $|\alpha_k|^2$ and, should that be the outcome of the measurement, the state would collapse to $|k\rangle$. For example, if we have 4 qubits in state $\frac{1}{\sqrt{2}}(|0\rangle + |15\rangle)$, we will obtain 0 (or 0000) with probability $1/2$ (in which case, the state would collapse to $|0\rangle = |0000\rangle$), and 15 (or 1111) with probability $1/2$, which would lead the state to collapse to $|15\rangle = |1111\rangle$.

To consider yet another example, if we have the state $\frac{1}{\sqrt{3}}(|001\rangle + |010\rangle + |100\rangle)$ and we measure all of its qubits, we will obtain 001, 010, or 100 with probability $1/3$ each. The state will collapse to the ket that corresponds to the obtained result.

Exercise 8.5

Compute the possible results and their probabilities when measuring the following states:

(a) $\frac{1}{\sqrt{2}}(|000\rangle - |111\rangle)$

(b) $\frac{1}{\sqrt{2}}|00101\rangle + \frac{i}{2}|01101\rangle - \frac{1}{2}|10101\rangle$

(c) $\frac{3}{5}|111000\rangle - \frac{4i}{5}|000111\rangle$

(d) $\frac{1}{2}(|0001\rangle - |0010\rangle + |0100\rangle - |1000\rangle)$

Describing what happens when a single qubit is measured in a multi-qubit system is rather laborious, but it follows the same logic that we applied for two-qubit systems. Imagine, then, that you have n qubits in some state $\sum_{j=0}^{2^n-1} \alpha_j |j\rangle$. If we measure the qubit in position l, we can only obtain 0 or 1 as a result. The probability of obtaining 0 should be the sum of all the probabilities of computational basis states that have 0 in their l-th position. And if the outcome of the measurement were 0, the state would collapse to a superposition in which only the basis states with 0 in position l remain, while keeping the amplitudes they had in the original state, but with a global normalization factor so that the resulting state is still normalized and valid.

More explicitly, let L_0 be the set of all computational basis states whose l-th bit is 0. Then, when measuring the l-th qubit of $\sum_{j=0}^{2^n-1} \alpha_j |j\rangle$, we will obtain 0 with probability

$$\sum_{j\in L_0} |\alpha_j|^2$$

and the state will collapse to

$$\sum_{j\in L_0} \frac{\alpha_j}{\sqrt{\sum_{j\in L_0} |\alpha_j|^2}} |j\rangle .$$

Similarly, if we denote by L_1 the set of all computational basis states whose l-th bit is 1, we will obtain 1 with probability

$$\sum_{j\in L_1} |\alpha_j|^2$$

and the state will collapse to

$$\sum_{j\in L_1} \frac{\alpha_j}{\sqrt{\sum_{j\in L_1} |\alpha_j|^2}} |j\rangle .$$

Some examples will help us drive all of this home. If we consider the state $\frac{1}{\sqrt{3}}(|001\rangle + |010\rangle + |100\rangle)$ and we measure its first qubit, we will obtain 0 with probability $2/3$: $1/3$ is contributed by $|001\rangle$, and $1/3$ by $|010\rangle$. If the outcome of the measurement is indeed 0, the state of the system will then collapse to

$$\frac{\sqrt{3}}{\sqrt{2}}\left(\frac{1}{\sqrt{3}}|001\rangle + \frac{1}{\sqrt{3}}|010\rangle\right) = \frac{1}{\sqrt{2}}(|001\rangle + |010\rangle).$$

Also, we will obtain 1 with probability $1/3$, and, if that is the outcome, the state will collapse to $|100\rangle$.

If we have the state

$$\frac{1}{4}|0001\rangle + \frac{1}{2}|0010\rangle + \frac{3}{4}|0100\rangle + \frac{\sqrt{2}}{4}|1000\rangle$$

and we measure its third qubit, we will obtain 0 with probability $1/16 + 9/16 + 2/16 = 12/16 = 3/4$. The state will, in that case, collapse to

$$\frac{2}{\sqrt{3}}\left(\frac{1}{4}|0001\rangle + \frac{3}{4}|0100\rangle + \frac{\sqrt{2}}{4}|1000\rangle\right) = \frac{1}{2\sqrt{3}}|0001\rangle + \frac{3}{2\sqrt{3}}|0100\rangle + \frac{\sqrt{2}}{2\sqrt{3}}|1000\rangle.$$

Additionally, we will obtain 1 with probability $1/4$, in which case the state will collapse to simply $|0010\rangle$.

Exercise 8.6

Determine the collapsed states and the probabilities of the results when:

(a) measuring the second qubit of $\frac{1}{\sqrt{3}}(|001\rangle + |010\rangle + |100\rangle)$

(b) measuring the fist qubit of $\frac{1}{4}|0001\rangle + \frac{1}{2}|0010\rangle + \frac{3}{4}|0100\rangle + \frac{\sqrt{2}}{4}|1000\rangle$

(c) measuring the third qubit of $\frac{1}{2}(|000\rangle + i|011\rangle - |101\rangle - i|110\rangle)$

Obviously, the measurement rules for systems with one and two qubits that we introduced in *Chapter 2* and *Chapter 5* are just particular cases of the measurement rules for n-qubit systems. Moreover, if you measure all the qubits of a multi-qubit system in sequence, one after the other, you will recover the probabilities that we described at the beginning of this section for the case in which all the qubits are measured at once. The computation is a little bit cumbersome, but not conceptually difficult. A good way to convince yourself of the result is to work out the case with 3 qubits, for instance.

And that is all we had to say about measuring systems with several qubits. By now, you know the drill: once we have learned how to measure multi-qubit systems, it is time to learn about multi-qubit gates. That will be the topic of the next section.

8.3 Multi-qubit gates and universality

As you've surely deduced, in order to transform n-qubit states, we need to apply quantum gates that are nothing more than unitary matrices. Since the vector that represents an n-qubit state has 2^n elements, our matrices in this case must be of size $2^n \times 2^n$. This may look daunting, but there are ways of obtaining these transformations from smaller ones, as we will study in this section. In fact, we will start by considering the simplest case of multi-qubit gates: those that can be constructed as tensor products.

8.3.1 Tensor product gates

In *Section 5.3.1*, we showed how to obtain two-qubit gates from a couple of one-qubit gates acting independently on each of the system qubits. Likewise, when we have n qubits, we can consider gates that are obtained by combining the individual action of smaller gates on different parts of the system. For instance, in a system with 4 qubits, we could have a one-qubit gate acting on the first qubit, a two-qubit gate acting on the second and third, and no action (or, what is the same, the identity gate) on the fourth.

In order to obtain the explicit mathematical expression of the matrix of such gates, we need to generalize the tensor product to matrices of any size. If we have a pair of matrices

$$A = \begin{pmatrix} a_{11} & \cdots & a_{1k} \\ \vdots & \ddots & \vdots \\ a_{j1} & \cdots & a_{jk} \end{pmatrix}, \qquad B = \begin{pmatrix} b_{11} & \cdots & b_{1m} \\ \vdots & \ddots & \vdots \\ b_{l1} & \cdots & b_{lm} \end{pmatrix},$$

of sizes $j \times k$ and $l \times m$, then their tensor product is the matrix of size $(jk) \times (lm)$ given by

$$A \otimes B := \begin{pmatrix} a_{11}B & \cdots & a_{1k}B \\ \vdots & \ddots & \vdots \\ a_{j1}B & \cdots & a_{jk}B \end{pmatrix} = \begin{pmatrix} a_{11}b_{11} & \cdots & a_{11}b_{1m} & \cdots & a_{1k}b_{11} & \cdots & a_{1k}b_{1m} \\ \vdots & \ddots & \vdots & \cdots & \vdots & \ddots & \vdots \\ a_{11}b_{l1} & \cdots & a_{11}b_{lm} & \cdots & a_{1k}b_{l1} & \cdots & a_{1k}b_{lm} \\ \vdots & \ddots & \vdots & \cdots & \vdots & \ddots & \vdots \\ a_{j1}b_{11} & \cdots & a_{j1}b_{1m} & \cdots & a_{jk}b_{11} & \cdots & a_{jk}b_{1m} \\ \vdots & \ddots & \vdots & \cdots & \vdots & \ddots & \vdots \\ a_{j1}b_{l1} & \cdots & a_{j1}b_{lm} & \cdots & a_{jk}b_{l1} & \cdots & a_{jk}b_{lm} \end{pmatrix}.$$

For instance, the tensor product of an X gate with a $CNOT$ gate would be

$$\begin{pmatrix} 0 & 1 \\ 1 & 0 \end{pmatrix} \otimes \begin{pmatrix} 1 & 0 & 0 & 0 \\ 0 & 1 & 0 & 0 \\ 0 & 0 & 0 & 1 \\ 0 & 0 & 1 & 0 \end{pmatrix} = \begin{pmatrix} 0 & 0 & 0 & 0 & 1 & 0 & 0 & 0 \\ 0 & 0 & 0 & 0 & 0 & 1 & 0 & 0 \\ 0 & 0 & 0 & 0 & 0 & 0 & 0 & 1 \\ 0 & 0 & 0 & 0 & 0 & 0 & 1 & 0 \\ 1 & 0 & 0 & 0 & 0 & 0 & 0 & 0 \\ 0 & 1 & 0 & 0 & 0 & 0 & 0 & 0 \\ 0 & 0 & 0 & 1 & 0 & 0 & 0 & 0 \\ 0 & 0 & 1 & 0 & 0 & 0 & 0 & 0 \end{pmatrix}.$$

A straightforward computation shows that if U_1 is an n_1-qubit gate, U_2 is an n_2-qubit gate, $|\psi_1\rangle$ is an n_1-qubit state, and $|\psi_2\rangle$ is an n_2-qubit state, then it holds that

$$(U_1 \otimes U_2)(|\psi_1\rangle \otimes |\psi_2\rangle) = (U_1 |\psi_1\rangle) \otimes (U_2 |\psi_2\rangle).$$

Moreover, $U_1 \otimes U_2$ is unitary and $(U_1 \otimes U_2)^\dagger = U_1^\dagger \otimes U_2^\dagger$. Hence, $U_1 \otimes U_2$ is an $(n_1 + n_2)$-qubit gate that acts as U_1 on the first n_1 qubits and as U_2 on the other n_2 qubits.

This allows us to create a good number of gates that act on n qubits by taking the tensor product of smaller gates, and it is quite useful in practice. However, not every multi-qubit gate can be obtained in this form. We give an example in the following section.

8.3.2 The Toffoli gate

The Toffoli gate, also called CCNOT for reasons that will become apparent in a moment, is a three-qubit gate whose action is defined by the way it transforms the computational basis states. It holds that

$$\mathrm{CCNOT}\left|000\right\rangle = \left|000\right\rangle, \qquad \mathrm{CCNOT}\left|001\right\rangle = \left|001\right\rangle, \qquad \mathrm{CCNOT}\left|010\right\rangle = \left|010\right\rangle,$$
$$\mathrm{CCNOT}\left|011\right\rangle = \left|011\right\rangle, \qquad \mathrm{CCNOT}\left|100\right\rangle = \left|100\right\rangle, \qquad \mathrm{CCNOT}\left|101\right\rangle = \left|101\right\rangle,$$
$$\mathrm{CCNOT}\left|110\right\rangle = \left|111\right\rangle, \qquad \mathrm{CCNOT}\left|111\right\rangle = \left|110\right\rangle.$$

As you can see, the input state only changes under one condition: when both the first and second qubits are 1. In that case, a NOT gate is applied on the third qubit. We say that the two first qubits control the application of the NOT gate and that is why the Toffoli gate is called the **controlled-controlled-NOT**, **CCNOT** or **CCX** gate. The first two qubits are called **control qubits** and the third one is the **target**. The representation of the CCNOT gate in a circuit is as follows:

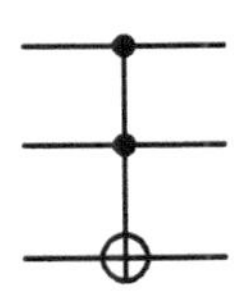

and its matrix is

$$\begin{pmatrix}
1 & 0 & 0 & 0 & 0 & 0 & 0 & 0 \\
0 & 1 & 0 & 0 & 0 & 0 & 0 & 0 \\
0 & 0 & 1 & 0 & 0 & 0 & 0 & 0 \\
0 & 0 & 0 & 1 & 0 & 0 & 0 & 0 \\
0 & 0 & 0 & 0 & 1 & 0 & 0 & 0 \\
0 & 0 & 0 & 0 & 0 & 1 & 0 & 0 \\
0 & 0 & 0 & 0 & 0 & 0 & 0 & 1 \\
0 & 0 & 0 & 0 & 0 & 0 & 1 & 0
\end{pmatrix}.$$

To learn more...

The Toffoli gate was introduced by Tommaso Toffoli in 1980. Its relevance comes from the fact that any classical Boolean gate (i.e., one that only takes bits as inputs and also returns bits as outputs) can be implemented by using the Toffoli gate and some auxiliary inputs. Since the Toffoli gate is reversible (in fact, it is unitary and is its own inverse), it follows that any Boolean function admits a reversible implementation.

Of course, you can also consider CCNOT gates in which the target qubit is the first one (or the second one) and the other qubits act as controls. The circuit representation and matrix will change accordingly.

An interesting property of the Toffoli gate is that it can't be obtained as the tensor product of gates that act on 1 and 2 qubits. In fact, you should try to prove it yourself, as we invite you to do in the following exercise.

Exercise 8.7

Prove that the CCNOT gate can't be obtained as the tensor product of two smaller gates.

All this may have led you to believe that, in addition to the one-qubit and two-qubit gates that we have studied so far, when designing quantum circuits, you would also need to consider some specific gates that act on 3 qubits, some that act on 4 qubits, and so on, as if every bump in the number of qubits could introduce some flashy new gates that had nothing to do with the ones used in smaller systems. Fortunately, this is not the case. In fact, the number of different gates that you need to have at your disposal in order to create arbitrary multi-qubit gates is surprisingly low, as we will show in the next section.

8.3.3 Gate universality

Although we know that the Toffoli gate can't be obtained as the tensor product of gates acting on fewer qubits, this does not mean that it can't be decomposed and constructed from just one-qubit and two-qubit gates. In fact, the circuit in *Figure 8.1* shows a way of building up the CCNOT gate from smaller ones. This can be useful, for instance, to implement the gate in actual quantum hardware in which only one- and two-qubit gates are available.

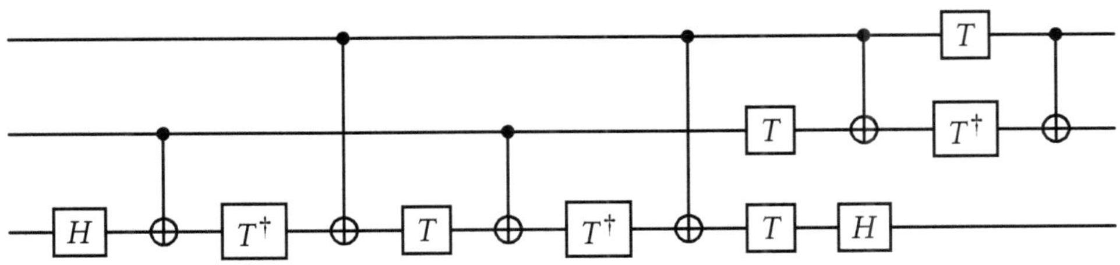

Figure 8.1: Implementing the Toffoli gate with one-qubit and two-qubit gates

To learn more...

You can prove that the circuit of *Figure 8.1* indeed implements the Toffoli gate by showing that it transforms all the computational basis states in the way that it should. It is not complicated, but it is tedious and boring. Rather than asking you to do it at this point, we will wait until *Chapter 10*, in which we will learn how to implement circuits with many qubits in Qiskit and we will rely on the computer to do the hard work for us.

In the circuit of *Figure 8.1*, we are using a fairly small number of different gates. We have the H, T, and $T^\dagger$ gates, which are all one-qubit gates, and the CNOT gate, which is a two-qubit gate. In fact, it is easy to see that $T^\dagger = T^7$, so we could also replace each occurrence of $T^\dagger$ with seven consecutive instances of T. Thus, only three different quantum gates, namely, T, H, and $CNOT$, are needed to implement the Toffoli gate.

This is far more than a happy coincidence. In fact, it can be proved that with just those three gates, you can "implement" any quantum gate on any number of qubits. Or, to be more precise, given any quantum gate U on n-qubits and any desired error bound ϵ, it is possible to create a circuit that only uses the T, H, and CNOT gates and implements a gate

$\tilde{U}$ (also on n-qubits) such that, for any n-qubit state $|\psi\rangle$, the distance between the states $U|\psi\rangle$ and $\tilde{U}|\psi\rangle$ is at most ϵ.

This means that if you use $\tilde{U}$ instead of U, you will not notice almost any difference, provided that ϵ is small enough. Of course, the smaller the ϵ, the bigger the circuit that you will need in order to implement $\tilde{U}$, but (at least in principle) it is always possible to achieve a precision as high as needed.

To learn more...

The distance between $U|\psi\rangle$ and $\tilde{U}|\psi\rangle$ is measured here as is usually done with regular vectors. That means that if $U|\psi\rangle = \sum_{j=0}^{2^n-1} \alpha_j |j\rangle$ and $\tilde{U}|\psi\rangle = \sum_{j=0}^{2^n-1} \beta_j |j\rangle$, then their distance is $\sqrt{\sum_{j=0}^{2^n-1} |\alpha_j - \beta_j|^2}$.

Since with $\{T, H, \mathrm{CNOT}\}$ we can approximate any given gate with as much precision as we want, we say that this set of gates is **universal**. There are other sets of gates that are also universal in a more strict sense. For example, if you take the set of all one-qubit gates plus the CNOT gate, you can implement any gate exactly, with no need for imperfect approximations (for mathematical proofs of all these facts, you can check [13]). The price to pay, though, is that you are no longer working with a finite set of gates.

In any case, not every decomposition of gates with the set $\{T, H, \mathrm{CNOT}\}$ needs to be an approximation. For instance, the Toffoli implementation in *Figure 8.1* is exact. And so are the ones that we invite you to work with in the following exercise.

Exercise 8.8

Prove the following gate equivalences:

(a) $S = T^2$.

(b) $Z = S^2$.

(c) $S^\dagger = S^3$.

(d) $T^\dagger = T^7$.

(e) $X = HZH$.

Other useful decompositions can be obtained when working with controlled gates. So far, we have only considered controlled versions of the NOT gate, namely, our beloved CNOT and the recently introduced CCNOT. But any quantum gate has its controlled (and multiply-controlled) version. In fact, if U is a quantum gate that acts on n-qubits, its controlled version CU is defined by

$$CU\left|0\right\rangle\left|\psi\right\rangle = \left|0\right\rangle\left|\psi\right\rangle, \qquad CU\left|1\right\rangle\left|\psi\right\rangle = \left|1\right\rangle(U\left|\psi\right\rangle),$$

where $\left|\psi\right\rangle$ is any computational basis state of n-qubits. The action on the rest of states is defined by linearity.

For instance, CZ acts on the basis states as follows:

$$CZ\left|00\right\rangle = \left|00\right\rangle, \qquad CZ\left|01\right\rangle = \left|01\right\rangle,$$
$$CZ\left|10\right\rangle = \left|1\right\rangle \otimes Z\left|0\right\rangle = \left|10\right\rangle, \qquad CZ\left|11\right\rangle = \left|1\right\rangle \otimes Z\left|1\right\rangle = -\left|11\right\rangle.$$

Exercise 8.9

Prove that if U is a unitary gate that acts on n-qubits, then CU is a unitary gate that acts on $(n+1)$-qubits and, in fact, $(CU)^{\dagger} = \mathrm{C}(U^{\dagger})$.

The definition of multiply-controlled gates is similar. For instance, if U is any quantum gate acting on n qubits, then its controlled-controlled version $\mathrm{CC}(U)$ is defined by

$$\mathrm{CC}U\left|00\right\rangle\left|\psi\right\rangle = \left|00\right\rangle\left|\psi\right\rangle, \qquad \mathrm{CC}U\left|01\right\rangle\left|\psi\right\rangle = \left|01\right\rangle\left|\psi\right\rangle,$$
$$\mathrm{CC}U\left|10\right\rangle\left|\psi\right\rangle = \left|10\right\rangle\left|\psi\right\rangle, \qquad \mathrm{CC}U\left|11\right\rangle\left|\psi\right\rangle = \left|11\right\rangle U\left|\psi\right\rangle,$$

where $\left|\psi\right\rangle$ is any computational basis state of n qubits. When $U = \mathrm{NOT}$, this coincides, of course, with the definition of the Toffoli gate.

Although seemingly very complicated, all these gates admit decompositions with just one-qubit and two-qubit gates. For all the mathematical details, we refer you to the book by Nielsen and Chuang [13].

That is all for this section. You now know all that you will need in order to understand the full power of quantum algorithms. We'll begin exploring them in earnest in the next chapter.

Summary

In this chapter, we have studied general systems with multiple qubits. We have learned how to describe their states, how to measure them one qubit at a time (or all at once!), and how to transform them with gates that act on several qubits at the same time. We have also discussed universality, a property of some sets of gates that allow us to decompose big quantum gates into simpler ones.

You are now fully equipped. You have all the mathematical knowledge required to understand complex quantum algorithms. Are you excited? Well, you should be, because starting in the next chapter, we are going to study problems that quantum computers can solve much, much faster than their classical counterparts. Buckle up, we are going to light speed!

9

The Full Power of Quantum Algorithms

Life is beautiful.
The only problem is that many people mistake beautiful for easy.

— Mafalda M.

So far, we have discussed several applications of quantum computing. These have shown us how superposition and entanglement can work together in quantum protocols, and, when discussing Deutsch's algorithm in *Section 6.3*, we also mentioned how the notion of interference can play a very significant role in them.

Nevertheless, all the protocols and algorithms that we have considered so far have only used, at most, two qubits, which is a very significant limitation. In this chapter, we are moving one step forward. We will be exploiting all that we have learned in *Chapter 8* to introduce more powerful algorithms that will use arbitrarily many qubits. This will prepare us to better understand the more sophisticated quantum algorithms that we will discuss later in the book.

The plan for the next few pages is the following. We will start easily yet boldly with a protocol that runs on three-qubit systems: quantum teleportation, which is, in a way, the dual of superdense coding. We will then introduce two quantum algorithms that may use an arbitrarily large amount of qubits; one of which (if not both!) is the younger sibling of our good old Deutsch's algorithm.

These are the topics that will be covered in this chapter:

- Quantum teleportation
- The Deutsch–Jozsa algorithm
- The Bernstein–Vazirani algorithm

After reading this chapter, you will have a solid understanding of the quantum teleportation protocol and the Deutsch–Jozsa and Bernstein–Vazirani algorithms. This will allow you to see how the building blocks of quantum algorithms work and can be used in practice.

Without further ado, let's get started. It's time to bring the fantasies of some science fiction aficionados to life. Let's talk about quantum teleportation!

9.1 Quantum teleportation

In *Section 6.1*, we introduced superdense coding: a protocol that enables us to send classical data (two bits) through quantum data (one qubit). In this section, we will take the opposite route, and we will discuss **quantum teleportation**: a protocol that will enable us to send quantum data (a qubit state) relying on the transmission of classical data (two classical bits). As you will soon see for yourself, the similarities between superdense coding and quantum teleportation go beyond the symmetry of their goals. Indeed, just like superdense coding, quantum teleportation will initially assume that both communicating parties each has a qubit of an entangled pair (in a Bell state). So, yes, entanglement is going to once again play a crucial role in this protocol.

Before we get to discussing the details of quantum teleportation, maybe it is good to sit down for a moment and reflect on the significance of what it promises to achieve. Imagine for a second that you have a qubit in some arbitrary state: this can be any state—one that is arbitrarily difficult to write down, or one in which the amplitudes may be the weirdest irrational numbers mankind has ever seen. By classical means, would there be a way for you to always be able to send to us its state with a fixed number of classical bits? Even if you always were able to fully describe the state of the qubit, the answer is in the negative, because it is impossible to send a pair of arbitrary complex amplitudes—with full accuracy—through a fixed or bounded number of bits. After all, a finite number of bits can only represent a finite amount of states, and the set of one-qubit states is uncountably[1] infinite.

What quantum teleportation promises to do is the following. Imagine that you have a qubit in your lab in an arbitrary state $|\psi\rangle$, and you may not even know what the state is. Assume further that you and us each have a qubit from a pair in a Bell state. With quantum teleportation, just by doing some quantum operations and measurements, and by sending us two bits, you will make our qubit from our pair get into state $|\psi\rangle$—no matter how far apart we are. Naturally, because of the no-cloning theorem, you will lose the state of the original qubit, because otherwise we would end up with two copies of the same state. After all, this is quantum teleportation, not quantum cloning! Now, this sounds impressive, doesn't it?

Important note

Quantum teleportation enables a party to send a qubit state to another party by just sending them two classical bits. The protocol assumes that they initially have one qubit each, and that the pair of qubits is in a Bell state.

Quantum teleportation is a powerful protocol and, for it, two qubits don't quite do the job; we need an extra one, so this is going to be our first three-qubit protocol. Later in the

[1]The word "uncountably" is not added for literary emphasis; it is an actual mathematical term. If you are intrigued, we strongly invite you to satisfy your curiosity by studying a tiny bit of logic and set theory!

chapter, we will explore other algorithms that can use an arbitrarily large number of qubits, but we wanted to start slow.

9.1.1 The details of quantum teleportation

Let us now get into the intricacies and details of quantum teleportation. First things first, we should fix what our initial situation is and what prerequisites we need.

In what follows, we will assume that Alice has a qubit in a certain (perhaps even unknown) state $|\psi\rangle$, and she wants to teleport this qubit state to Bob's lab. In addition to this, as mentioned earlier, we will also require Alice and Bob to each have a qubit from a pair that is in a Bell state; in particular, we will assume that this particular pair is in the Bell state

$$\frac{1}{\sqrt{2}}(|00\rangle + |11\rangle).$$

Nevertheless, and as in superdense coding, any Bell state would do, although the choice of a different Bell state would require some minor adaptations, as we will later discuss.

In accordance with the conventions that we have been adopting through this book, in this pair of qubits, we will assume that the first qubit is Alice's and the second one is Bob's.

We will now bring Alice's qubit in state $|\psi\rangle$ and the entangled pair of qubits together into a unified three-qubit system, in which the first qubit will be Alice's $|\psi\rangle$ qubit, and the two other qubits will be those of the entangled pair. Thus, the initial state of this three-qubit system would be

$$|\psi\rangle \otimes \frac{1}{\sqrt{2}}(|00\rangle + |11\rangle).$$

Exercise 9.1

In the initial state that we have considered for our three-qubit system:

(a) If a measurement of the first qubit yielded a 0, what would the state of the system collapse to?

(b) If a measurement of the second qubit yielded a 0, what would the state of the system collapse to?

(c) Clearly, the state can be written as the product of the state of the first qubit and the state of the last two qubits. What implications does this have for measurements?

This is our initial setup. Let's now carry out the steps that will lead the state $|\psi\rangle$ to be teleported to Bob's lab.

The first step in the protocol is for Alice to apply a CNOT gate controlled by its original $|\psi\rangle$ qubit and targeted to her qubit from the entangled pair, and then to follow that by an application of the Hadamard gate on her original qubit. This wording may be confusing, so—since a picture is worth a thousand words—let us better visualize this with a quantum circuit. These are the gates that Alice should be applying on her qubits:

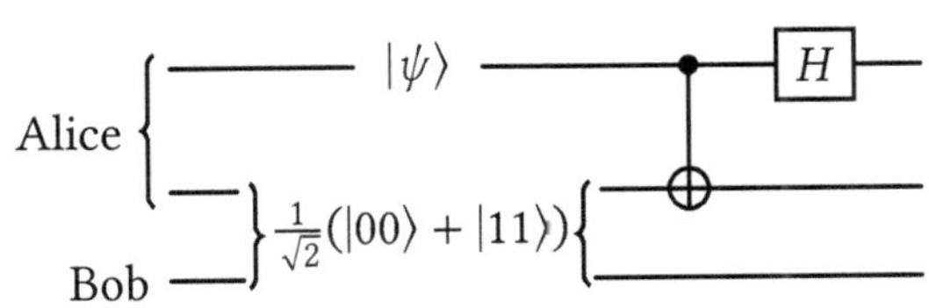

The obvious question now is: how do these gates modify the state of the system? Let's find out.

In what follows, we will assume that $|\psi\rangle = \alpha|0\rangle + \beta|1\rangle$. With this in mind, it can be readily checked that the action of the CNOT gate leaves the state of the system as

$$\alpha|0\rangle \otimes \frac{1}{\sqrt{2}}(|00\rangle + |11\rangle) + \beta|1\rangle \otimes \frac{1}{\sqrt{2}}(|10\rangle + |01\rangle),$$

which can be further simplified as

$$\frac{1}{\sqrt{2}}\left(\alpha(|000\rangle + |011\rangle) + \beta(|110\rangle + |101\rangle)\right).$$

Thus, the subsequent application of a Hadamard gate on the first qubit will transform the state of the system into

$$\frac{1}{2}\left(\alpha(|000\rangle + |100\rangle + |011\rangle + |111\rangle) + \beta(|010\rangle - |110\rangle + |001\rangle - |101\rangle)\right).$$

Exercise 9.2

Using the linearity of quantum gates and the properties of the bra-ket notation, show—if you haven't already!—that the result of applying the CNOT operation in the scenario that we have just considered is indeed

$$\alpha\,|0\rangle \otimes \frac{1}{\sqrt{2}}(|00\rangle + |11\rangle) + \beta\,|1\rangle \otimes \frac{1}{\sqrt{2}}(|10\rangle + |01\rangle).$$

Is it easier to compute these results using the bra-ket notation or the 8×8 matrices that represent three-qubit quantum gates?

Alright, so far, so good. Now our next step will be for Alice to measure the two qubits she has in her possession, that is, the first two qubits of our system. As we learned in *Chapter 8*, if the outcome of the measurement is 00, the state of the whole system will collapse to

$$|C_{00}\rangle := \alpha\,|000\rangle + \beta\,|001\rangle = |00\rangle \otimes (\alpha\,|0\rangle + \beta\,|1\rangle) = |00\rangle \otimes |\psi\rangle,$$

which means that Alice's qubits will be in the state $|00\rangle$ and Bob's qubit will be in the state $|\psi\rangle$—just as we wanted! In this case, we have successfully teleported the state $|\psi\rangle$ to Bob and there's nothing else we have to do.

Now what happens if get a different measurement outcome? For 01, the state of the system will collapse to

$$|C_{01}\rangle := \alpha\,|011\rangle + \beta\,|010\rangle = |01\rangle \otimes (\alpha\,|1\rangle + \beta\,|0\rangle),$$

which is… oddly close to what we want, but we're not quite yet there! In this situation, Alice's qubits are in the state $|01\rangle$ (which is fair enough, nothing particularly interesting here), but Bob's qubit is in the state $\alpha |1\rangle + \beta |0\rangle$, which is like $|\psi\rangle$ except for the fact that the amplitudes are flipped. Nevertheless, this can be very easily fixed: all we would have to do is apply an X gate on Bob's qubit, and thus its state would turn out to be

$$X(\alpha |1\rangle + \beta |0\rangle) = \alpha |0\rangle + \beta |1\rangle = |\psi\rangle .$$

So, if Alice's measurement is 01, she just has to relay this information to Bob and he will know he has to apply an X gate to his qubit and, in that way, he will own his very own copy of the precious state $|\psi\rangle$.

The two remaining outcomes of Alice's measurement leave us in a similar situation. If the outcome of her measurement is 10, the state collapses to

$$|C_{10}\rangle := \alpha |100\rangle - \beta |101\rangle = |10\rangle \otimes (\alpha |0\rangle - \beta |1\rangle).$$

So, in this case, in order to retrieve $|\psi\rangle$, Bob would need to apply a Z gate to his qubit as

$$Z(\alpha |0\rangle - \beta |1\rangle) = \alpha |0\rangle + \beta |1\rangle .$$

Lastly, if Alice's measurement yields 11 as a result, the state of the system will be

$$|C_{11}\rangle := \alpha |111\rangle - \beta |110\rangle = |11\rangle \otimes (\alpha |1\rangle - \beta |0\rangle),$$

and, in this case, Bob will have to apply the gates X and Z (in that order) to retrieve $|\psi\rangle$.

This is pretty much the story. In this final step, we need Alice to measure her qubits and send these results to Bob. After receiving those results—if we collect and simplify the conclusions we have obtained—Bob must perform the following operations sequentially in order to complete the teleportation process:

- If the last bit of Alice's measurement is 1, he must apply an X gate to his qubit
- If the last bit of Alice's measurement is 1, he must apply a Z gate to his qubit

This procedure is represented in quantum circuits as follows:

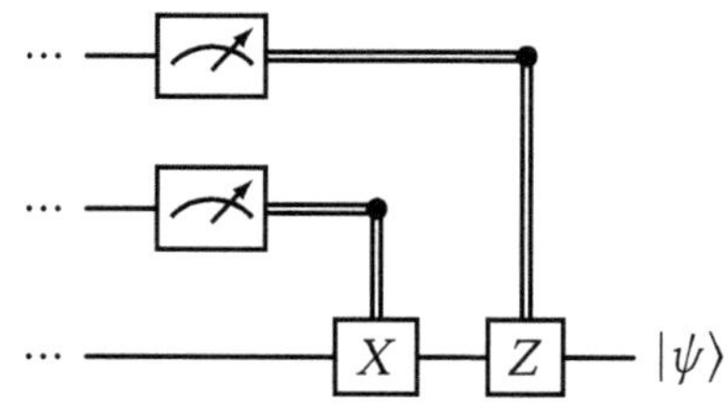

This way of writing down the circuit can seem cryptic, but it's actually kind of natural. In a way, it is as if the bit from the measurement operation were controlling the application of the X and Z gates, in the same way that a control qubit operates in a controlled gate. That's why we have connected the measurement operation to the quantum gate, just as we do with controlled quantum gates. Nevertheless, unlike in controlled gates, this operation is controlled by classical information (bits), and we represent this using a double wire. In general, in a quantum circuit, double wires are used to represent classical information, whereas normal wires always represent quantum information.

Important note

The quantum circuit that summarizes the operations needed to make quantum teleportation work is the following:

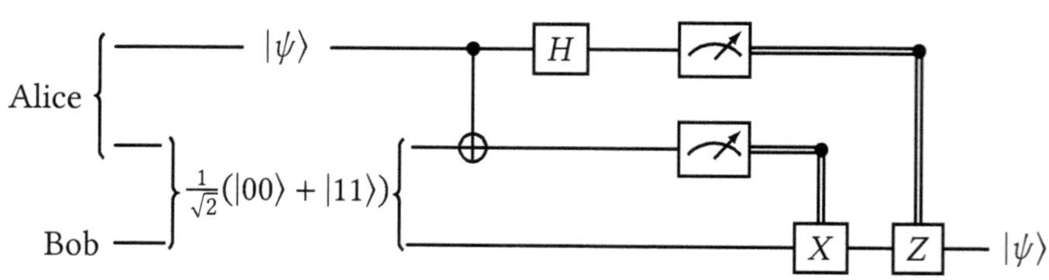

Earlier in this section, we mentioned that while we were going to use the Bell state $\frac{1}{\sqrt{2}}(|00\rangle + |11\rangle)$ for the entangled pair, any other Bell state would do the trick. In fact, a different choice of Bell state would only mean that Bob would have to apply different combinations of gates upon receiving the results from Alice's measurement. The following exercise will give you the chance to work through this.

Exercise 9.3

Consider a quantum teleportation setup analogous to the one we have discussed, but having the state of the entangled paired shared by Alice and Bob initialized to $\frac{1}{\sqrt{2}}(\left|01\right\rangle + \left|10\right\rangle)$. If Alice's measurement yields 00, what gates should Bob apply to his qubit in order to retrieve $\left|\psi\right\rangle$?

And that's quantum teleportation. As an additional remark, if you want to trace back the origin of this protocol, you may wish to know that it was first introduced in the article "Teleporting an unknown quantum state via dual classical and Einstein-Podolsky-Rosen channels" [59].

Now, that was a lot of information to absorb, so let's zoom out and look at the big picture.

9.1.2 An overview of quantum teleportation

Throughout this section, we have discussed quantum teleportation and seen how it enables the transfer of a quantum state from one qubit to another. In the setup that we discussed, we have considered two characters: Alice and Bob. Alice was in possession of one qubit in the state $\left|\psi\right\rangle$ that we wanted to teleport, and, in addition, Alice and Bob each had a qubit from an entangled pair in a Bell state. From this point, Alice only had to perform a CNOT and Hadamard operations on the two qubits under her control and then measure them. Then she needed to send the two-bit measurement result to Bob and, with that information, Bob could apply some simple transformations on his qubit in order to set its state to $\left|\psi\right\rangle$.

Before moving on to other things, a few important remarks are in order about this protocol:

- Quantum teleportation **does not violate the no-cloning theorem**. Notice that the only way for Alice to be able to send the state $\left|\psi\right\rangle$ to Bob is by losing access to that state herself. At the end of the execution of the protocol, Bob's qubit is in state $\left|\psi\right\rangle$, but both of Alice's qubits are in a state from the computational basis determined by the outcome of her measurement.

- Quantum teleportation **works at any distance**. No matter how far apart Alice and Bob may be, entanglement still works—and the protocol still works. This has been demonstrated experimentally, teleporting a state through a distance of 1400 km [60].
- Quantum teleportation **does not allow information to be sent faster than the speed of light**. Keep in mind that, in order for Bob to have access to $|\psi\rangle$, he needs to first wait to receive Alice's two bits telling him which gate operations he needs to perform. Those two bits, being the classical information that they are, cannot travel faster than the speed of light. The general statement of this fact is known as the **no-communication theorem** or the **no-signaling principle**.

To learn more...

The ideas behind quantum teleportation can go far beyond the mere teleportation of states. As we will mention in *Chapter 14*, a protocol known as **quantum gate teleportation** draws on some ideas from the quantum teleportation protocol in order to enable the fault-tolerant implementation of quantum gates.

That's all there is to the quantum teleportation of a qubit. We can now move on to discuss our first quantum algorithm that will use an arbitrarily large number of qubits, and it's an algorithm that will certainly sound familiar. Remember Deutsch's algorithm from *Chapter 6*? Well, he has grown up and he wants to pay a visit!

9.2 The Deutsch–Jozsa algorithm

Back in *Chapter 6*, we introduced Deutsch's algorithm, which enabled us to detect whether a one-bit Boolean function was constant or balanced using a single call to its oracle. In this section, leveraging the power of multi-qubit systems, we are going further: we are about to introduce an algorithm that will enable us to complete an analogous task for Boolean functions on any number of bits.

This algorithm is, as you may have guessed, the **Deutsch–Jozsa** algorithm, which was introduced by David Deutsch and Richard Jozsa in their paper "Rapid solution of problems by quantum computation" [58]. Actually, as we have already mentioned, Deutsch's algorithm is just a particular case of the Deutsch–Jozsa algorithm for two-qubit systems (and hence for one-bit Boolean functions).

Before we can discuss this shiny new algorithm, we first need to understand what it is capable of accomplishing and, thus, we need to go beyond the definition of one-bit Boolean functions that we introduced in *Chapter 6*. So, let's go big. In general, an n-bit Boolean function is a function that takes as an argument n bits and returns a single bit as output. For example, the function f, defined as

$$f(0,0) = 0, \qquad f(0,1) = 1, \qquad f(1,0) = 1, \qquad f(1,1) = 0$$

would be a two-bit Boolean function. Now, Boolean functions are said to be **constant** if they are, well, constant: if they map all possible inputs to the same bit. The function f above is not constant because $f(0,0) = 0$ but $f(0,1) = 1$. In addition, a Boolean function is said to be **balanced** if exactly half of its inputs are mapped to 0 and the other half gets mapped to 1. The function f above is, for example, balanced.

It is important to highlight that some Boolean functions may be neither constant nor balanced. For example, the two-bit Boolean function

$$f(0,0) = 0, \qquad f(0,1) = 1, \qquad f(1,0) = 1, \qquad f(1,1) = 1$$

is not constant (because, for example, $f(0,0) \neq f(0,1)$), but it isn't balanced either, because one of its inputs gets mapped to 0, whereas three get mapped to 1—neither of those are half of the inputs!

Exercise 9.4

Write down all the two-bit balanced and constant functions.

With this, we know exactly the kind of functions we are going to be working with and how they can be classified. Now, when we studied Deutsch's algorithm, we saw how quantum computers used oracles in order to evaluate those functions. Let's discuss those oracles in a little more detail and see how they can be used for general Boolean functions.

9.2.1 Oracles, in all their glory

As we have mentioned on countless occasions, one of the defining features of quantum gates is that they are reversible (or invertible): when you apply a quantum gate, you should always be able to apply another one that undoes its action. This seemingly innocent condition can prove to be hard to swallow if we are accustomed to classical computing. For instance, this requirement forbids the existence of a quantum AND gate, since the AND operation is not reversible; if an AND gate returns a 0, it's impossible to tell whether it came from the input 00, 01, or 10.

This need for reversibility would make it impossible to construct a quantum gate that would directly implement an n-bit Boolean function f, for $n > 1$. By that, we mean a gate that would simply take n qubits in computational basis states (for $n > 1$) as input and return the (computational basis) output of f on those inputs, while discarding the inputs altogether.

Exercise 9.5

In the previous statement, we assumed that $n > 1$. For which *one-bit* Boolean functions f does there exist a one-qubit gate G_f that directly implements f? This would mean that $G_f(|0\rangle) = |f(0)\rangle$ and $G_f(|1\rangle) = |f(1)\rangle$.

Given these constraints, oracles are a very convenient abstraction. Generalizing what we had for one-bit Boolean functions, given any n-bit Boolean function f, an oracle for f is an $(n+1)$-qubit gate O_f, which acts on the computational basis states as follows. If x is any sequence of n bits and b is a single bit, consider the $(n+1)$-dimensional computational basis state $|x, b\rangle$. The oracle O_f transforms this state as $O_f |x, b\rangle = |x, b \oplus f(x)\rangle$, where $\oplus$ denotes the XOR operation (or addition modulo 2). Graphically,

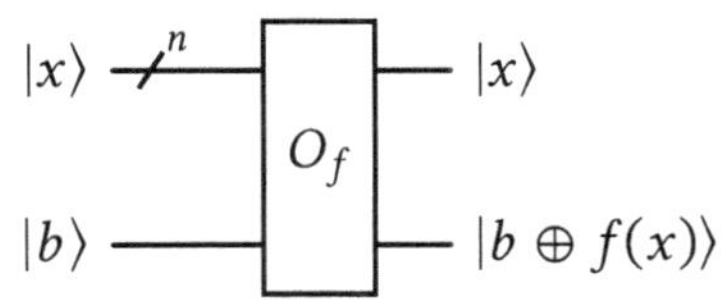

where we are using the tiny stick with the n above it to denote that the first wire actually represents a bundle of n qubits.

Naturally, from their definition on the computational basis, the action of oracles is extended by linearity to all possible states.

There are a couple of neat things about oracles, the first of which being that, for any Boolean function f, its corresponding oracle is always a valid quantum gate: the transformation it induces is always unitary. In addition to this, oracles provide us with a very convenient way of measuring how many times a certain quantum algorithm calls a given function.

To learn more...

In order to prove that the oracle of a Boolean function is, indeed, a unitary operation and hence a valid quantum gate, it suffices to notice that it maps an orthonormal basis (the computational basis) into another orthonormal basis (which is, again, the computational basis, even if with a different ordering of its elements). Should you want to better understand this, these notions are covered in *Appendix A*.

So, now that we have introduced all the definitions that we need, we can finally state what the Deutsch–Jozsa algorithm promises to deliver: it will discern whether a function is constant or balanced with a single call to its oracle.

Important note

For any natural number n, consider an n-bit Boolean function f that is constant or balanced. Using a single evaluation of the oracle of f, the Deutsch–Jozsa algorithm can determine, in a single execution, whether f is constant or balanced.

Notice that there is a significant difference in the prerequisites of Deutsch's algorithm and Deutsch–Jozsa. When we considered one-bit Boolean functions, we always knew that they would be either constant or balanced, so we could formulate our problem for any arbitrary function. Here, we are requiring that the oracle that we are provided be specifically that of a constant or balanced function. That is, if we are given some function that doesn't fit into these two categories, we are not claiming to be able to detect it; actually, if an oracle for such a function is used, the algorithm will just return a random output. For this reason, this kind of problem is called a **promise problem**: we need to be "promised" that the input will satisfy the required constraints.

With that, we are ready to introduce the Deutsch–Jozsa algorithm.

9.2.2 The magic behind Deutsch–Jozsa

How does the Deutsch–Jozsa algorithm work? You won't be surprised to learn that it works just like Deutsch's algorithm, but with more qubits, to the extent that—as we have hinted already—Deutsch's algorithm is just the Deutsch–Jozsa algorithm for $n = 1$. Thus, and in spite of any déjà vu that we may encounter, let us describe, step by step, what Deutsch–Jozsa is all about.

In what follows, we will consider an n-bit Boolean function f. To use this algorithm on f, we need a system with $n + 1$ qubits. The first n of these must be initialized to $|0\rangle$, while the last one must be initialized to $|1\rangle$; of course, if all qubits are initialized to $|0\rangle$, we could just apply an X gate on the last qubit and move on with our lives. With this setup, our first step will be (surprise, surprise) to apply a Hadamard gate on each of the qubits, as follows:

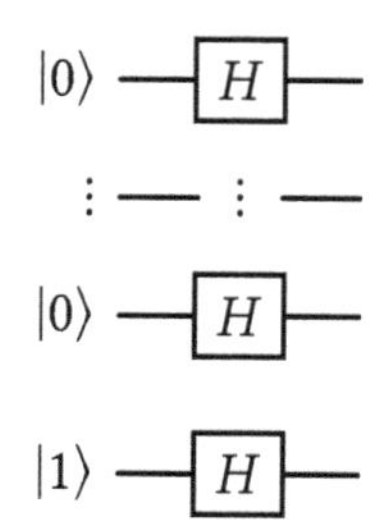

The state of the system right after this operation will be

$$\left(\frac{1}{\sqrt{2}}(|0\rangle + |1\rangle)\right) \otimes \cdots \otimes \left(\frac{1}{\sqrt{2}}(|0\rangle + |1\rangle)\right) \otimes \left(\frac{1}{\sqrt{2}}(|0\rangle - |1\rangle)\right),$$

which, by expanding the first n tensor products and collecting all the $\frac{1}{\sqrt{2}}$ factors, can be rearranged as

$$\frac{1}{\sqrt{2^{n+1}}}(|0\rangle + |1\rangle + \cdots + |2^n - 1\rangle) \otimes (|0\rangle - |1\rangle) = \frac{1}{\sqrt{2^{n+1}}}\left(\sum_{x=0}^{2^n-1} |x\rangle\right) \otimes (|0\rangle - |1\rangle).$$

Notice that this means that the first n qubits are in a balanced superposition of all the computational basis states; that is, all the kets formed with binary strings of length n (which we have decided to write as decimal numbers, for brevity). To convince yourself that this equality is indeed true, you can test it for small n; and, if you want to verify it rigorously (which we strongly recommend!), you can go back to *Exercise 8.2.*

Now that we're in full superposition, our next step will be to apply the oracle of f, so our circuit thus far would look like this:

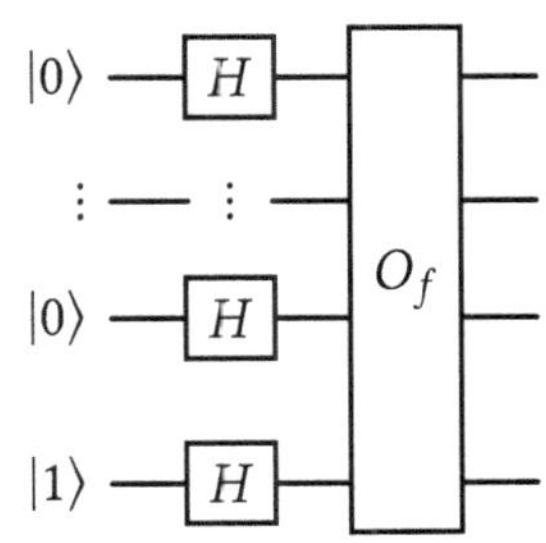

The application of the oracle will leave the state of our system as

$$\left(\frac{1}{\sqrt{2}}\right)^{n+1}\left(\sum_{x=0}^{2^n} |x\rangle\right) \otimes (|0 \oplus f(x)\rangle - |1 \oplus f(x)\rangle) = \sum_{x=0}^{2^n} \frac{(-1)^{f(x)}}{\sqrt{2^{n+1}}} |x\rangle (|0\rangle - |1\rangle),$$

where we have used the fact that, for any bit b, we have

$$|0 \oplus b\rangle - |1 \oplus b\rangle = (-1)^b(|0\rangle - |1\rangle).$$

Remember that we proved this in *Section 6.3*, just in case you needed a reminder!

We are almost done! All that is left to do is apply a Hadamard gate on each of the first n qubits, as follows:

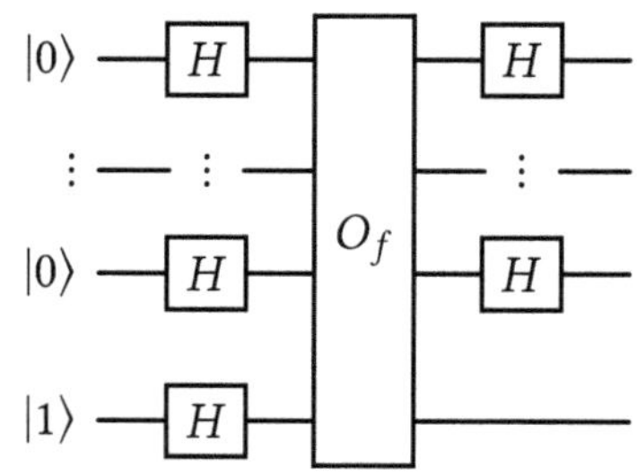

Figuring out what the state will be after this operation is slightly trickier than in the cases that we considered previously. To get started, consider an n-qubit state from the computational basis, $|x_1 \cdots x_n\rangle$; here, we are assuming that x_i are bits (so, if $x_1 = 0$, that means that the first qubit is in state $|0\rangle$). The action of n Hadamard gates (each acting on one of the n qubits) will be

$$\begin{aligned}
&(H \otimes \cdots \otimes H)|x_1 \cdots x_n\rangle = H|x_1\rangle \otimes \cdots \otimes H|x_n\rangle \\
&= \left(\frac{1}{\sqrt{2}}(|0\rangle + (-1)^{x_1}|1\rangle)\right) \otimes \cdots \otimes \left(\frac{1}{\sqrt{2}}(|0\rangle + (-1)^{x_n}|1\rangle)\right) \\
&= \left(\frac{1}{\sqrt{2}}((-1)^{x_1 \cdot 0}|0\rangle + (-1)^{x_1 \cdot 1}|1\rangle)\right) \otimes \cdots \otimes \left(\frac{1}{\sqrt{2}}((-1)^{x_n \cdot 0}|0\rangle + (-1)^{x_n \cdot 1}|1\rangle)\right) \\
&= \frac{1}{\sqrt{2^n}} \sum_{y_1,\ldots,y_n=0}^{1} (-1)^{x_1 y_1 + \cdots + x_n y_n} |y_1 \cdots y_n\rangle . \\
&= \frac{1}{\sqrt{2^n}} \sum_{y_1,\ldots,y_n=0}^{1} (-1)^{x_1 y_1 \oplus \cdots \oplus x_n y_n} |y_1 \cdots y_n\rangle .
\end{aligned}$$

Incidentally, regarding the last equality, notice that, in general, given any two bits x and y, we have $(-1)^{x+y} = (-1)^{x \oplus y}$. Feel free to run a proof by induction on n if you want to (rigorously) verify the remaining equalities (we love induction, and we hope that you do as well). Again, it is at least a good idea to work out the computation for small n to get an intuition of what is going on with those tensor products!

In the sequel, for any pair of natural numbers x and y (between 0 and $2^n - 1$) with binary representations $x_1 \cdots x_n$ and $y_1 \cdots y_n$, respectively, we will use $x \odot y$ to denote $x_1 y_1 \oplus \cdots \oplus x_n y_n$. Having introduced this notation, we can apply the result above to deduce that the state of our system after the application of the Hadamard gates will be

$$\sum_{x=0}^{2^n-1} \frac{(-1)^{f(x)}}{\sqrt{2^{n+1}}} ((H \otimes \cdots \otimes H)\left|x\right\rangle)(\left|0\right\rangle - \left|1\right\rangle) = \sum_{x=0}^{2^n-1} \frac{(-1)^{f(x)}}{\sqrt{2^{n+1}}} \left(\sum_{y=0}^{2^n-1} \frac{(-1)^{x \odot y}}{\sqrt{2^n}} \left|y\right\rangle \right) (\left|0\right\rangle - \left|1\right\rangle),$$

which can be rearranged and simplified as

$$\sum_{x=0}^{2^n-1} \sum_{y=0}^{2^n-1} \frac{(-1)^{f(x)+x \odot y}}{\sqrt{2^{2n+1}}} \left|y\right\rangle (\left|0\right\rangle - \left|1\right\rangle) = \sum_{y=0}^{2^n-1} \sum_{x=0}^{2^n-1} \frac{(-1)^{f(x)+x \odot y}}{2^n} \left|y\right\rangle \otimes \frac{1}{\sqrt{2}}(\left|0\right\rangle - \left|1\right\rangle).$$

And that's all we need, for here comes the moment of truth. The last and final step of the Deutsch–Jozsa algorithm is to perform a measurement of the first n qubits. If we do so, the probability of obtaining the outcome 0 in all n qubits will be

$$\left| \sum_{x=0}^{2^n-1} \frac{(-1)^{f(x)+x \odot 0}}{2^n} \right|^2 = \left| \sum_{x=0}^{2^n-1} \frac{(-1)^{f(x)}}{2^n} \right|^2.$$

Clearly, if f is constant (and $(-1)^{f(x)}$ takes the same value for all x), this will add up to 1; that would be an example of **constructive interference**. On the contrary, if f is balanced (and $(-1)^{f(x)} = 1$ for half of the values of x, and $(-1)^{f(x)} = -1$ for the other half), the sum will add up to 0; in this case, we would find a **destructive interference**.

Exercise 9.6

Show that, indeed, the probability of obtaining $0 \cdots 0$ in the final measurement of the Deutsch–Jozsa algorithm is the one we have claimed.

In summary, we have shown that, if the function f is constant, the final measurement of the algorithm will always return the outcome $0 \cdots 0$, whereas if f is balanced, at least one of the bits in this measurement is guaranteed to be 1.

Important note

Given oracle O_f for an n-bit Boolean function f, if we run a single shot of the circuit

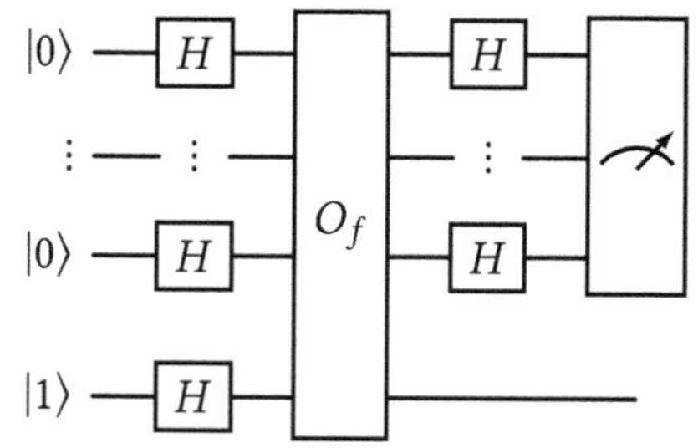

the measurement on the first n qubits will return the outcome 0 on each and every qubit if the function is constant, and it will return 1 on at least one qubit if the function is balanced. This is the Deutsch–Jozsa algorithm.

As we mentioned in *Section 6.3*, this algorithm was introduced by David Deutsch and Richard Jozsa in their paper "Rapid solution of problems by quantum computation" [58]. And, as promised in *Section 6.3*, we have seen how this algorithm generalizes Deutsch's algorithm to Boolean functions with an arbitrarily large number of inputs.

If we were to use a classical computer to solve the problem that the Deutsch–Jozsa algorithm tackles for an n-bit Boolean function f, we would need $2^n/2 + 1$ evaluations of f in the worst case. Think about it as follows: if you have evaluated f on half of the inputs and have always gotten the same output, can you truly tell that f is constant or balanced? For all you know, it could be constant, but it could also be balanced! It might just be that f takes a different value in the other half of inputs you haven't yet evaluated.

Thus, the Deutsch–Jozsa algorithm is one of those algorithms that perfectly illustrates the power and might of quantum algorithms. It solves a problem (deterministically!) with a single evaluation of the oracle of a function, while any deterministic classical algorithm might require $2^{n-1} + 1$ evaluations of the function. Notice the difference: we are comparing one evaluation to an exponentially growing number of evaluations. This is huge.

Nevertheless, it might be best not to get too excited just yet, for Deutsch–Jozsa is a brilliant algorithm for a problem that no one asked to solve. However, the ideas behind it will power some of the very useful algorithms that we will introduce later in the book!

Before moving on to other matters, it could be a good idea to reflect on some of the abstract ingredients that make the Deutsch–Jozsa algorithm clearly outperform any classical counterpart. And, along the way, we may find some opportunities to further clarify some common myths about quantum computers.

9.2.3 The truth about quantum parallelism

"Quantum computers are faster because they can use superposition to evaluate all possible solutions at once". Such a catchy and inspiring statement. We have heard it. You have heard it. We've all heard it, and that doesn't make it any less misleading. As you can probably tell by now, designing a quantum algorithm doesn't appear to be as simple as just putting some qubits in a state in superposition and just moving with the flow. Indeed, as we have mentioned on countless occasions already, quantum algorithms are powered by an intricate balance with many elements at play, and superposition is just one of them. By now, you have had plenty of opportunities to see how entanglement also plays a major role in the design of quantum protocols, and we will now also discuss how **interference**—as neglected as it may be in those cool, popular science videos—plays a crucial role in the design of quantum algorithms.

Generalizing a tiny bit, most quantum algorithms follow the same structure:

1. Get all the qubits in a state of **superposition** (with Hadamard gates all over the place).
2. Get some transformations done, potentially relying on **entanglement**.
3. Rely on **interference** to extract the information that we want from the resulting state.

The Deutsch–Jozsa algorithm (and Deutsch's algorithm) follows exactly this structure. Once all the qubits are in a state of superposition, the oracle of the Boolean function transforms the state, and then some Hadamard gates generate the necessary interference, bringing us to the final state that we want to measure.

To learn more...

Interference is a physical phenomenon that occurs when two waves are combined, resulting in a new wave. The short reason as to why this is relevant in quantum mechanics is the wave-particle duality, which is one the core concepts in quantum mechanics and—very vaguely speaking—states that quantum-mechanical objects are both particles and waves at the same time. This idea is behind some of the weird things that qubits can do. Or is it a consequence of it? Maybe it's best not to go down philosophical rabbit holes just now.

For our purposes, it may suffice to know that interference manifests itself every time a quantum gate leads to the addition or subtraction of some of the amplitudes of a state. If you are interested in learning more about the physics behind interference, the book *Quantum Physics of Atoms, Molecules, Solids, Nuclei, and Particles* [17] by Eisberg and Renick might be a good place to get started.

In any case, you may be pleased to know that Deutsch's algorithm has a very interesting physical analogue, which is the **Mach-Zehnder interferometer**. This device is used to detect whether an object induces a phase difference in photons (in quantum computing terms, whether it applies a certain "gate" on them!). In particular, in the interferometer, photons go in and they go through beam splitters (which are kind of Hadamard gates). Then, each of the beams goes through different routes, and one of them goes through the object that may induce the phase shift (this is the oracle). Finally, the beams meet again at a beam splitter (again, similar to a Hadamard gate) and the interference that they produce there gives the relevant information about the phase difference that they had. Of course, this is a vague analogy: we are not claiming that a Mach-Zehnder interferometer actually implements Deutsch's algorithm!

As you have seen, getting the qubits in a state of superposition was the easy part of the algorithm. Most of the magic is in the two other ingredients!

Keep this general structure in mind, because you will see it is used time and again in most quantum algorithms.

After this clarification, we can move on and turn our attention to another algorithm that is very similar in nature to the Deutsch–Jozsa algorithm. Actually, it is so similar that you almost won't be able to tell them apart from each other!

9.3 The Bernstein–Vazirani algorithm

You may have heard the expression "killing two birds with one stone". In this chapter, we are doing exactly that—at least metaphorically (we love birds). Remember the Deutsch–Jozsa algorithm? Now we are going to use it to kill two problems with one (quantum) shot. Get ready to see the Bernstein–Vazirani algorithm: Deutsch–Jozsa with a new outfit!

Assume that you are given an n-bit Boolean function f, and you are promised that, for some number b with binary representation $b_1 \cdots b_n$, the output of the function is computed as

$$f(x_1, \dots, x_n) := x_1 b_1 \oplus \cdots \oplus x_n b_n.$$

Under these hypotheses, classically, how many evaluations of f would you need to perform in order to find b? You would need at least n: one for each bit of the binary representation of b. You could find b_1 as $f(1, 0, \dots, 0)$, b_2 as $f(0, 1, 0, \dots, 0)$, and so on and so forth. Now, if you had access to a quantum computer… how many evaluations of f would you need to perform? As it turns out, if you have an oracle for f, the Bernstein–Vazirani algorithm can get the job done with a single evaluation. Just a single shot! How would this algorithm work? Well, get ready for a surprise.

The Bernstein–Vazirani algorithm works exactly as the Deutsch–Jozsa algorithm. It only differs in the assumptions on the Boolean function that it considers, but the circuit it uses is a carbon copy. Indeed, assume that we run the circuit from the Deutsch–Jozsa algorithm for a function f that satisfies the hypothesis of the Bernstein–Vazirani algorithm. Recall that the circuit is the following:

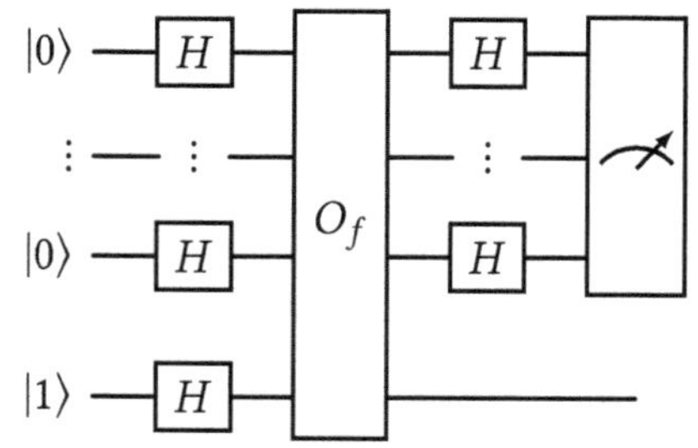

Right before performing the measurement operations, according to our study of the Deutsch–Jozsa algorithm, the state of the system will be the following:

$$\sum_{y=0}^{2^n-1}\sum_{x=0}^{2^n-1}\frac{(-1)^{f(x)+x\odot y}}{2^n}\left|y\right\rangle\otimes\frac{1}{\sqrt{2}}(\left|0\right\rangle-\left|1\right\rangle).$$

This means that the probability of measuring any y (any sequence of bits that would represent a number y between 0 and 2^n-1) will be

$$\left|\sum_{x=0}^{2^n-1}\frac{(-1)^{(x\odot b)+(x\odot y)}}{2^n}\right|^2,$$

where we have used the fact that $f(x)=x\odot b$. Setting $y=b$ makes this expression equal to 1, which means that the measurement of the n qubits is guaranteed to return $b_1,\ldots,b_n$, hence revealing the bit string that we were looking for!

Important note

If we run the Deutsch–Jozsa algorithm on an n-bit Boolean function f such that, for a fixed sequence of bits $b_1,\ldots,b_n$, we always have

$$f(x_1,\ldots,x_n)=x_1b_1\oplus\cdots\oplus x_nb_n,$$

then the final measurement of the n first qubits of the circuit is guaranteed to return the bits $b_1, \ldots, b_n$. This is the Bernstein–Vazirani algorithm.

To learn more…

On its own, the Bernstein–Vazirani algorithm offers a speed-up over its classical counterparts (one evaluation instead of n), but it's not as impressive as the one that the Deutsch–Jozsa algorithm offered for its own problem (one evaluation instead of $2^n/2 + 1$).

Nevertheless, it is possible to create a recursive version of the Bernstein-Vazirani problem that yields a quantum algorithm with a superpolynomial advantage over any classical counterpart [61].

The Bernstein–Vazirani algorithm was introduced by Ethan Bernstein and Umesh Vazirani in their paper "Quantum Complexity Theory" [61]. That's all we have to say about this algorithm. Now, let's wrap things up!

Summary

In this chapter, we have introduced our first protocols that were built on systems with more than two qubits.

We first discussed quantum teleportation, a protocol that, analogous to superdense coding, enabled two communicating parties to send quantum information (in the form of a qubit state) through the communication of classical data (two bits) thanks to the power of quantum entanglement. All of this, of course, was built on the prerequisite that the two communicating parties shared some qubits from an entangled pair.

We then introduced the Deutsch–Jozsa algorithm, which generalizes Deutsch's algorithm and works for Boolean functions with inputs of arbitrary dimension. Along the way, we made some remarks about the true nature of quantum parallelism, highlighting the role that entanglement and interference play in most quantum algorithms.

We concluded by exploring, through the Bernstein-Vazirani algorithm, how the very same circuit used in the Deutsch–Jozsa algorithm could be applied to solve different problems.

In the following chapter, we will see how all of these protocols and algorithms can be coded into Qiskit, and we will learn how to master all the possibilities it offers to handle systems with arbitrarily many qubits.

10

Coding with Many Qubits in Qiskit

Software and cathedrals are much the same—
first we build them, then we pray.

— Samuel T. Redwine

In the last couple of chapters, we have introduced systems with arbitrarily many qubits and we have seen some of the eye-opening things that can be done with them. In this chapter, we are putting all of that knowledge into practice. We already know how to use Qiskit to work with one or two qubits, and it is time for us to break free of every limit and construct circuits with any number of qubits.

The contents of this chapter are the following:

- Working with many qubits in Qiskit
- Quantum teleportation

- The Deutsch–Jozsa algorithm
- The Berstein–Vazirani algorithm

By the end of this chapter, you will know how to construct any quantum circuit in Qiskit—regardless of its size. You will also be able to implement all the protocols and algorithms that we discussed in *Chapter 9*; namely, the quantum teleportation protocol and the Deutsch–Jozsa and Bernstein-Vazirani algorithms. In this regard, we will also go through how to implement a quantum oracle for *any* Boolean function.

This will be our last foundational chapter on Qiskit. Are you ready for a ride?

10.1 Working with many qubits in Qiskit

To get started, we will import the `QuantumCircuit` class from the Qiskit package—nothing particularly exciting just yet:

```
from qiskit import QuantumCircuit
```

As you are about to find out, working with several qubits is not at all different from working with a couple of them. All it takes is instantiating circuits with a larger number of qubits and… well, everything else just grows in size accordingly. That's pretty much it! There's no secret beyond that.

While having more qubits does not change much of how Qiskit works, going beyond two qubits enables us to use some new features from the framework, the most obvious of which is being able to access new quantum gates.

In *Chapter 8*, we introduced the Toffoli gate. This is a three-qubit gate in which two qubits act as controls and one acts as a target, in such a way that—informally speaking—an X gate is applied on the target when the controls are in state $|1\rangle$. To use this gate in Qiskit, we only have to call the `ccx` method, whose name comes from the fact that the Toffoli gate is a doubly controlled X gate. We create and draw a five-qubit circuit using this gate in the following piece of code:

```
# Create a qubit with 5 qubits:
qc = QuantumCircuit(5)
# Apply the Toffoli gate:
qc.ccx(0,2,3)
qc.draw("mpl")
```

In the call to `ccx`, the first two arguments (`0,2`) identify the control qubits, and the third one (`3`) specifies the target qubit. Upon running the preceding code, you will get the following representation:

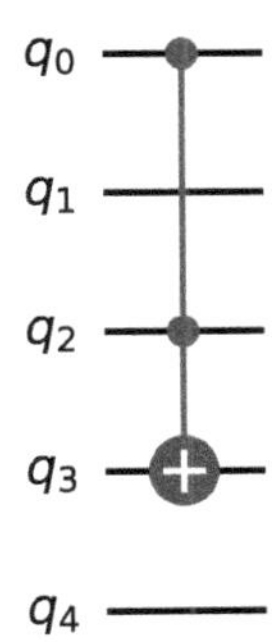

In addition to the Toffoli gate, there are plenty of other useful multi-qubit quantum gates, and we will now introduce one of them. Given a system with $n + 1$ qubits, the **multi-controlled NOT gate** controlled by the first n qubits and targeting the last qubit is an $(n + 1)$-qubit gate, which takes any computational basis state $|x_1\rangle \cdots |x_n\rangle |b\rangle$ and transforms it into

$$|x_1\rangle \cdots |x_n\rangle |b \oplus ((x_1) \cdot (x_2) \cdots (x_n))\rangle ,$$

where $(x_1) \cdot (x_2) \cdots (x_n)$ denotes the multiplication of the bits $x_1, \ldots, x_n$. For example, for $n = 3$, this gate would take $|1\rangle |1\rangle |1\rangle |0\rangle$ to $|1\rangle |1\rangle |1\rangle |1\rangle$, but it would keep $|1\rangle |0\rangle |1\rangle |0\rangle$ unchanged (since the second control qubit is in state $|0\rangle$). Informally speaking, this multi-controlled NOT gate works—on computational basis states—by applying an X on the target qubit if and only if the control qubits are all in state $|1\rangle$.

It is worth pointing out that the CNOT and Toffoli gates can be viewed as particular cases of the multi-controlled NOT gate. They would correspond, respectively, to multi-controlled NOT gates with one and two control qubits.

The multi-controlled NOT gate can be used in Qiskit with the method `mcx`, which takes two arguments: the first must be a list specifying the indices of the control qubits, and the second must be the index of the target qubit. In the following example, we add to `qc` a multi-controlled NOT gate targeting the third qubit, and controlled by the first, fourth, and fifth qubits:

```
qc.mcx([0,3,4],2)
qc.draw("mpl")
```

The representation that we get after performing this operation is the following:

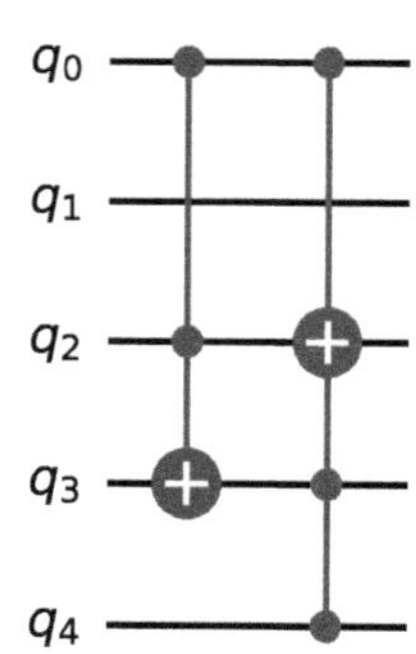

Notice how the control qubits are represented by thick dots, the target qubit is denoted by the ⊕ symbol, and all of the qubits are joined by a vertical bar. Also, keep in mind that the preceding representation also includes the Toffoli gate that we applied previously, as we have used the same circuit object.

Exercise 10.1

Create a quantum circuit with six qubits and apply a multi-controlled NOT gate that does the following:

- Targets the fourth qubit
- Is controlled by the first, third, fifth, and sixth qubits

These are all the quantum gates that we will need for now. Let's dive a little bit deeper into how quantum circuits are internally represented within Qiskit.

10.1.1 Using registers

Back in *Chapter 7*, we introduced registers, and we will now devote a few paragraphs to reviewing what they are and seeing how they behave on systems with many qubits.

Registers are groups of qubits or (classical) bits, and quantum circuits operate on a collection of quantum registers (storing qubits) and classical registers (storing bits). Indeed, when we call `QuantumCircuit(num_q, num_c)`, we are simply instantiating a quantum circuit with a quantum register named `q` storing `num_q` qubits, and with a classical register named `c` having `num_c` bits. And, while that is the default, it is always possible to add new registers to a quantum circuit—without going any further, the `measure_all` method adds a classical register by the name of `meas`.

For example, if we call `measure_all` on our previous circuit, we can see how a new classical register has been added to our circuit.

Consider the following piece of code:

```
qc.measure_all()

print("Quantum registers:", qc.qregs)
print("Classical registers:", qc.cregs)
```

Remember that, in every quantum circuit object, the lists of quantum and classical circuits are stored in the attributes `qregs` and `cregs` respectively. Upon running the preceding code, this is the output that we get:

```
Quantum registers: [QuantumRegister(5, 'q')]
Classical registers: [ClassicalRegister(5, 'meas')]
```

If we want to add our own quantum and classical registers to a circuit, all we need to do is create some `QuantumRegister` or `ClassicalRegister` objects, add them as attributes to our quantum circuit for easier access, and include them in the circuit through the `add_register` method, as we illustrate in the following example:

```
from qiskit import QuantumRegister, ClassicalRegister

qc.new_qreg = QuantumRegister(1, name = "new")
qc.new_creg = ClassicalRegister(10, name = "new_c")
qc.add_register(qc.new_qreg, qc.new_creg)
```

Notice how, when creating the registers, we are specifying how many bits or qubits they hold, and we are also giving the register a name. After running this, we can print the lists of registers and draw our circuit:

```
print("Quantum registers:", qc.qregs)
print("Classical registers:", qc.cregs)
qc.draw("mpl")
```

The console output of this instruction will be the following:

```
Quantum registers: [QuantumRegister(5, 'q'), QuantumRegister(1, 'new')]
Classical registers: [ClassicalRegister(5, 'meas'),
                      ClassicalRegister(10, 'new_c')]
```

And this will be the representation of our new circuit:

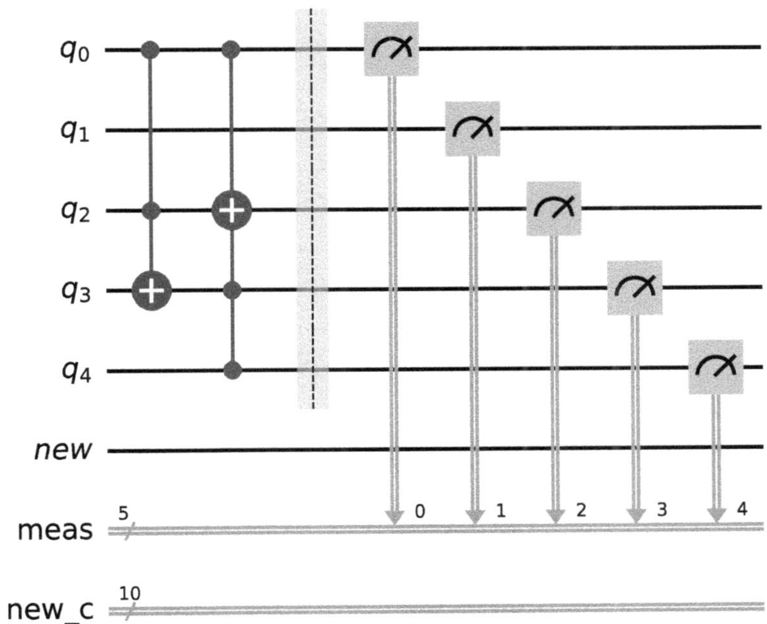

Having clarified how to use registers in our new general setting, we can now pay off a debt that we created for ourselves a couple of chapters ago.

10.1.2 How to verify the equivalence of circuits

Back in *Subsection 8.3.2*, when we first came to know the Toffoli gate, we presented a possible implementation of it in terms of more elementary gates. However, we didn't prove that the circuit we provided actually implemented the Toffoli gate. It is time for us to fix that, and we will do it with the help of Qiskit. To begin, we will prepare a Qiskit circuit with our implementation of the Toffoli gate. This can be done as follows:

```
tof = QuantumCircuit(3)
tof.h(2)
```

```
tof.cx(1,2)
tof.tdg(2)
tof.cx(0,2)
tof.t(2)
tof.cx(1,2)
tof.tdg(2)
tof.cx(0,2)
tof.t(1)
tof.t(2)
tof.cx(0,1)
tof.h(2)
tof.t(0)
tof.tdg(1)
tof.cx(0,1)
tof.draw("mpl") # Draw it to check it!
```

Notice that we are using the `t` method to apply a T gate (as we discussed in *Chapter 4*), and we are using the `tdg` (short for "T dagger") method to apply the inverse of the T gate. The representation that we get after defining this circuit is the following:

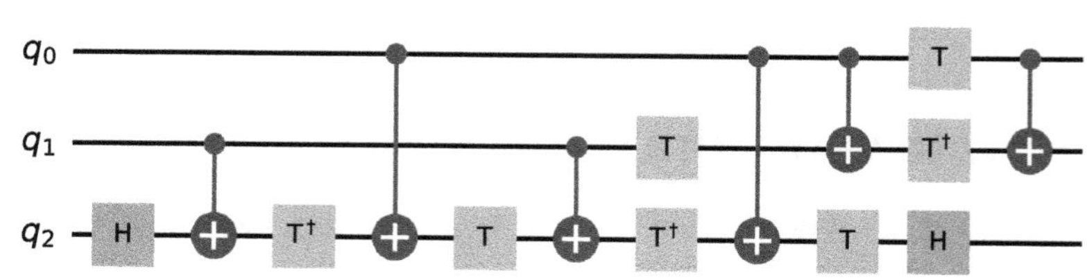

This matches the construction that we provided in *Subsection 8.3.2.*

With this, we have a Qiskit circuit that contains what we presume to be an implementation of the Toffoli gate, and now is when Qiskit comes to the rescue. Qiskit provides a way of computing which unitary operation a given quantum circuit implements. This can be accessed via the `Operator` class, which takes a quantum circuit as an initialization argument and processes it in order to extract the operator that it implements.

Therefore, in order to verify whether the action of our circuit is equivalent to that of the Toffoli gate, we just have to get an `Operator` object from our circuit and from the Toffoli gate, and compare them! We can do this with the following piece of code:

```
from qiskit.quantum_info import Operator
real_tof = QuantumCircuit(3)
real_tof.ccx(0,1,2) # This circuit has the actual Toffoli gate.

# Check if the operators are equal:
print(Operator(tof) == Operator(real_tof))
```

Upon running this, we get that, indeed, both operators are equal. And that's one of the many ways in which Qiskit can make our lives easier!

If you actually look into the `Operator` objects that Qiskit constructs, you will find some 8×8 matrices with the entries that would correspond to the coordinate matrix of the Toffoli gate. Keep in mind, however, that, instead of displaying a zero, some entries may display values such as `1.11022302e-16+0.j`; the reason for these "zero-like yet not zero" values lies in the fact that Qiskit performs its computations using floating-point arithmetic, so the results it gives may not always be exact.

To learn more...

Comparing two results after performing computations with floating-point numbers is a risky business, because operations on floating-point numbers are not exact. Qiskit takes this into account and only compares operator objects for approximate equality.

That pretty much covers it in terms of how to use Qiskit with many qubits. Let's now try to program the protocols that we introduced in *Chapter 9*. We will begin with quantum teleportation!

10.2 Quantum teleportation

The quantum teleportation protocol aims to "teleport" a state from one qubit $|\psi\rangle$ to another and, in order to do that, it relies on the following circuit:

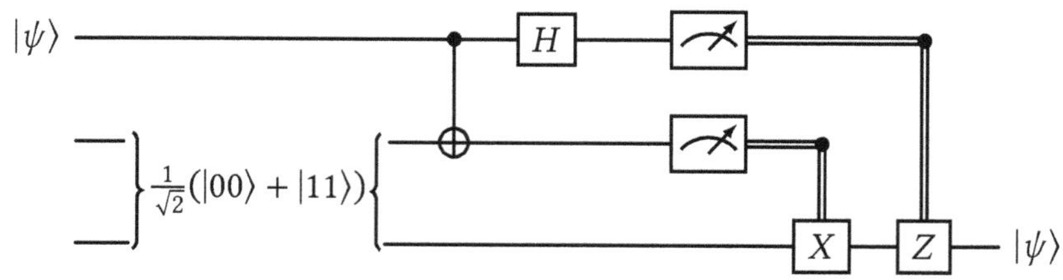

The state $|\psi\rangle$ meant to be teleported should be the initial state of the first qubit and, as we showed in *Chapter 9*, at the end of the execution, the third qubit will be in state $|\psi\rangle$.

To implement this protocol in Qiskit, we will first implement its circuit, ignoring the state $|\psi\rangle$, that is, assuming it to be $|0\rangle$; we will store this circuit in a global variable, `qt`.

To get started, we will initialize the circuit and get the two bottom qubits in a Bell state using a Hadamard gate followed by a CNOT gate, as usual:

```
qt = QuantumCircuit(3)
qt.h(1)
qt.cx(1,2)
```

Then we can apply a CNOT gate on the first two qubits and a Hadamard gate on the first one, following the specification:

```
qt.cx(0,1)
qt.h(0)
```

With that, we have reached the point at which we need to perform some measurements. To store the results from the measurements, we will include a classical register named `"aux"` with two bits. Then we will measure each of the first two qubits and store the results on their corresponding bits:

```
qt.aux = ClassicalRegister(2, name = "aux")
```

```
qt.add_register(qt.aux)


qt.measure(0, qt.aux[0])
qt.measure(1, qt.aux[1])
```

Notice that we could've also performed the measurements with the instruction

```
qt.measure([0,1], [qt.aux[0], qt.aux[1]])
```

When we specify the qubits and bits on which the measurement is meant to operate, we can send them as lists, or—if we only wish to measure one qubit (at a time)—provide their direct values without the need to create a list.

At this point, we have to perform some operations conditioned by the results of the measurements. If the result of the measurement on the second qubit was 1 (and, therefore, if the second bit in aux is 1), we must apply an X gate on the third qubit. Similarly, if the measurement on the first qubit was 1, we have to apply a Z gate on the third qubit. To do this, we can use the `if_test` method as follows:

```
# Apply X on qubit 2 if aux[1] == 1
with qt.if_test((qt.cregs[0][1], 1)):
    qt.x(2)


# Apply Z on qubit 2 if aux[0] == 1
with qt.if_test((qt.cregs[0][0], 1)):
    qt.z(2)
```

Observe how, to perform operations conditional on a certain bit object, `bit` being `1`, we only have to put those operations inside a `with qc.if_test((bit, 1))` block. In general, given any classical bit object `bit` (like `qt.cregs[0][1]` above) from a circuit `circuit`, any circuit instructions in a block `with circuit.if_test((bit,1))` will only be executed if `bit` stores the value `1`—and this will, of course, be evaluated at runtime, immediately before the first instruction inside the block. If instead we used `with qc.if_test((b,0))`, the circuit instructions within the block would only be executed if `b` stores `0` at runtime.

The use of `if_test` enables us to embed classical control operations directly into quantum circuit objects, fitting all the operations in a single job that can then be submitted and executed; this allows us to evaluate classical conditions while the circuit is being executed on real hardware. Notice how, if we had to evaluate this locally (and thus use several jobs), we wouldn't be able to implement the quantum teleportation protocol, as, in IBM's interface, it is not possible to share the state of a quantum circuit between different jobs.

To learn more...

We will not be using the `if_test` method further, but we should mention that one could implement more complex constructions such as if-else conditionals. It is also possible to implement other flow control structures like "for" loops, "while" loops, or "switch" statements using the `for_loop`, `while_loop`, and `switch` methods, respectively. All of these tools are fully explained in the Qiskit documentation, should you ever need to use them.

Thus far, we have completed our implementation of the quantum teleportation circuit. However, in order to be able to test the protocol, we will also add a measurement operation on the last qubit, which will send its results to a new classical register that we will label as `final`:

```
qt.final = ClassicalRegister(1, name = "final")
qt.add_register(qt.final)
qt.measure(2,qt.final)
```

Of course, this measurement will not reveal everything about the state (as we have discussed on numerous occasions), but at least it will give us some information about it—namely, it will reveal the module of its amplitudes. As we mentioned in *Chapter 2*, if we wanted to fully determine the state using measurements, we would have to resort to quantum tomography techniques [21].

If we draw the circuit that we have built using `qt.draw("mpl")`, we will get the following output:

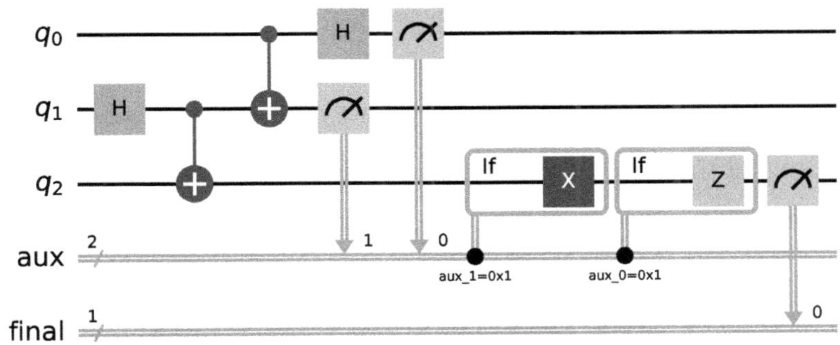

Now that we have our circuit ready, we should go ahead and test it. For this, we will first initialize a simulator backend and a sampler object:

```
from qiskit_aer import AerSimulator
from qiskit_ibm_runtime import SamplerV2 as Sampler

backend = AerSimulator(seed_simulator = 18620123)
sampler = Sampler(backend)
```

With these objects, we can implement a wrapper function that will take as input any one-qubit circuit preparing a state $|\alpha\rangle$ and will apply the quantum teleportation protocol on that state (composing $|\alpha\rangle$ before our `qt` circuit on our first qubit):

```
def quantum_teleportation(state_preparation):
    # Compose the state_preparation circuit with qt.
    # We use front = True for state_preparation to go before qt.
    circuit = qt.compose(state_preparation, front = True)
    # Get the results.
    job = sampler.run([circuit])
```

```
    return job.result()[0].data.final
```

The preceding function composes the circuits, runs the sampler, and returns a dictionary with the measurement counts—that should be more than enough for us to test our implementation of the quantum teleportation protocol. Notice how, in the definition of the function, we have used the `compose` method. This takes as an argument another circuit and returns a brand new circuit in which the instructions of the argument circuit are added, by default, at the end of the original circuit. If we instead want the instructions of the argument circuit to be added at the beginning of the original circuit, we can use the optional `front = True` parameter, as we did above. It is important to note that, when using the `compose` method on two circuits as `original.compose(other)`, Qiskit assumes that `other` does not have more qubits or bits than `original`, and creates a new circuit object appending the m qubits and n bits of `other` to the first m qubits and n bits of `original`.

To begin with a very simple test, we will teleport the state $|0\rangle$. For this, we only need to pass an empty state preparation circuit (remember that all circuits are initialized to $|0\rangle$ by default, so an empty circuit will suffice to prepare that state):

```
result_0 = quantum_teleportation(QuantumCircuit(1))
print(result_0.get_counts())
```

After running the preceding code, we get the expected result:

```
{'0': 1024}
```

Let's now consider a slightly more sophisticated example and teleport the $|+\rangle$ state. Its state preparation circuit only needs to include a Hadamard gate, so we can run the following lines of code:

```
state_plus = QuantumCircuit(1)
state_plus.h(0)

result_plus = quantum_teleportation(state_plus)
print(result_plus.get_counts())
```

This gives us an even distribution of measurement outcomes, which is perfectly consistent with the state $|+\rangle$:

```
{'1': 513, '0': 511}
```

Exercise 10.2

In the case that we are considering, we expect the final state to be $|+\rangle$, but the final measurement that we perform cannot guarantee us whether the final state is indeed $|+\rangle$. All we know for certain is that the real amplitudes of the final state have modulus $1/\sqrt{2}$—but the state could still be $|-\rangle$, for example!

Tweak the preceding circuit so that the measurement operation that we perform can enable us to determine that the final state is $|+\rangle$, and no other.

Hint: Consider using a Hadamard gate.

Now it's time for you to try to teleport a different quantum state!

Exercise 10.3

Use our implementation of the quantum teleportation protocol to teleport the state $|1\rangle$.

And that's all there is to implementing the quantum teleportation protocol! In the next section, we will see how the Deutsch–Jozsa algorithm can be run on Qiskit.

10.3 The Deutsch–Jozsa algorithm

Now that we've put our Qiskit skills into practice with quantum teleportation, it's time to reach for new heights. In this section, we will implement the Deutsch–Jozsa algorithm.

As we saw in *Chapter 9*, the Deutsch–Jozsa algorithm only involves the execution of a rather simple circuit. Indeed, the trickiest bit for us to be able to recreate may not lie in the implementation of the algorithm itself... but in the construction of oracles!

When we studied the Deutsch–Jozsa algorithm, we took the oracle as a given. However, if we want to be able to simulate and test the algorithm fully by ourselves, we will have to figure out this whole business of building quantum oracles. That will be the whole purpose of the following subsection.

10.3.1 Let's build some oracles

In this subsection, we will present a small procedure that will enable us to construct a quantum oracle for any Boolean function. The implementations that this procedure will yield will, by no means, be optimal—but they will be good enough for our purposes. Sometimes perfection can be sacrificed for convenience.

To get started nice and easily, let's consider a very simple three-bit Boolean function f, which we will define to take $f(111) = 1$, and to take the value 0 for any inputs different from 111 (that is, $f(000) = \cdots = f(110) = 0$). Clearly, the following four-qubit circuit will implement an oracle for this function:

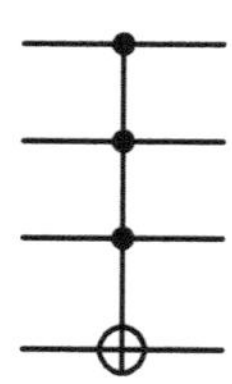

Certainly, for any computational basis state $|x\rangle|y\rangle|z\rangle \otimes |b\rangle$, the state will not change unless $x = y = z = 1$, in which case it will be $|x\rangle|y\rangle|z\rangle \otimes |b \oplus 1\rangle$. Therefore, for any input computational basis state of the aforementioned form, the output state will be

$$|x\rangle|y\rangle|z\rangle \otimes |b \oplus f(x, y, z)\rangle,$$

exactly as required for an oracle computing f!

Let's now try to tackle a slightly more sophisticated example. Consider the function g, which takes the value 0 except for the input 110, for which $g(110) = 1$. Implementing an oracle for this function is a little bit more complicated—but not too much! We can simply take this circuit:

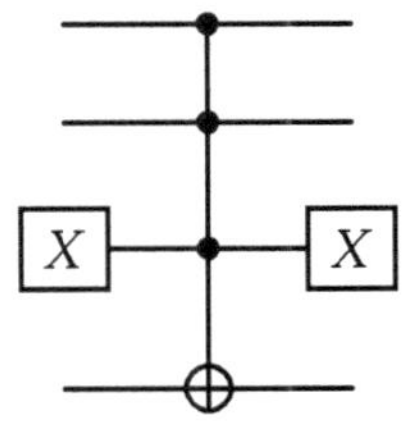

It is very easy to verify that, for any state of the form $|x\rangle|y\rangle|z\rangle \otimes |b\rangle$, the control bits will all be equal to one if and only if $x = 1$, $y = 1$, and $z = 0$. Therefore, the state of the last qubit will flip if and only if $(x, y, z) = (1, 1, 0)$, and the first three qubits will suffer no change whatsoever (notice that the X gate in the third gate is inverted by the X gate at the end). In summary, this circuit will transform the aforementioned state into

$$|x\rangle|y\rangle|z\rangle \otimes |b \oplus g(x, y, z)\rangle,$$

thus implementing an oracle for g.

In the two examples that we have considered, you may have been able to identify a pattern. Given any n-bit string $s = s_1 \cdots s_n$, it is very straightforward to construct an oracle for the n-bit function σ_s that takes the value 1 on s and 0 on any other value. As you may have already deduced on your own, the $(n + 1)$-qubit circuit that would implement such an oracle would only need to perform these operations:

1. For every bit $k = 1, \ldots, n$, apply an X gate on qubit k if $s_k = 0$.
2. Apply a multi-controlled X gate targeting the last qubit with controls on the first n qubits.
3. Repeat step 1.

In case it isn't obvious why this would actually implement an oracle for σ_s, we can sketch a short proof. Assume that the input state to the circuit described is a computational basis state of the form $|x_1\rangle \cdots |x_n\rangle \otimes |b\rangle$. After all the X gates from step 1 have been applied, this state will be as follows:

$$|x_1 \oplus (s_1 \oplus 1)\rangle \cdots |x_n \oplus (s_n \oplus 1)\rangle \otimes |b\rangle.$$

Notice how we have used the fact that applying an X gate in the computational basis state $|x\rangle$ transforms it into $|x \oplus 1\rangle$; also keep in mind that $a \oplus 1$ is 1 if $a = 0$, and is 0 when $a = 1$. At this point, the first n qubits will be equal to 1 if and only if $s_k = x_k$ for every $1 \leq k \leq n$—that's because $x_k \oplus s_k = 0$ if and only if $x_k = s_k$. Consequently, the application of the multi-controlled X gate in step 2 will bring the state to

$$|x_1 \oplus (s_1 \oplus 1)\rangle \cdots |x_n \oplus (s_n \oplus 1)\rangle \otimes |b \oplus \sigma_s(x_1, \ldots, x_n)\rangle .$$

Finally, the action of the X gates in step 3 will transform the state into

$$|x_1\rangle \cdots |x_n\rangle\, |b \oplus \sigma_s(x_1, \ldots, x_n)\rangle ,$$

showing that our circuit, as claimed, implements an oracle for σ_s.

We've just seen how oracles for functions of the form σ_s—taking null values for all inputs except s—are very easy to construct, but how could this help us build an oracle for any general function? Well, consider an arbitrary n-bit function f and let $\{s_1, \ldots, s_n\}$ be the set of n-bit strings s for which $f(s) = 1$. It is trivial that

$$f(y) = \sigma_{s_1}(y) \oplus \cdots \oplus \sigma_{s_n}(y),$$

because only one of the terms on the right-hand side can be different from zero for a given y.

And here comes the cool bit: an oracle for this function can be implemented as the mere concatenations of the oracles for $s_1, \ldots, s_n$. To see why, consider an n-bit string $x = x_1 \cdots x_n$ and a bit b. If we apply those oracles in sequence on the input state $|x_1\rangle \cdots |x_n\rangle \otimes |b\rangle$, the output state will simply be as follows:

$$|x_1\rangle \cdots |x_n\rangle \otimes \left|b \oplus \sigma_{s_1}(x) \oplus \cdots \oplus \sigma_{s_n}(x)\right\rangle = |x_1\rangle \cdots |x_n\rangle \otimes |b \oplus f(x)\rangle .$$

To illustrate this way of constructing oracles, consider the three-bit Boolean function f defined as $f(x) = 0$ for all x except $f(000) = 1$, $f(100) = 1$, and $f(111) = 0$. An oracle for this function could be constructed as follows:

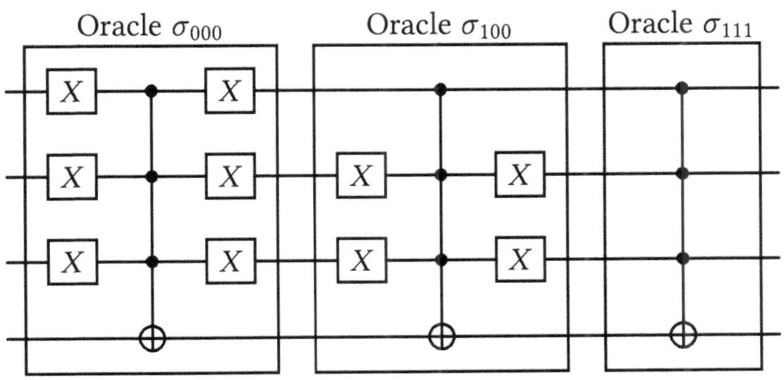

We can now implement this procedure for constructing oracles in a `build_oracle` function. This function will take as input a list of binary strings for which a function takes the value 1, and it will return a quantum circuit implementing an oracle for it. For example, if we wanted to build an oracle for a four-bit Boolean function that only takes the value 1 on the inputs 0000 and 0001, we would pass the argument `["0000", "0001"]` to the `build_oracle` function.

To construct such a function and implement our procedure, we can use the following:

```
def build_oracle(strings_one):
    # If the function is never 1, the oracle is the identity.
    # Hence, we return an empty circuit.
    if len(strings_one) == 0:
        return QuantumCircuit()

    # Number of bits that the function takes as input:
    n = len(strings_one[0])

    qc = QuantumCircuit(n+1)
    for x in strings_one:
        # Find the positions in the string x where the bit is 0.
```

```
        # For this, we find the list of indices i such that x[i]=='0'.
        bits_zero = []
        for i in range(len(x)):
            val = x[i]
            if val == '0':
                bits_zero.append(i)

        # Step 1 in our construction.
        for bit in bits_zero:
            qc.x(bit)

        # Step 2.
        qc.mcx(list(range(n)), n)

        # Step 3.
        for bit in bits_zero:
            qc.x(bit)

    return qc
```

The preceding code is just a fairly direct translation of our algorithm into Python. First, we perform a basic input check: if the list we have been given is empty, that means that no string takes the value 1, and hence the oracle will just be an empty circuit. Then, we extract the length of the strings (assuming, of course, that all of them have the same length), and we iterate over them. For every string, we apply a CNOT gate on the qubits corresponding to the bits with value 0 (as in Step 1). Then, we apply a multi-controlled NOT gate (as in step 2), and, lastly, we apply the NOT operations from step 3.

For example, in order to test this function on the three-bit f function that we considered before, we can run the following instruction:

```
build_oracle(["000", "100", "111"]).draw("mpl")
```

This displays the following output, which perfectly matches our previous circuit:

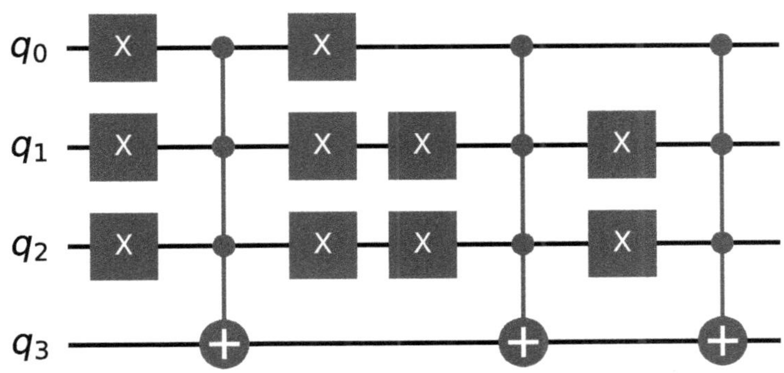

With `build_oracle`, we can effortlessly construct oracles for any Boolean function—although, as we've mentioned, these may be *far* from optimal. Let's now see how we can use it within the Deutsch–Jozsa algorithm.

10.3.2 Implementing the Deutsch–Jozsa algorithm

The circuit behind the Deutsch–Jozsa algorithm couldn't be more straightforward. Given an oracle for an n-bit Boolean function, it is an $(n + 1)$-qubit circuit with the bottom qubit initialized to $|1\rangle$ consisting of the following:

- The application of a Hadamard gate on each qubit
- The action of the oracle
- The application of a Hadamard gate on the first n qubits
- The measurement of the first n qubits

Taking as an argument an oracle circuit, `oracle`, we can implement the circuit for the Deutsch–Jozsa algorithm as follows:

```
def DJ(oracle):
    # Number of qubits in the circuit (same as the oracle).
    # If we are working with an n-bit function, nqubits = n + 1.
    nqubits = oracle.num_qubits

    # Create a circuit with (nqubits) qubits and (nqubits-1) bits.
    # Remember we are only going to measure the first (nqubits-1) qubits!
    qc = QuantumCircuit(nqubits, nqubits - 1)

    # Initialize the bottom qubit to |1>.
    qc.x(nqubits - 1)

    for i in range(nqubits):
        qc.h(i)
    qc.barrier()
    qc = qc.compose(oracle)
    qc.barrier()
    for i in range(nqubits - 1):
        qc.h(i)
        qc.measure(i,i)
    return qc
```

And that's it! To test our implementation, let us first consider the simplest of examples: a constant three-bit Boolean function that always returns zero (clearly, the circuit for its oracle will be empty). To prepare the Deutsh–Jozsa circuit, we only have to run the following:

```
oracle = QuantumCircuit(4)
dj = DJ(oracle)
dj.draw("mpl")
```

With this, we get the following visualization of the circuit, which matches the specification:

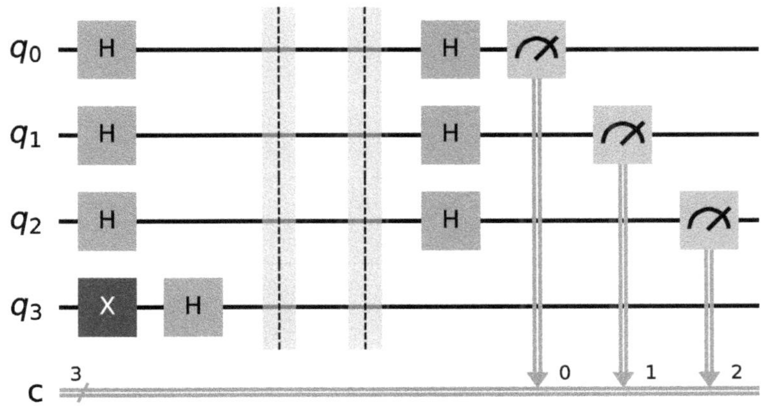

Now we can run this circuit with our sampler, keeping in mind that a single shot suffices for the Deutsch–Jozsa algorithm:

```
job = sampler.run([dj], shots = 1)
result = job.result()[0].data.c
print(result.get_bitstrings())
```

Since the function that we used was constant, we get, as expected, the measurement outcome `['000']`.

We can now try out a more elaborate example, with a balanced three-bit Boolean function. In order to construct the oracle of a Boolean function, we only have to specify a function that takes the value 1 at four distinct inputs (since there are a total of eight possible three-bit strings). Thus, let's run the following piece of code:

```
# Oracle of a balanced function (takes the value 1 with four inputs):
oracle = build_oracle(["000", "101", "100", "110"])
# Prepare the D-J circuit:
```

```
dj = DJ(oracle)
# Run it!
job = sampler.run([dj], shots = 1)
result = job.result()[0].data.c
print(result.get_bitstrings())
```

After running this code, we get the outcome `['010']`. As expected, since our function was balanced, the measurement does not yield `000`.

Exercise 10.4

Construct the oracle for another balanced function and verify that, indeed, our implementation of the Deutsch–Jozsa algorithm works on it.

That is one way in which the Deutsch–Jozsa algorithm can be implemented using Qiskit. To conclude this chapter, we will explore how the Bernstein–Vazirani algorithm can be run.

10.4 The Bernstein–Vazirani algorithm

The circuit used in the Bernstein–Vazirani algorithm is exactly identical to the one used in Deutsch–Jozsa, so we don't have to write any new lines of code. All we need to do is consider a Boolean function defined as

$$f_b(x_1, \ldots, x_n) := x_1 b_1 \oplus \cdots \oplus x_n b_n$$

for an n-bit binary string b. According to our discussion in *Chapter 9*, if we run the Deutsch–Jozsa algorithm with the oracle for f_b, the final measurement will return exactly b.

Let us then consider the function f_b, defined as above, for $b = 101$. A quick computation reveals that the only values for which f_b takes the value 1 are 001, 011, 100, and 110. Thus, in order to retrieve b, we would only need to run the Deutsch–Jozsa algorithm with the following oracle:

```
oracle = build_oracle(["001", "011", "100", "110"])
```

Thus, we can proceed as follows:

```
dj = DJ(oracle)
job = sampler.run([dj], shots = 1)
result = job.result()[0].data.c
print(result.get_bitstrings())
```

And, as expected, the outcome from the measurement is, precisely, `['101']`.

Exercise 10.5

Try the Bernstein–Vazirani algorithm on a different f_b function for a different binary string b.

That was our last algorithm for this chapter, so let us wrap things up!

Summary

In this chapter, we have discussed how to work with arbitrarily many qubits in Qiskit, and we have learned how to implement all the multi-qubit quantum algorithms that we have considered so far.

We began the chapter with a brief discussion of some of the Qiskit features that we had yet to cover. We mentioned how the Toffoli gate can be used in Qiskit circuits, and we also took the chance to introduce multi-controlled NOT gates. From that point, we briefly covered the use of quantum and classical registers within Qiskit, and we saw how Qiskit enables us to compute the operator that a quantum circuit implements. In connection with that last point, we were able to verify the correctness of an implementation of the Toffoli gate that we had introduced earlier in the book.

We then implemented the quantum teleportation protocol and, in the process, we learned how (classical) conditionals can be embedded right into Qiskit circuits. We also tested our implementation on a couple of simple cases.

After that, we dived into the Deutsch–Jozsa algorithm, first discussing how to implement any quantum oracle (albeit with a not necessarily efficient method) in a quantum computer, and how to then do it on Qiskit. With that foundation, we were able to implement the Deutsch–Jozsa algorithm and test it on a few functions. From that point, it was trivial to also implement the Bernstein–Vazirani algorithm.

That's all we have to say about Qiskit… for now. We still have a bit of coding to do, and it's going to be lots of fun. But before we do that, we must introduce the stars of the show. In the following chapters, we will explore some of the quantum algorithms to which we can attribute a lot of the attention that quantum computing gets nowadays. Are you ready to learn about them? We are!

Part 4

The Stars of the Show: Main Quantum Algorithms

This is the part of the book where all the concepts that we have introduced and studied in detail in the previous chapters are going to fall into place to form beautiful, intricate, and useful methods, including the famous Shor's and Grover's algorithms.

In the following chapters, you will learn how to use quantum computers to factor big integers efficiently (something that constitutes a serious threat to current cryptographic protocols) and to accelerate searching over unsorted collections of data. Moreover, you will understand the principles behind these algorithms, namely the quantum Fourier transform and amplitude amplification. And, of course, you will learn how to program these methods using Qiskit.

This part includes the following chapters:

11

Finding the Period and Factoring Numbers

The sun rises and the sun sets, and hurries back to where it rises.

— Ecclesiastes 1:5

This is one of the most important chapters of this book, for it introduces what many consider to be the most relevant quantum algorithm ever conceived: Shor's algorithm for integer factorization. This was one of the first quantum algorithms to not only show an advantage over any known classical algorithms, but to do it for a problem with enormous practical significance. Indeed, Shor's algorithm has strong implications regarding the security of certain encryption protocols that are used on a daily basis on the Internet. For all this, it's difficult to overstate the importance of Shor's algorithm—and we shall study it in all the detail that it deserves.

To this extent, we will introduce a new tool (the quantum Fourier transform) and we will explain how many of the quantum phenomena we have been discussing up until this point—including superposition, entanglement and interference—play a central role in the inner workings of the method. In addition, we will turn our attention to phase estimation, a problem with many applications in science and engineering, and we will show how it can be easily addressed with the tools introduced for Shor's factoring algorithm.

The topics covered in this chapter are the following:

- The prime importance of prime factors
- Shor's algorithm
- Preparing a periodic sequence
- Finding the period with the quantum Fourier transform
- Quantum phase estimation

After reading this chapter, you will know why factoring big integers is an important mathematical problem that lies at the heart of some widely-used cryptographic methods. You will also have an idea as to why this problem is hard to solve with classical computers, and understand why it becomes easy once you have access to a powerful-enough quantum computer. You will also see why the study of certain periodic sequences is a viable approach to integer factorization, and how the quantum Fourier transform can help with it. Staying faithful to our practical approach, you will also be able to implement the quantum Fourier transform in a quantum circuit that only uses a polynomial number of gates, all of them acting on one or two qubits at a time. Finally, you will also know how all these tools relate to the problem of phase estimation, and how to solve it using quantum algorithms.

11.1 The prime importance of prime factors

Most students in primary and secondary school are familiar with integer factorization. Part of their math classes is devoted to solving problems like "find the prime factors of 42" or "express 500 as the product of powers of prime numbers"[1].

[1] The answers to these questions, by the way, are that the prime factors of 42 are 2, 3 and 7, and that 500 can be written as $2^2 \cdot 5^3$). Just in case your arithmetic was a little bit rusty!

In fact, you may remember solving that kind of problem yourself, and it is very likely that you were taught to use a rather crude but effective method called **trial division**. It consists in trying to divide the number to be factored by each prime number in ascending order. For instance, for 42, you would first try with the smallest prime number, which is 2. Since 2 divides 42, you would write 2 down as one of the factors, and then repeat the process with $42/2 = 21$. In this case, 2 does not divide 21, but the next prime number, 3, does. At this point, you would write down 3 as another factor of 42 and proceed with $21/3 = 7$. You would then start over and try all the primes in order: $2, 3, 5 \ldots$, with no success until finding that 7 indeed divides 7 (not very surprising) and concluding the procedure because you have already reached a prime number. In the end, you would have that $42 = 2 \cdot 3 \cdot 7$. Has this brought back any nice memories from your eleven-year-old self?

There are several things to note about this problem and this particular way of solving it. First of all, trial division seems to be a straightforward and natural approach to factorization. But is it efficient, or is there a better way? We will have a lot to say about this topic in this chapter but, spoiler alert, trial division is not a particularly good approach. If you learned about integer factorization in school, you probably used it to find the least common multiple of several numbers in order to express fractions with a common denominator and operate on them. Well, let us tell you a secret: you do not need to find the factors of two numbers to compute their least common multiple. In fact, this may work reasonably well for numbers with small factors, but once they get big, it becomes highly impractical. You can perform the operation much faster with something called Euclid's algorithm[2].

Important note

When working with integers, the size of the problem is the number of digits of the numbers we are considering. For instance, if the input to an algorithm is 12345, then the size of the input is 5, not 12345. Alternatively, we can consider that the input size is the number of bits required to describe the number. For instance, if

[2]Technically, Euclid's algorithm is used to compute the greatest common divisor d of two numbers a and b. But once that you find it, you can obtain their least common multiple by simply computing $a \cdot b$ and dividing it by d.

the input is 129, then the number in binary is 1000001 and the input size is 7. The number of bits needed to write down a number is $\log_2 10$ (or about 3.32) times its number of digits, hence—from the point of view of computational complexity—both ways of describing the input size are equivalent as they are related by a constant factor (for more on this, please consult *Appendix C*).

This is a reasonable choice because when we are working with integers (for instance, to perform arithmetical operations such as addition, multiplication, or powers) we operate individually on the digits (or the bits, if we are using binary code as computers do). Thus, we want to assess how the running time of the algorithm grows when we have more digits. And, of course, we want this growth to be moderate.

But numbers grow much more rapidly than their number of digits. For instance, when we move from 10^3 to 10^6, the number has grown 1000 times, but its size in digits has only been increased by 3. In fact, when we increase the number of digits by one, the actual numbers grow by a factor of 10. An integer value is exponential in its number of digits!

It turns out that trial division requires iterating over many of the numbers smaller than a given integer N. This method is, thus, exponential in the number of digits of N, and cannot be used with big integers. However, Euclid's algorithm runs in polynomial time relative to the number of digits, making it practical (see *Introduction to algorithms* [62] for all the details).

All this now begs the question of why on Earth would anyone want to study the prime factors of an integer. Well, it turns out that prime numbers have an enormous significance for mathematicians that study number theory, because prime numbers are kind of the "atoms" of numbers and can tell us a lot about their properties. But even for people who are not into number theory, prime factors are of uttermost importance, even if they're unaware of their very existence. The **RSA** protocol, introduced in the late 1970s by Rivest, Shamir, and Adleman [63], is one of the most widely used procedures for securing digital communications and encrypting data, and it uses keys that are obtained from the product of two big prime numbers. Here's the deal: if you were able to easily factor any integer, you could break this protocol and compromise the security of the Internet! So, yeah, you have to thank prime factors for that pizza you ordered online.

Now that we know how high the stakes are, let's go back to our previous question: what is the fastest way of finding the prime factors of an integer with a classical algorithm? Well, nobody knows for sure, but we certainly know that trial division is too slow. There exist other, more advanced methods such as the **general number field sieve** [64], which is currently the most efficient algorithm for big numbers, but their running time is still almost exponential. In fact, as proved in [65], [66], the running time of the general number field sieve for integers of n bits is

$$O(e^{cn^{1/3}\log n^{2/3}})$$

for some positive constant c. This means that the time is bounded from above by a constant times $e^{cn^{1/3}\log n^{1/3}}$ (see *Appendix C* for all the details on the O notation), which grows very rapidly with n. This is not an isolated case, for the vast majority of experts firmly believe that there cannot exist an algorithm that factors integers efficiently on a classical computer.

And here is where things become interesting. In the mid 1990s, Peter Shor shook the scientific world by proposing a quantum algorithm able to factor integers efficiently. This method [67], makes it possible to find factors of a big integer in time that is

$$O(n^2 \log n \log\log n),$$

which is much, much, *much* faster than $O(e^{cn^{1/3}\log n^{2/3}})$, and makes it practical to solve the factorization problem if one has access to a sufficiently powerful quantum computer. It goes without saying that this development made the interest in quantum computing grow tremendously and its impact can still be felt today, with a lot of effort invested in creating

post-quantum cryptographic protocols (also called **quantum-safe protocols**), such as those recently standardized by NIST [68], [69]. These new methods will allow us to migrate from cryptographic protocols that can be broken with quantum computers, such as RSA, and keep our information secure.

To learn more...

RSA is an **asymmetric** or **public-key cryptographic protocol**. One of the advantages of this kind of procedure is that they simplify key distribution and that they enable other applications such as the digital signing of messages in addition to encryption. For this reason, although they are usually slower than symmetric-key protocols (such as the one-time pad that we studied in *Chapter 3*), they are a fundamental part of today's Internet security.

There exist other asymmetric cryptographic protocols, such as those based on elliptic curves, that are widely used today, but they are also vulnerable to attacks using quantum computers. In fact, Peter Shor gave an algorithm to break those methods in the same paper in which he introduced his quantum factoring algorithm! To learn more about public key cryptography and how it is affected by quantum computers, we recommend checking out the book by Katz and Lindell [30].

Now that we know that factorization is a very important problem (and that trial division is not the smartest way to solve it!), we will devote the rest of the chapter to explaining how Shor's algorithm can be used to find primes that divide integers, starting with an overview of how the method works.

11.2 Shor's algorithm

By now, you are probably intrigued about how Shor's algorithm can be so much faster than classical algorithms at factoring numbers. Does it use some incredibly complicated quantum circuit or what? Well, in fact, you may be surprised to learn that most of the operations required to factor integers with Shor's algorithm are run on classical computers! This is what we call a **hybrid method**, because part of the process is carried out on a classical machine, and the rest on a quantum computer. The reason for this decision is that a good part of the computations involved in Shor's algorithm can be run efficiently on classical computers, so we don't really need to resort to quantum processors to carry them out. We will only need quantum computers for a specific part of the process that would be extremely costly on a classical device.

That's enough of a preamble. Let's see for ourselves what the different parts of the procedure are. Imagine that you want to factor a big integer N. Shor's algorithm instructs you to perform the following steps:

1. Check whether N is a power of a prime number. That is, check whether there exist a prime number p and an exponent b such that $N = p^b$. If this is the case, you are done: p is a factor of N.

2. Otherwise, choose an integer a between 2 and $N - 1$ uniformly at random. Check if a and N have a common factor $m > 1$. In that case, you are done: m is the factor you were looking for.

3. Find the period of a modulo N. That is, find r such that a^r leaves 1 as a remainder when divided by N. In symbols, we write

   $$a^r \equiv 1 \mod N$$

   and we say that a^r is congruent to 1 modulo N or, simply, that a^r is 1 modulo N. For more about modular arithmetic, please refer to *Appendix A*.

4. If r is odd, go back to step 2 and choose a different a.

5. Compute

   $$x = a^{\frac{r}{2}} + 1 \bmod N$$

   and

   $$y = a^{\frac{r}{2}} - 1 \bmod N$$

6. If $x = 0$ or $y = 0$, go back to step 2 and choose a different a. Otherwise, compute

$$p = \gcd(x, N)$$

and

$$q = \gcd(y, N),$$

where gcd stands for the greatest common divisor. Then, p and q will be non-trivial factors (that is, factors other than 1 and N) of N.

There are many things to unpack here, so let's go bit by bit. We'll begin by analyzing why the algorithm works and how we can perform each of its steps efficiently.

11.2.1 Analysis of the method

You surely have noticed that the algorithm that we have just introduced can end at steps 1 or 2 if $N = p^b$ or if you find a factor m of N. This is pretty unlikely, but, hey, if you do get lucky, you can call it a day and go celebrate.

A little more problematic is the fact that steps 4 and 5 can bring you back to step 2 to choose a different integer a. This sounds like trouble. Could it be the case that you would have to choose many different a's before you find one that works? Fortunately, that is not so. As Shor proved in his seminal paper [67], the probability of a not working when chosen uniformly at random is at most $1/2^{k-1}$, where k is the number of different odd prime factors of N. But N is not a power of a prime (we would have detected that in step 1) and we can safely assume that it is not even (otherwise, we don't need any computation whatsoever, let alone a quantum one, to know that 2 is a factor of N). Thus, N will have at least two different odd prime factors (that is, $k \geq 2$) and the probability of failure is at most $1/2$. If you repeat the process a fixed number of times, say 100 times, the probability of not finding a suitable value of a will be at most $1/2^{100}$, which is negligibly small.

So, with just a little patience, we will find a value of a that can work for our purposes. But, why is that enough to find a non-trivial factor of N? Well, notice that

$$x \cdot y = (a^{\frac{r}{2}} + 1)(a^{\frac{r}{2}} - 1) = a^r - 1 \equiv 1 - 1 = 0 \mod N,$$

since $a^r \equiv 1 \mod N$. This means that N divides $x \cdot y$, so any prime factor of N divides either x or y. Now, if all the factors of N divided x, then x would be a multiple of N and $\gcd(x, N)$ would be just N, but that situation (and the similar case $\gcd(y, N) = N$) is excluded at the beginning of step 6. Thus, p and q are both different from N and they are both different from 1. Why? Simple. If, for instance, $p = \gcd(x, N) = 1$, then no prime factor of N could divide x, but N divides xy, so $q = \gcd(y, N) = N$, which we just ruled out. Moreover, both p and q divide N, so we have succeeded in finding us some non-trivial factors of N. Hurray!

We now need to explain in a little bit more detail how we perform each of the steps and we also need to show that they can be carried out (either with a classical or a quantum computer) in polynomial time in the number of bits of N. Let's explore each of them separately.

The computations of step 1 rely on the fact that there exist classical algorithms that can efficiently check whether a given integer N is a prime number or not. This efficient primality testing can be carried out deterministically with something such as the AKS algorithm [70] or, with a probability of error as small as needed, with the much more efficient, but probabilistic, Rabin-Miller test [71] (see *Introduction to algorithms* [62] for a thorough treatment of this latter method). Now, assume that there exists a prime p such that $N = p^s$ for some integer $s \geq 1$. In this case, $s = \log_p N$. Thus, the biggest s can be is $\log_2 N$. In this way, you can just compute the j-th root of N (which can be done efficiently) for every integer $1 \leq j \leq \log_2 N$ and, if it is an integer, check whether it is a prime number with either the AKS algorithm or the Rabin-Miller test. You will have to do this at most $\log_2 N$ times, which is the number of bits of N—so this is certainly efficient.

In step 2, after selecting a, you need to check if it has a common factor with N, and that's easy. You can just compute the greatest common divisor of a and N by using Euclid's algorithm, which, as we have already mentioned, is efficient. Of course, you will also use it for the gcd computations in step 6.

Step 5 can also be performed efficiently on a classical computer, but with a small catch. When raising a to $r/2$, you should *not* compute $a^2 = a \cdot a$, $a^3 = a \cdot a^2$, $a^4 = a \cdot a^3$ ... and so on until you reach $a^{\frac{r}{2}}$—that would take an exponential amount of time on the number of bits of r. Instead, you compute $a^2 = a \cdot a$, $a^4 = a^2 \cdot a^2$, $a^8 = a^4 \cdot a^4$ ... and then you multiply together the powers that you need to obtain $a^{\frac{r}{2}}$. Let's show how to do this with an example.

Imagine that r is 54 and, hence, $r/2 = 27$. To raise some integer a to 27, compute a^2, a^4, a^8 and a^{16}. You stop there because the next power would be a^{32}, which is already bigger than a^{27}. Then, notice that $27 = 16 + 8 + 2 + 1$ and compute $a^{16} \cdot a^8 \cdot a^2 \cdot a = a^{16+8+2+1} = a^{27}$. Something similar can be done for any r with just $O(\log r)$ multiplications or, to put it in another way, a number of multiplications that is proportional to the size in bits of r, which is exactly what we wanted! There is an additional small detail, though. After each multiplication, you should compute the remainder of the result when divided by N. That is, you should work modulo N all the time. Otherwise, the numbers could grow very big and the operations would no longer be efficient.

So, we have seen that steps 1, 2, 5, and 6 can be carried out efficiently with classical algorithms. And, obviously, step 4 just requires a simple division (or, even easier, checking if the last bit of r is 1). But what about step 3? It turns out that, for a classical computer, this would be the real bottleneck of the algorithm. Indeed, there is no known classical algorithm that can perform this computation efficiently and it is widely believed that such an algorithm simply does not exist. This is where quantum computers can kick in! Throughout the rest of the chapter, we will explain how a clever quantum circuit can allow us to obtain the period of a, thus completing all the steps that we need in order to find a non-trivial factor of N. It will be quite a ride, we assure you.

But before we get to that, let's see some examples of how to perform all the steps in Shor's algorithm with some small numbers for which we can carry out the computations with just pen and paper.

11.2.2 A simple example

Imagine that you want to factor the integer 21. Not a very challenging task, sure, but it will help us understand how Shor's algorithm works while keeping the computations simple. So please bear with us while we go through each of the steps of the procedure.

First, we need to check if 21 is the power of a prime number, and it obviously is not, because we know it is the product of two different primes (namely 3 and 7). But if you wanted to do everything by the book, you would need to compute the roots of 21 up to order $\log_2 21$ and check if any of them is a prime. In our case, $\log_2 21$ is less than 5, so we would need to compute $\sqrt{21}$, $\sqrt[3]{21}$, and $\sqrt[4]{21}$. With the help of a calculator, you can easily see that none of them is even an integer.

Now, we need to select some integer a between 2 and 20 at random. Let's say that we get $a = 4$ in our first try[3]. The greatest common divisor of 4 and 21 is 1, so we were not lucky in this step (in general, we very rarely will), and we need to go to step 3 and find the period of 4 modulo 21. That is, we need to find r such that $a^r = 1 \bmod 21$. There is no shortcut for this computation (unless you have a quantum computer!). But in this case, since 21 is small, we can just try all the values for r, starting with $r = 2$, and quickly find out that, indeed, $4^3 = 64$, which is congruent to 1 modulo 21. Alas, r is odd, so it does not work for us and we need to select[4] a different value for a.

Let's say that we now select $a = 3$. Well, in this case, $\gcd(3, 21) = 3$ and we have already found a factor of 21. That was cool, right? But when N is big (say, thousands of bits long) and it only has a handful of factors, this kind of situation will be very unlikely. We just wanted to show that this might happen from time to time.

[3]Well, we confess that the choice was not completely random. We selected this value for didactic purposes, and we will keep selecting different values for a again at, ehem, "random" to illustrate all the aspects of the algorithm.

[4]At "random", of course.

Let's then select[5] a more interesting value for a. Let's try with $a = 5$. If we compute the gcd of 21 and 5, we readily find that they have no common factors, so we can proceed to find the period of 5. In this case, $r = 6$ is the smallest positive integer such that $5^6 \equiv 1 \mod 21$, as you can easily check for yourself. Since 6 is even, we can go to step 5 and compute $x = 5^3 + 1 \mod 21$ and $y = 5^3 - 1 \mod 21$. These values happen to be $x = 0$ and $y = 19$, so unfortunately we need to try our luck again with a different a. The problem here is that $5^3 + 1 = 126$, which is a multiple of 21, so the gcd of 126 and 21 is 21, which is a *trivial* factor of 21. We do not need any computation to know that 21 divides 21, do we? Similarly, $5^3 - 1 = 124$ and $\gcd(124, 21) = 1$, which is again a trivial factor of 21.

Finally, let's show what happens if we obtain $a = 2$ when selecting values at random. Obviously, $\gcd(2, 21) = 1$ so we need to compute the period of 2 modulo 21. It is easy to check that $2^6 = 64 \equiv 1 \mod 21$. Then, we compute $x = 2^3 + 1 = 9$ and $y = 2^3 - 1 = 7$. The gcd of 9 and 21 is 3, which is a factor of 3, the other being $7 = \gcd(7, 21)$.

These are all the different situations that you can find when applying this factorization algorithm. We were so "lucky" that we obtained values for a that perfectly illustrated each of them! What were the odds, huh?

Exercise 11.1

Check what happens if you select values for a that range from 6 to 20 when trying to factor 21.

Exercise 11.2

Apply the factorization algorithm to $N = 15$ with every possible value of a from 2 to 14.

With this example, we have a clearer intuition as to how Shor's algorithm works. And thus, we are now ready to tackle the most difficult part of the method: figuring out how to use a quantum computer to find periods of numbers. As you will soon learn, it all starts with the

[5]We mean "choose at random", obviously.

creation of some peculiar periodic sequences, such as the ones that we will study in the next section.

11.3 Preparing a periodic sequence

As we have seen in the previous section, we can find a factor of an integer N if, given an arbitrary integer a between 2 and $N-1$, we are able to find its period: that is, an integer r such that $a^r \equiv 1 \mod N$. No efficient classical algorithm has been found for this task and it is a widespread belief that it is impossible to do it with classical computers. In this section and the next one, however, we will show how quantum computers can enable us to solve this problem in polynomial time on the size of N and a. Let's get to it!

11.3.1 Introducing a very periodic state

The first thing that we will need to do is prepare a periodic sequence that will include all the values a^r for r from 0 to $2^m - 1$, for a certain value of m that we will discuss later. In fact, we will use a quantum computer to prepare the state

$$\frac{1}{\sqrt{2^m}} \sum_{j=0}^{2^m-1} |j\rangle|a^j \mod N\rangle,$$

where the first register[6] has m qubits and the second has as many qubits as the length of N in bits.

This may seem a little bit mysterious, so let's give an example to clarify what type of state we will be working with and why we are interested in it. Imagine that $N = 15$, $a = 2$, and $m = 3$. Then the state that we have just mentioned becomes

$$\frac{1}{\sqrt{8}} \left(|0\rangle|1\rangle + |1\rangle|2\rangle + |2\rangle|4\rangle + |3\rangle|8\rangle + |4\rangle|1\rangle + |5\rangle|2\rangle + |6\rangle|4\rangle + |7\rangle|8\rangle\right),$$

because $2^4 = 16 \equiv 1 \mod 15$, $2^5 = 32 \equiv 2 \mod 15$, $2^6 = 64 \equiv 4 \mod 15$, and $2^7 = 128 \equiv 8 \mod 15$.

[6]A quantum register is just a group of qubits that represents a certain value or shares a common function.

Notice how the sequence of values in the second register is periodic: they go 1, 2, 4, 8 and then they repeat again in the same order. This property is key in making apparent the reason why we have chosen this state. Imagine that we measure the second register: what results can we obtain? Well, there are four different possibilities, each with probability $1/4$:

$$\frac{1}{\sqrt{2}}\left(|0\rangle|1\rangle + |4\rangle|1\rangle\right), \quad \frac{1}{\sqrt{2}}\left(|1\rangle|2\rangle + |5\rangle|2\rangle\right),$$
$$\frac{1}{\sqrt{2}}\left(|2\rangle|4\rangle + |6\rangle|4\rangle\right), \quad \frac{1}{\sqrt{2}}\left(|3\rangle|8\rangle + |7\rangle|8\rangle\right).$$

Do you notice anything special about these states? The difference between the values on the first register of each of them is 4, which is exactly the period of 2 modulo 15 (and the period of the initial sequence in the second register). This is the value that we want to compute! And this is no coincidence. In fact, if you measure the second register of $\frac{1}{\sqrt{2^m}}\sum_{j=0}^{2^m-1}|j\rangle|a^j \bmod N\rangle$, it will collapse to a certain value $|k\rangle$ and every value $|j\rangle$ in the first register will satisfy $a^j \equiv k \mod N$. Thus, if you take any $|j_1\rangle$ and $|j_2\rangle$ values on the first register such that $j_1 > j_2$, it will hold that

$$a^{j_1-j_2} = \frac{a^{j_1}}{a^{j_2}} \equiv 1 \mod N,$$

because both a^{j_1} and a^{j_2} are equal to k modulo N. Notice that this means that $j = j_1 - j_2$ is a positive integer such that $a^j \equiv 1 \mod N$.

Exercise 11.3

Take $N = 21$, $a = 8$, and $m = 3$. If you measure the second register of $\frac{1}{\sqrt{2^m}}\sum_{j=0}^{2^m-1}|j\rangle|a^j \bmod N\rangle$ and obtain 1 as a result, what are the possible values for the first register?

What is $8^j \bmod 21$ when j is the difference of two such values? What if the result of the measurement is 8?

The jackpot appears to be at our fingertips, because a number j such that $a^j \equiv 1 \mod N$ is exactly what we need for our factorization algorithm... But we should not get too excited yet. We're still facing a small problem: how can we get j_1 and j_2? If, after measuring the second register we measure the first one, we will certainly obtain some value j such that $a^j \equiv k \mod N$. But then the state will collapse to $|j\rangle|k\rangle$, losing all the information about the other values in the first register. And if we start all over again with $\frac{1}{\sqrt{2^m}}\sum_{j=0}^{2^m-1}|j\rangle|a^j \mod N\rangle$ and we measure the second register, there is no guarantee at all that we will obtain k again. So close, yet so far!

Don't worry. It will require some extra work, but we will be able to obtain the information that we need from the state. That, however, will have to wait until the next section. Now that we know why the state $\frac{1}{\sqrt{2^m}}\sum_{j=0}^{2^m-1}|j\rangle|a^j \mod N\rangle$ is important, we should concentrate on studying how to prepare it with a quantum circuit.

11.3.2 Quantum circuit to obtain the periodic sequence

Preparing the state $\frac{1}{\sqrt{2^m}}\sum_{j=0}^{2^m-1}|j\rangle|a^j \mod N\rangle$ will be much easier than it seems a first glance. In fact, it all relies on one little trick that we have applied again and again, and on an arithmetical operation that we will use as a controlled quantum gate. Wait and see!

Since we want to prepare a state whose first register has all the values from 0 to $2^m - 1$, it should come as no surprise that we will begin by applying a column of Hadamard gates on our first quantum register, which will have m qubits. The second quantum register will have n qubits, where n is the number of bits required to write N, and it will be initialized to state $|1\rangle$. Notice that this is the decimal value 1, so if you expand it in binary, it will have $n - 1$ leading zeroes and then a solitary one at the end. It is easy to obtain this value in the second register: we initialize all qubits to $|0\rangle$, as usual, and then we apply an X gate on the least significant qubit of the register. With this, plus the application of the H gates on the first register, we now have the state

$$\frac{1}{\sqrt{2^m}}\sum_{j=0}^{2^m-1}|j\rangle|1\rangle.$$

As you can see, this gives the first part of the state that we need. We are almost there! To get the correct states on the second register, notice that we just need a quantum gate that takes $|j\rangle|1\rangle$ to $|j\rangle|a^j \bmod N\rangle$. Does this look familiar? It should! It's almost like the oracles that we all know and love, but, instead of adding the result, now we are multiplying it (modulo N); that is, we are multiplying the result of $a^j \bmod N$ times the content of the second register (which is 1). In fact, this idea of accumulating multiplications on the second register allows us to decompose the operation a little bit further. Let's see how.

Imagine that, for each l, you can implement a quantum gate $U_{a^{2^l}}$ that acts as follows on computational basis states:

$$U_{a^{2^l}}|b\rangle = |b \cdot a^{2^l} \bmod N\rangle.$$

With that kind of gate, we can easily implement the transformation from $|j\rangle|1\rangle$ to $|j\rangle|a^j \bmod N\rangle$. In fact, we can do it with the circuit represented in *Figure 11.1*. Notice that we control the application of a $U_{a^{2^l}}$ with the $(l+1)$-th most significant bit of the first register. Suppose that the first register contains $|j\rangle$ and the expansion of j in binary is $j_{m-1}j_{m-2}\ldots j_0$. If j_l is 0, then we are not applying any gate from the $(l+1)$-th qubit, which is the same as multiplying by $1 = a^{j_l 2^l}$. If j_l is 1, then we are multiplying the second register by $a^{2^l} = a^{j_l 2^l}$. All things considered, we are multiplying (modulo N) the content of the second register by

$$a^{j_{m-1}2^{m-1}} a^{j_{m-2}2^{m-2}} \cdots a^{j_0 2^0},$$

which is exactly equal to a^j because

$$j = j_{m-1}2^{m-1} + j_{m-2}2^{m-2} + \cdots + j_0 2^0.$$

This might be a little difficult to see from an abstract point of view, so lets go through an example to make things more concrete. Imagine that $m = 4$ and we consider $j = 5$. Then, the binary expansion of j is 0101. When we apply the control gates as in *Figure 11.1* with 0101 on the upper (first) register, then we are applying only the gates $U_{a^{2^2}}$ and $U_{a^{2^0}}$. In total, we are multiplying the second register by $a^{2^2+2^0} = a^5$ modulo N, as expected. If, for instance, j is 9, its binary expansion is 1001 and we apply the gates $U_{a^{2^3}}$ and $U_{a^{2^0}}$, which is the same as multiplying by $a^{2^3+2^0} = a^9$ modulo N.

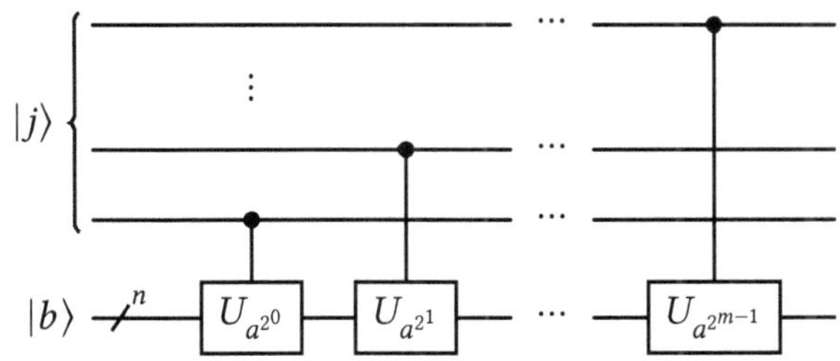

Figure 11.1: Circuit to compute modular exponentiation. It maps $|j\rangle|b\rangle$ *to* $|j\rangle|b \cdot a^j \bmod N\rangle$

Thus, the circuit in *Figure 11.1* takes $|j\rangle|b\rangle$ to $|j\rangle|b \cdot a^j \bmod N\rangle$ and, by linearity, takes the state

$$\frac{1}{\sqrt{2^m}} \sum_{j=0}^{2^m-1} |j\rangle|1\rangle$$

to the state

$$\frac{1}{\sqrt{2^m}} \sum_{j=0}^{2^m-1} |j\rangle|a^j \bmod N\rangle,$$

exactly as we need.

All this depends, though, on the very existence of the quantum gate $U_{a^{2^l}}$ for every $l \geq 0$. And there is a very subtle detail here: is this even a *reversible* operation? It should take $|b\rangle$ to $|b \cdot a^j \bmod N\rangle$ but, for instance, if $N = 9$, $a = 3$ and $j = 2$, then $a^j \bmod N = 0$ and the operation cannot be reversed. Are we in trouble, then? Well, not quite. If we have reached this point in the application of the method, it's because a has no factors in common with N; otherwise, we would have successfully terminated the application of the algorithm in step 2. Under this condition, it can be proved that multiplication by a^j is always reversible modulo N (for more on this, please refer to *Appendix A*), so we are safe.

In fact, $U_{a^{2^l}}$ is an arithmetical operation that can be implemented with a polynomial number of classical gates and, consequently, also with a polynomial number of quantum gates. For some of the quantum circuits and techniques proposed for implementing modular exponentiation, please check [72]–[76].

Important note

With the following circuit,

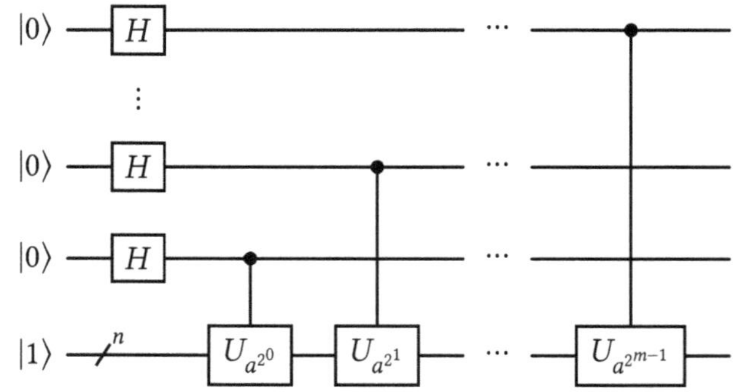

we can prepare the state $\frac{1}{\sqrt{2^m}}\sum_{j=0}^{2^m-1}|j\rangle|a^j \bmod N\rangle$, provided that a and N have no common factors. This state includes a periodic sequence that will help us find a number $k > 0$ such that $a^k \equiv 1 \mod N$. Notice how we have used two of the key quantum phenomena to our advantage: superposition with the H gates and then entanglement between $|j\rangle$ and $|a^j \bmod N\rangle$.

With all this, we now know how to prepare the state that we wanted. Next, we will learn how to extract the period information from it. Get ready, because we are going to introduce one of the most important operations in quantum computing: the quantum Fourier transform.

11.4 Finding the period with the quantum Fourier transform

We have reached the final and crucial part of method: how to obtain information about the periodicity of the sequence that we have prepared. For this kind of task, there is a well-known tool in the classical domain: the **discrete Fourier transform** or **DFT** (see *Introduction to algorithms* [62] for a complete explanation of this operation). However, given the size of the state we are working with (which has a number of terms that is exponential in m), it is impossible in practice to apply classical methods to the corresponding sequence.

Here is where quantum computing comes to the rescue, with a much faster version of the DFT. Let's introduce the quantum Fourier transform!

11.4.1 QFT FTW!

The **quantum Fourier transform** or **QFT** is the quantum version of the DFT and has very similar applications and properties. The QFT on m qubits is defined as the linear transformation that takes the computational basis state $|j\rangle$ to

$$\frac{1}{\sqrt{2^m}} \sum_{k=0}^{2^m-1} e^{\frac{2\pi ijk}{2^m}} |k\rangle,$$

where i is the imaginary unit (that is, $i^2 = 1$). This certainly can be a lot to process in one go, so let's try and examine it more carefully.

One of the most striking elements in the definition of the QFT is the appearance of those $e^{\frac{2\pi ijk}{2^m}}$ values (for a refresher on complex exponentials and Euler's formula, please check *Appendix A*). Numbers of the form $e^{\frac{2\pi is}{n}}$, with s and n integers, are extremely important in many areas of mathematics (number theory and group theory, for example) and they receive a special name: they are called **roots of unity**. The reason for this particular denomination is that it clearly holds that

$$\left(e^{\frac{2\pi is}{n}}\right)^n = e^{\frac{2\pi is}{n}\cdot n} = e^{2\pi is} = \left(e^{2\pi i}\right)^s = 1^s = 1,$$

because $e^{2\pi i} = 1$. Thus, these numbers are certainly roots of 1, because when you raise them to n you get exactly 1. In this particular case, we say that they are **nth roots of unity**, for obvious reasons.

Roots of unity also have some beautiful geometric properties. For instance, when represented in the complex plane, they all lie in the circle of radius 1, starting with 1 (which is, of course, always a root of 1) and then the remaining roots can be found by rotating an angle $2\pi/n$ counterclockwise until returning again to 1. There are exactly n different nth

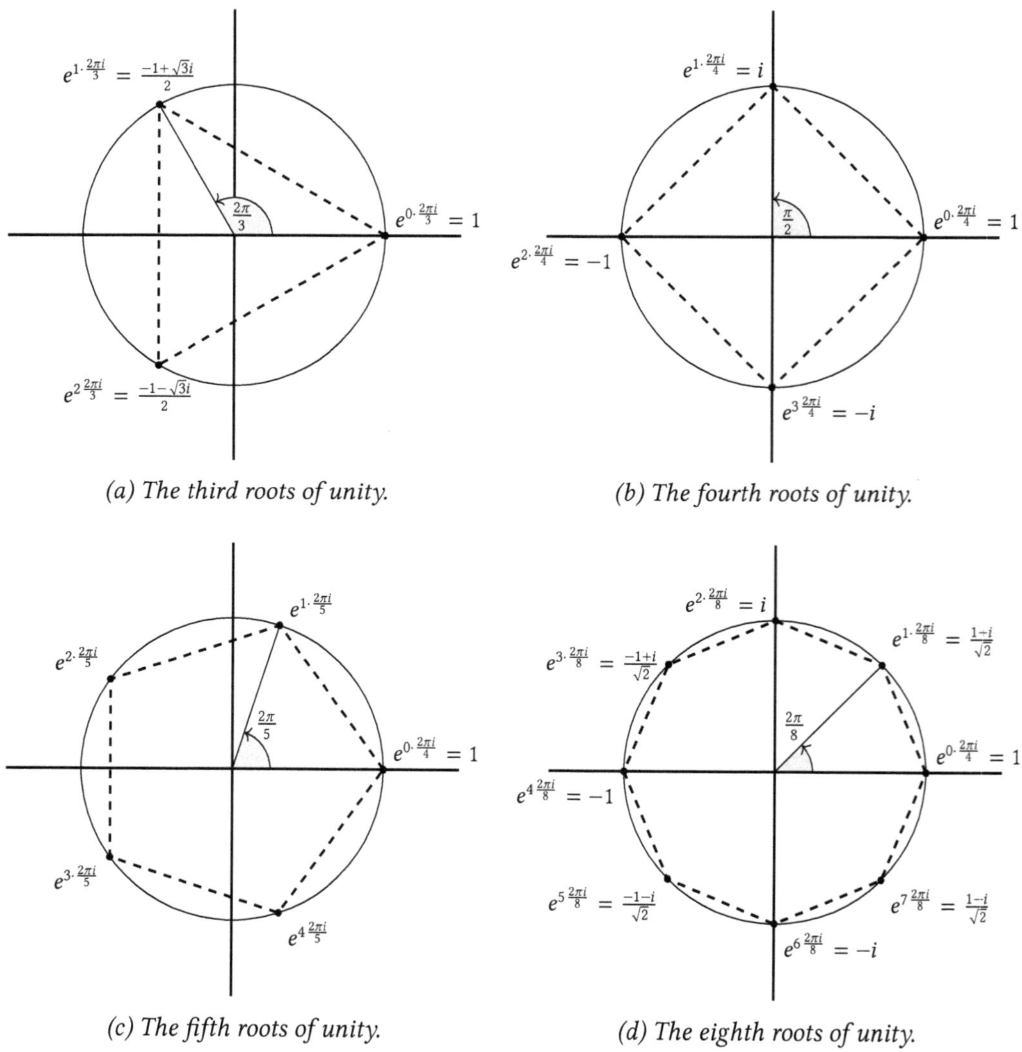

(a) The third roots of unity.

(b) The fourth roots of unity.

(c) The fifth roots of unity.

(d) The eighth roots of unity.

Figure 11.2: The nth roots of unity for $n = 3, 4, 5, 8$

roots of unity, which are explicitly

$$1 = e^{0\cdot\frac{2\pi i}{n}}, e^{\frac{2\pi i}{n}}, e^{\frac{4\pi i}{n}}, e^{\frac{6\pi i}{n}}, \ldots, e^{\frac{2(n-1)\pi i}{n}},$$

because $e^{\frac{2n\pi i}{n}} = e^{2\pi i}$ which is, again, 1 and thus the values start repeating for $s \geq n$. These points form a regular polygon of n sides. See *Figure 11.2* for some examples of this.

For convenience, it is common to represent

$$\omega_n = e^{\frac{2\pi i}{n}},$$

and then the n different nth roots of unity are simply

$$1 = \omega_n^0, \omega_n, \omega_n^2, \ldots, \omega_n^{n-1}.$$

Another important (and useful) property of roots of unity can be found when computing sums of their powers. Take any integer s that is not a multiple of n and consider the sum of the values $1, \omega_n^s, \omega_n^{2s}, \ldots, \omega_n^{(n-1)s}$. The result will always be 0 because this is a geometric progression of ratio $\omega_n^s \neq 1$ and the sum is then given by

$$\begin{aligned}\frac{1-\omega_n^s\cdot\omega_n^{(n-1)s}}{1-\omega_n^s} &= \frac{1-\omega_n^{ns}}{1-\omega_n^s} = \frac{1-(\omega_n^n)^s}{1-\omega_n^s}\\ &= \frac{1-1^s}{1-\omega_n^s} = \frac{1-1}{1-\omega_n^s} = 0,\end{aligned}$$

where we have used that $\omega_n^n = 1$. This is a simple but very important property that we will use several times through the remainder of this chapter.

Important note

If we define ω_n to be $e^{\frac{2\pi i}{n}}$, where $n \geq 2$ is an integer, then the following n values are all the different nth roots of unity:

$$1, \omega_n, \omega_n^2, \ldots, \omega_n^{n-1}.$$

Moreover, for any integer s that is not a multiple of n, it holds that

$$1^s + \omega_n^s + \omega_n^{2s} + \cdots + \omega_n^{(n-1)s} = 0.$$

However, if we take s to be kn for some integer k, then $\omega_n^s = \omega_n^{kn} = (\omega_n^n)^k = 1^k = 1$, and then

$$1^s + \omega_n^s + \omega_n^{2\cdot s} + \cdots + \omega_n^{(n-1)s} = 1 + 1 + 1 + \cdots + 1 = n.$$

Now, let's try to look more closely at what the QFT does as a linear transformation. For that, let's explicitly write the QFT matrix for some simple cases. Remember that the QFT on m qubits takes $|j\rangle$ to

$$\frac{1}{\sqrt{2^m}} \sum_{k=0}^{2^m-1} \omega_{2^m}^{jk} |k\rangle.$$

Consequently, when $m = 1$, the QFT takes $|0\rangle$ to

$$\frac{1}{\sqrt{2}} (|0\rangle + |1\rangle)$$

and $|1\rangle$ to

$$\frac{1}{\sqrt{2}} (|0\rangle - |1\rangle),$$

because $\omega_2^0 = 1$ and $\omega_2^1 = e^{\pi i} = -1$. This means the matrix associated to the QFT in this case is

$$\begin{pmatrix} \frac{1}{\sqrt{2}} & \frac{1}{\sqrt{2}} \\ \frac{1}{\sqrt{2}} & -\frac{1}{\sqrt{2}} \end{pmatrix} = \frac{1}{\sqrt{2}} \begin{pmatrix} 1 & 1 \\ 1 & -1 \end{pmatrix}.$$

Wait! We know this matrix! This is just an old friend of ours: the Hadamard gate. It is always reassuring to meet an old acquaintance, isn't it? Maybe that QFT is not as scary as it seemed.

What about the case when $m = 2$? Notice that, in general, the amplitude of $|k\rangle$ on QFT$|j\rangle$ is exactly $\omega_{2^m}^{jk}/\sqrt{2^m}$, so the matrix is then

$$\frac{1}{2}\begin{pmatrix} 1 & 1 & 1 & 1 \\ 1 & i & -1 & -i \\ 1 & -1 & 1 & -1 \\ 1 & -i & -1 & i \end{pmatrix}.$$

Exercise 11.4

Check that the expression that we have given for the QFT matrix on 2 qubits is correct.

In general, the elements of QFT matrices will all be roots of unity (which is obvious, since, by construction, all of them are integer powers of roots of unity). More interestingly, QFT matrices are always symmetric and each row (and column) is a sequence of nth roots of unity for a fixed n. In fact, it is easy to see that a row of the QFT matrix for m qubits has as its elements

$$1, \omega_{2^m}^{k}, \omega_{2^m}^{2k}, \ldots, \omega_{2^m}^{(2^m-1)k},$$

for a certain non-negative integer k. If k and 2^m have no common factors, these elements will be exactly the 2^mth roots of unity in a certain order. This is what happens with the second and fourth column of the matrix for the two-qubit QFT that we have shown above. On the other hand, if $\gcd(k, 2^m) = l > 1$, then the sequence $1, \omega_{2^m}^{k}, \omega_{2^m}^{2k}, \ldots, \omega_{2^m}^{(2^m-1)k}$ will consist of l repetitions of the $(2^m/l)$th roots of unity. In the case of the two-qubit QFT, the third row of the matrix shows exactly this behavior, but you can easily check that this is the case too for any other number of qubits.

Exercise 11.5

What are the different sequences of roots of unity that appear as the rows of the matrix for the QFT on 3 qubits?

The final property of the QFT that we want to highlight in this section has to do with its inverse, which is usually called (not at all surprisingly) the **inverse quantum Fourier transform**. We will explicitly show later in this chapter that the QFT can be implemented with a quantum circuit and, thus, it is unitary and hence invertible. However, we can show it now by just analyzing its matrix. The (j, k) entry of the m-qubit QFT matrix is, as we already know, $\frac{1}{\sqrt{2^m}} e^{\frac{2\pi ijk}{2^m}}$. Then, its conjugate transpose will have $\frac{1}{\sqrt{2^m}} e^{\frac{-2\pi ikj}{2^m}}$ as its (j, k) entry, because the conjugate of e^{ix} is e^{-ix} when x is real (see *Appendix A* for a proof of this property). Then, to show that the QFT is unitary, we only need to show that the linear transformation L that takes $|k\rangle$ to

$$\frac{1}{\sqrt{2^m}} \sum_{l=0}^{2^m-1} e^{\frac{-2\pi ikl}{2^m}} |l\rangle$$

is its inverse.

With all that we know, this is quite easy to prove. In fact, if we first apply the QFT to $|j\rangle$, we know that we obtain

$$\frac{1}{\sqrt{2^m}} \sum_{k=0}^{2^m-1} e^{\frac{2\pi ijk}{2^m}} |k\rangle.$$

If we now apply L to this state, by linearity we get

$$\frac{1}{\sqrt{2^m}} \sum_{k=0}^{2^m-1} \left(e^{\frac{2\pi ijk}{2^m}} \frac{1}{\sqrt{2^m}} \sum_{l=0}^{2^m-1} e^{\frac{-2\pi ikl}{2^m}} |l\rangle \right) = \frac{1}{2^m} \sum_{l=0}^{2^m-1} \left(\sum_{k=0}^{2^m-1} (e^{\frac{2\pi i}{2^m}})^{k(j-l)} \right) |l\rangle.$$

But, from our discussion of the sum of powers of the roots of unity, we already know that $\sum_{k=0}^{2^m-1} (e^{\frac{2\pi i}{2^m}})^{k(j-l)}$ is equal to 2^m if $j = l$, and it is 0 if $j \neq l$ (because both j and l are less than 2^m, so $j - l$ can never be a multiple of 2^m). Hence, this latter state simplifies to just $|j\rangle$! This shows that L is the inverse quantum Fourier transform, and hence the QFT is unitary.

All this is so important that we want to state it succinctly and explicitly, while, at the same time, renaming some of the indices to have more symmetric, pleasant and aesthetic expressions. This gives rise to the remarkable fact that we highlight in the following note.

Important note

The quantum Fourier transform is a unitary operation that takes $|j\rangle$ to

$$\frac{1}{\sqrt{2^m}}\sum_{k=0}^{2^m-1} e^{\frac{2\pi ijk}{2^m}}|k\rangle.$$

Its inverse, called the inverse quantum Fourier transform, is the unitary operation that takes $|j\rangle$ to

$$\frac{1}{\sqrt{2^m}}\sum_{k=0}^{2^m-1} e^{-\frac{2\pi ijk}{2^m}}|k\rangle.$$

Wow, this was quite a wild ride, wasn't it? So many beautiful mathematical facts from just one linear transformation! But you ain't seen nothing yet, as Bachman-Turner Overdrive like to sing. Next, we will show how we can use the QFT to find information about periods in number sequences. Let's roll!

11.4.2 The QFT applied to periodic sequences

We now have all the ingredients to complete the quantum part in Shor's algorithm. We know how to create a periodic sequence with information about the modular powers of a and we know it all about the quantum Fourier transform. Let's see how to combine the two of them to obtain the period that we need to factor that big integer N.

Let's start with something simple. Imagine that we have a state on two qubits like

$$\frac{1}{\sqrt{2}}\left(|0\rangle + |2\rangle\right),$$

which has period 2. We are going to apply to it the inverse quantum Fourier transform (or IQFT), so let's explicitly write its matrix for the case of 2 qubits. Remember that the IQFT takes a computational basis state $|j\rangle$ on two qubits to

$$\frac{1}{2}\sum_{k=0}^{3} e^{-\frac{\pi ijk}{2}}|k\rangle = \frac{1}{2}\sum_{k=0}^{3}(-i)^{jk}|k\rangle.$$

Consequently, its matrix is

$$\frac{1}{2}\begin{pmatrix} 1 & 1 & 1 & 1 \\ 1 & -i & -1 & i \\ 1 & -1 & 1 & -1 \\ 1 & i & -1 & -i \end{pmatrix}.$$

Thus, the result of applying it to $\frac{1}{\sqrt{2}}(|0\rangle + |2\rangle)$ is

$$\frac{1}{2}\begin{pmatrix} 1 & 1 & 1 & 1 \\ 1 & -i & -1 & i \\ 1 & -1 & 1 & -1 \\ 1 & i & -1 & -i \end{pmatrix}\begin{pmatrix} \frac{1}{\sqrt{2}} \\ 0 \\ \frac{1}{\sqrt{2}} \\ 0 \end{pmatrix} = \begin{pmatrix} \frac{1}{\sqrt{2}} \\ 0 \\ \frac{1}{\sqrt{2}} \\ 0 \end{pmatrix}.$$

Notice how the first and third coefficients in the resulting vector are obtained, respectively, by multiplying the rows $\begin{pmatrix} 1 & 1 & 1 & 1 \end{pmatrix}$ and $\begin{pmatrix} 1 & -1 & 1 & -1 \end{pmatrix}$ times the original vector. Since the vector of the state $\frac{1}{\sqrt{2}}(|0\rangle + |2\rangle)$ only has non-zero elements at the first and third positions, and in the selected matrix rows, those positions are equal, we obtain positive interference and the resulting value is non-zero. However, in the second and fourth rows of the IQFT matrix, the elements at positions first and third are opposite of each other. Consequently, they interfere negatively and the net result is 0.

Similarly, if we consider the state $\frac{1}{\sqrt{2}}(|1\rangle + |3\rangle)$, which also has period 2, the new state after applying the IQFT to it will be

$$\frac{1}{2}\begin{pmatrix} 1 & 1 & 1 & 1 \\ 1 & -i & -1 & i \\ 1 & -1 & 1 & -1 \\ 1 & i & -1 & -i \end{pmatrix}\begin{pmatrix} 0 \\ \frac{1}{\sqrt{2}} \\ 0 \\ \frac{1}{\sqrt{2}} \end{pmatrix} = \begin{pmatrix} \frac{1}{\sqrt{2}} \\ 0 \\ -\frac{1}{\sqrt{2}} \\ 0 \end{pmatrix}.$$

Again, we have positive interference when multiplying by rows that repeat after 2 elements and negative interference in the other cases, where the rows don't have period 2.

All this is a manifestation of a more general fact. Indeed, any m-qubit state of the form

$$\frac{1}{\sqrt{2^{(m-l)}}}\sum_{t=0}^{2^{(m-l)}-1}\left|c+t\cdot 2^{l}\right\rangle,$$

for some integers l between 1 and $m-1$, and c between 0 and 2^l-1, has period 2^l. Particular cases include the states $\frac{1}{\sqrt{2}}(|0\rangle+|2\rangle)$ and $\frac{1}{\sqrt{2}}(|1\rangle+|3\rangle)$ that we have just considered (with $m=2$, $l=1$, and $c=0,1$). Other examples include the three-qubit states

$$\frac{1}{\sqrt{2}}\left(|1\rangle+|5\rangle\right),$$

where $l=2$ and $c=1$, and

$$\frac{1}{2}\left(|0\rangle+|2\rangle+|4\rangle+|6\rangle\right),$$

where $l=1$ and $c=0$.

These are (some of) of the states that we learned to prepare in *Section 11.3* and they all have period 2^l. In fact, their only non-zero amplitudes appear every 2^l elements. If we apply to them the IQFT, the j-th amplitude of the resulting vector will be given by the multiplication of the j-th row of the IQFT matrix times the vector of the periodic quantum state. Since the (j,k) entry of the IQFT matrix is

$$\frac{1}{\sqrt{2^m}}e^{-\frac{2\pi ijk}{2^m}},$$

the resulting amplitude is

$$\frac{1}{\sqrt{2^m}}\frac{1}{\sqrt{2^{m-l}}}\sum_{t=0}^{2^{(m-l)}-1}e^{-\frac{2\pi ij(c+t2^l)}{2^m}},$$

because the only non-zero amplitudes of the state we are transforming can be found at the positions indexed by $c+t2^l$.

We can rewrite the expression as

$$\frac{1}{\sqrt{2^{2m-l}}}\sum_{t=0}^{2^{(m-l)}-1} e^{-\frac{2\pi i j(c+t2^l)}{2^m}} = \frac{e^{-\frac{2\pi i jc}{2^m}}}{\sqrt{2^{2m-l}}}\sum_{t=0}^{2^{(m-l)}-1} e^{-\frac{2\pi i jt}{2^{m-l}}}.$$

Since the exponential inside the sum is $\omega_{2^{m-l}}^{-tj}$, we know that the whole sum will be 0 in most cases. In fact, it will only be non-zero when j is a multiple of 2^{m-l}. This means that, if we apply the IQFT to one of those states and then we measure, we will obtain as a result an integer of the form

$$\frac{k2^m}{2^l}.$$

With a similar, but slightly more complicated derivation, it can be shown that, if you apply the IQFT to one of the periodic states that we discussed in *Section 11.3*, it is highly probable that a measurement will yield an integer that is close to

$$\frac{k2^m}{r},$$

where r is the period of the state and k is an integer. Remember that our goal is to determine r, because that is the value that we need in Shor's algorithm. To that extent, we can repeat the process of preparing the periodic state several times, applying the IQFT, and measuring. In this way, we will obtain several values that, with high probability, will be close to

$$\frac{k_1 2^m}{r}, \frac{k_2 2^m}{r}, \ldots, \frac{k_t 2^m}{r},$$

for some integers $k_1, k_2, \ldots, k_t$. From them, you can use some classical post-processing (for instance, with the help of continued fractions—see the book by Nielsen and Chuang [13] for a detailed treatment) and (finally!) recover r. Throughout the whole process, m can be set to be about $2n$, where n is the size of N in bits, the number of circuit executions and measurements needed is polynomial in n, and the whole post-processing part can be carried out in polynomial time in n. For all the details, feel free to check the original paper by Shor [67].

Important note

The quantum part of Shor's algorithm involves running the following circuit:

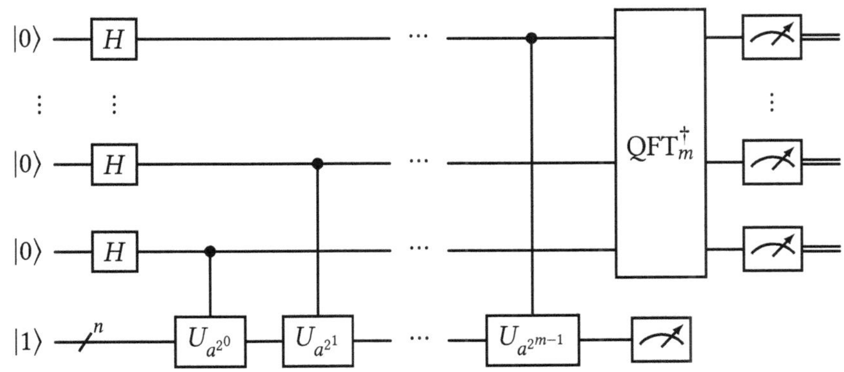

Here, $QFT_m^\dagger$ is the inverse Quantum Fourier transform on m qubits. The bottom register has n qubits, where n is the size in bits of the integer N that we want to factor.

From the measurement results of this circuit, by using a polynomial (in n) number of classical post-processing steps, with high probability we can recover r such that $a^r \equiv 1 \mod N$.

We are almost done with our study of Shor's algorithm. The only thing that remains is showing how to implement the (inverse) quantum Fourier transform as a circuit. We will take care of that right now.

11.4.3 Implementing the quantum Fourier transform

We are now going to learn how to construct a quantum circuit that implements the powerful quantum Fourier transform. You may be thinking that, given that the expression for the QFT is kind of complicated, its circuit will be big and cumbersome. However, as you will soon see, we can follow a recursive structure that will make it easy to obtain the circuit for the QFT on $m + 1$ qubits from the circuit for the m-qubit QFT. Let's then start from the simplest case and build our way up!

As we mentioned in *Section 11.4.1*, the matrix for the QFT on 1 qubit is exactly the same as the one for the Hadamard gate. Thus, the one-qubit QFT can be simply implemented with the following circuit:

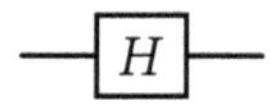

For the two-qubit case, we need to introduce a new quantum operation: the **SWAP gate**. As its name suggests, the SWAP gate exchanges the states of two qubits in a product state. That is, it takes $|\psi_1\rangle|\psi_2\rangle$ to $|\psi_2\rangle|\psi_1\rangle$. In particular, it acts on the computational basis states leaving $|00\rangle$ and $|11\rangle$ unchanged, and swapping $|01\rangle$ and $|10\rangle$. Hence, its matrix is

$$\begin{pmatrix} 1 & 0 & 0 & 0 \\ 0 & 0 & 1 & 0 \\ 0 & 1 & 0 & 0 \\ 0 & 0 & 0 & 1 \end{pmatrix}.$$

Obviously, this matrix is unitary. In fact, it is equal to its conjugate transpose and it is its own inverse, because if you swap the same things twice in a row, you return to the initial situation. Implementing it in terms of other gates that we already know is quite easy. We let you check it with the following exercise.

Exercise 11.6

Check that the SWAP gate can be implemented with the following circuit:

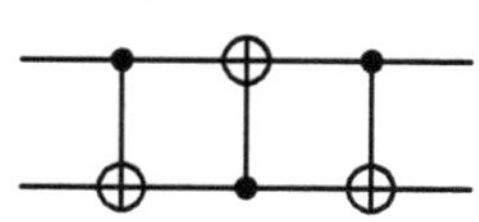

Check also that you can, alternatively, use the following circuit:

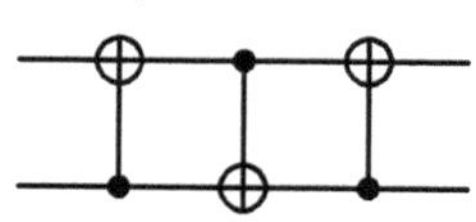

With this, we can now give the circuit for the two-qubit QFT, which is the following:

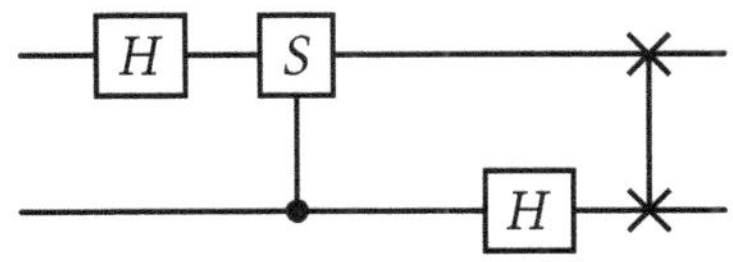

To check that this circuit works as expected, let's see how it acts in the computational basis state $|j\rangle = |x\rangle|y\rangle$, where $x, y = 0, 1$ and $j = 2x + y$. After the first H gate, the state becomes

$$\frac{1}{\sqrt{2}}\left(|0\rangle + (-1)^x|1\rangle\right)|y\rangle.$$

Then, when we apply the controlled-S gate, we obtain

$$\frac{1}{\sqrt{2}}\left(|0\rangle + (-1)^x i^y|1\rangle\right)|y\rangle.$$

Now, with the H gate on the bottom qubit, we get

$$\frac{1}{2}\left(|0\rangle + (-1)^x i^y|1\rangle\right)\left(|0\rangle + (-1)^y|1\rangle\right) =$$
$$= \frac{1}{2}\left(|00\rangle + (-1)^y|01\rangle + (-1)^x i^y|10\rangle + (-1)^{x+y} i^y|11\rangle\right).$$

With the final SWAP gate, the result is

$$\frac{1}{2}\left(|00\rangle + (-1)^x i^y|01\rangle + (-1)^y|10\rangle + (-1)^{x+y} i^y|11\rangle\right)$$
$$= \frac{1}{2}\left(|00\rangle + i^{2x+y}|01\rangle + i^{2y}|10\rangle + i^{2x+3y}|11\rangle\right).$$

Now, notice that $j = 2x + y$ and $i^4 = 1$. Then, we have

$$i^j = i^{2x+y},$$
$$i^{2j} = i^{4x+2y} = i^{4x} i^{2y} = (i^4)^x i^{2y} = i^{2y},$$
$$i^{3j} = i^{6x+3y} = i^{4x} i^{2x+3y} = i^{2x+3y}.$$

Hence, we can rewrite the final state as

$$\frac{1}{2}\left(i^{0\cdot j}|00\rangle + i^{j}|01\rangle + i^{2j}|10\rangle + i^{3j}|11\rangle\right) = \frac{1}{2}\left(i^{0\cdot j}|0\rangle + i^{j}|1\rangle + i^{2j}|2\rangle + i^{3j}|3\rangle\right),$$

which is exactly the action of the two-qubit QFT on the $|j\rangle$ state.

With a similar but longer computation, it can be checked that the three-qubit QFT can be implemented by the following circuit:

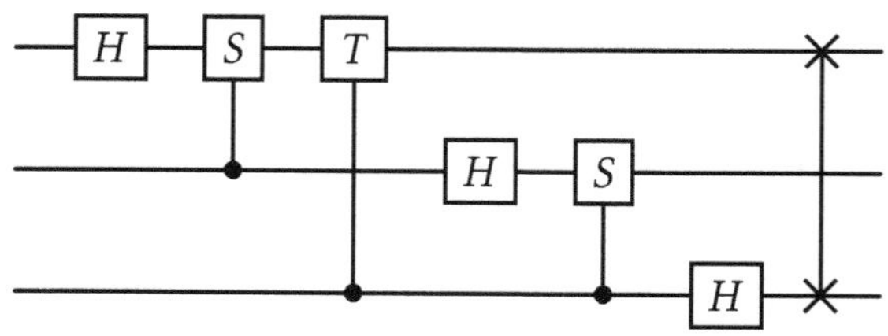

Notice how this builds upon the two-qubit QFT circuit. In fact, if we remove the bottom qubit and all the gates that interact with it, what we obtain is exactly the two-qubit circuit just before the SWAP gate. In a similar way, we can add an additional qubit with controlled gates targeting the other qubits to obtain the circuit for the four-qubit QFT. From that, adding a new qubit, we get the five-qubit QFT, and so on.

To give the general construction for the m-qubit QFT circuit, it is useful to remember a family of quantum gates that we introduced back in *Section 2.3.1*: phase gates. The phase gate $P(\theta)$ has matrix

$$P(\theta) = \begin{pmatrix} 1 & 0 \\ 0 & e^{i\theta} \end{pmatrix},$$

where θ is a real value. Particular cases include $Z = P(\pi)$, $S = P(\pi/2)$, and $T = P(\pi/4)$. To lighten up the notation a little bit, we will denote $P_k = P(\pi/2^k)$. With this, we are ready to give the circuit for the m-qubit QFT, which we highlight in the following important note.

Important note

The circuit for the m-qubit QFT is as follows:

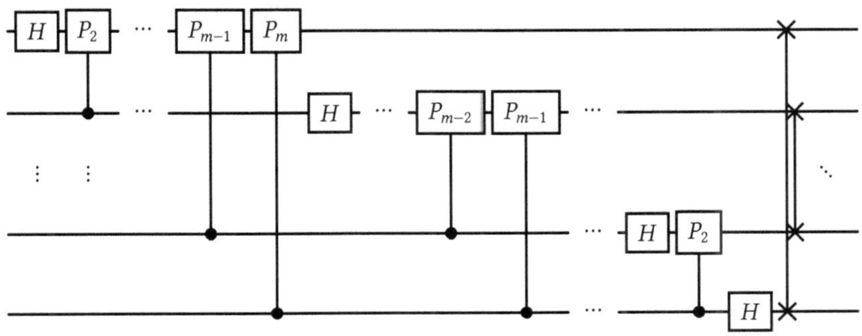

Here, we have denoted $P_k = P(\pi/2^k)$. Notice how the number of gates is $O(m^2)$.

The circuit for the inverse quantum Fourier transform can be readily obtained from this one. We just need to read the circuit from right to left, substituting each gate with its inverse (the SWAP and Hadarmard gates are their own inverses, while the inverse of $P(\pi/2^k)$ is $P(-\pi/2^k)$).

That's it! You now know all that you need in order to understand, implement, and apply Shor's factoring algorithm. We will close this chapter by making an observation about the techniques that we have used in the construction of the quantum circuit for the period estimation, and how they can be applied in problems of a more general nature.

11.5 Quantum phase estimation

An important problem in many areas of mathematics, science, and engineering is the determination of the eigenvalue associated to a certain eigenvector of a linear transformation. To fully understand what we are talking about, let's take some time to introduce some definitions (for a more thorough treatment, please refer to *Appendix A*).

Imagine that you have a linear transformation L from a vector space (over the complex numbers) into itself. We say that a vector $v \neq 0$ is an **eigenvector** of L if there exists a complex number λ such that $Lv = \lambda v$. In this case, we say that λ is the **eigenvalue** associated to the eigenvector v.

When L is a unitary operation, it can be proved (see *Appendix A*) that each of its eigenvalues is of the form $e^{2\pi i\theta}$ for some real number θ between 0 and 1. The problem of **phase estimation** is: given U, a unitary matrix, and $|\psi\rangle$, one of its eigenvectors, obtain an approximation of the eigenvalue λ such that $U|\psi\rangle = \lambda|\psi\rangle$. Since we know that $\lambda = e^{2\pi i\theta}$ for some real θ, this is equivalent to estimating θ, which is sometimes referred to as the phase of λ. This is the reason why this problem is called *phase* estimation.

To learn more...

Phase estimation is important, for instance, in physics and chemistry for its connections to problems in the study of materials and chemical compounds. For more on this, you can check out our book *A Practical Guide to Quantum Machine Learning and Quantum Optimization: Hands-on Approach to Modern Quantum Algorithms* [16].

Phase estimation is also central in other quantum algorithms, such as the HHL method for solving linear systems of equations [2] and the algorithm for quantum counting that we will study in *Chapter 12*.

With the techniques that we have studied in this chapter, we have all the elements to easily introduce **quantum phase estimation** (usually abbreviated as **QPE**), one of the most popular methods to solve the phase estimation problem. In fact, the circuit for QPE is as follows:

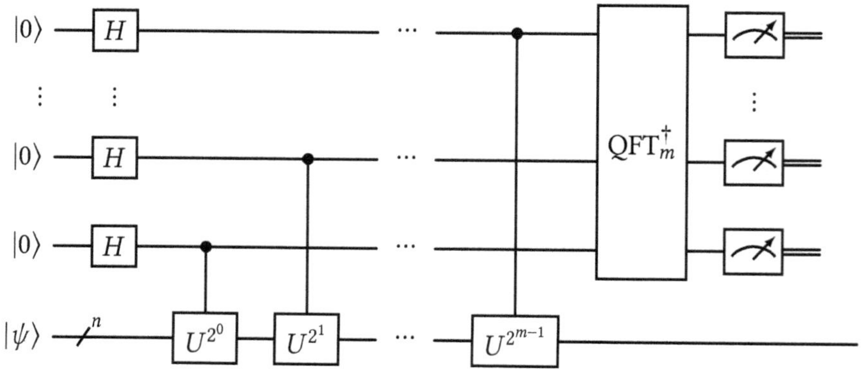

Sound familiar? This is so similar to the circuit used in Shor's algorithm that we could use it for playing "spot the difference"! Albeit similar, the two circuits are not identical. Notice that, in this case, the bottom register is initialized to the eigenvector $|\psi\rangle$ instead of $|1\rangle$, that the bottom register is not measured, and that the controlled operations are powers of the unitary U for which we want to estimate the phase of one of its eigenvalues.

With an analysis like the one that we performed in *Section 11.3.2*, we can easily see that we start with $|0\rangle|\psi\rangle$ and that the state just before the application of the inverse quantum Fourier transform is

$$\frac{1}{\sqrt{2^m}}\sum_{j=0}^{2^m-1} e^{2\pi i j\theta}|j\rangle|\psi\rangle.$$

The state in the second register never changes, so we can safely ignore it. Let's then focus on the first register only. If θ is of the form $l/2^m$ for some integer l, then this state is exactly the result of applying the m-qubit QFT to $|l\rangle$. Hence, when we apply the inverse quantum Fourier transform, we will obtain exactly l as the measurement result and, from that, we can easily recover θ as $l/2^m$. In a more general situation, with high probability, we will obtain l such that $l/2^m \approx \theta$. Increasing m, we can make this approximation as precise as desired. In fact, if the controlled powers of U can be implemented with a polynomial number of gates in the size of $|\psi\rangle$, it can be shown that this is an efficient method for phase estimation. So neat and convenient!

Exercise 11.7

How many qubits would you need to determine exactly the phase of the eigenvector $|1\rangle$ of the one-qubit gate S with QPE? What about $|1\rangle$ and T? Obtain the corresponding quantum circuits.

You may now be wondering what is the connection between QPE and Shor's algorithm. It's a legitimate question, so let's address it before ending this section. It can be proved that the eigenvalues of the modular multiplication unitary U_a are of the form $e^{\frac{2\pi i k}{r}}$, where k is an integer and r is the period of a modulo N. This can be easily seen because we know that U_a^r is the identity, since we are multiplying times a^r, which is 1 modulo N. Then, if λ is an eigenvalue of U_a and $|\psi\rangle$ is an associated eigenvector, it follows that

$$|\psi\rangle = U_a^r|\psi\rangle = U_a^{r-1}U_a|\psi\rangle = \lambda U_a^{r-1}|\psi\rangle = \cdots = \lambda^{r-1}U_a|\psi\rangle = \lambda^r|\psi\rangle,$$

from where we deduce that $\lambda^r = 1$ and, hence, λ is an rth root of unity. That is, $\lambda = e^{\frac{2\pi i k}{r}}$ for some integer k. Notice that, in this case, θ would be k/r.

Moreover, the state $|1\rangle$, which is the initial state of the second register in the quantum circuit of Shor's algorithm, can be written as a uniform superposition of all the eigenvalues of U_a. By linearity, if we interpret Shor's algorithm circuit as an instance of QPE, we will obtain, with high probability, an integer l such that $l/2^m = \theta = k/r$. That is, an integer of the form $k2^m/r$. This coincides with our earlier analysis of the method.

We find it extremely pleasant that, in a chapter devoted to periodicity, we have found ourselves returning to a previous step in our (long) journey, so let's end it on this happy note.

Summary

In this chapter, we have studied in detail one of the most important quantum algorithms: Shor's method for integer factorization.

We started by remarking on the importance of prime numbers and their central role in some cryptographic protocols in wide use today. Then, we took a first view at Shor's algorithm, describing the different steps in the method and giving some simple examples of how it is applied. As we showed, most of these steps can be carried out efficiently with classical algorithms, but finding the period of an integer would require the use of a quantum computer when the integer to factor is big.

For this reason, we then turned our attention to the creation of a certain type of periodic states with the help of quantum circuits that rely on the use of superposition and entanglement. To extract information about the period from these states, we introduced the mighty quantum Fourier transform. The study of this operation took us to wild and gorgeous places. We considered some beautiful complex numbers called roots of unity, we marveled once again with the power of interference, and we designed clever circuits to implement the QFT and its inverse.

Finally, we strayed a bit from the path to have a bird's-eye view of our discoveries, and noticed that the techniques that we have mastered can also be used to solve the problem of phase estimation. We then traced our steps back and connected this new problem to the task of finding factors of big integers.

This has been an arduous but joyful exploration. You may rest for a bit—you have earned it. But don't relax just yet. The journey is far from over! The next chapter will take us to another peak in our expedition. Get ready, for we are about to introduce the famous Grover's algorithm.

12

Searching and Counting with a Quantum Computer

When you go in search of honey, you must expect to be stung by bees.

— Joseph Joubert

In the previous chapter, we went through the wonders and mysteries that lie behind Shor's algorithm, perhaps the most famous quantum algorithm to have ever existed. Leveraging on the power of the quantum Fourier transform, we uncovered how quantum computers can efficiently factorize integer numbers and effectively threaten some of our current (public-key) cryptographic protocols. That's... quite a lot, and we are aware that we have left the bar too high. But rest assured: we are still hiding a few interesting things up our sleeve.

In this chapter, we are going to discuss another quintessential quantum algorithm: Grover's algorithm. Sadly (or perhaps fortunately!), this one will not allow us to blackmail our way to world dominance by breaking public-key cryptographic protocols. Nevertheless, it is still a powerful and exciting tool that might have lots of beneficial real-world applications. The problem that Grover's algorithm will address is that of searching through an unsorted list.

This search task might seem trivial to the unfamiliar eye, but that could not be further from the truth! It is a problem in which quantum computers can break boundaries that would be unthinkable for classical computers. Thus, to better contextualize what Grover's algorithm aims to achieve—and why that is worth our awe—we will devote the first section of this chapter to analyzing the problem of searching through an unsorted list. Once we are done with that introduction, we will discuss Grover's algorithm in all its might. And, lastly, to conclude the chapter, the quantum Fourier transform will make one last epic reappearance.

To put it succinctly, these are the contents of this chapter:

- Searching in an unsorted list
- Grover's algorithm
- Counting with the quantum Fourier transform

That might be enough of an introduction. For now, let us humble ourselves and quantumly look for needles in haystacks!

12.1 Searching in an unsorted list

In the Digital Age™, few people would consider searching to be a complicated computational task: we just take it for granted. We can almost instantly find files in our computers and we can even search through the whole 64 zetabytes of the Internet with just a few keystrokes using any search engine. Moreover, thanks to new advancements in artificial intelligence, searching has taken off to new heights. Just to give an example, some photo management applications now enable you to look for pictures in your digital photo albums by providing mere textual descriptions of their content.

It is fair to say that we live in the golden age of searching. Yet, as ubiquitous and widely used as search algorithms may be, few people (excluding computer scientists, of course) actually know how those algorithms work. Most would actually believe that, when looking for some files on their computer, their system just inspects each file one by one, in whichever order they may be arranged, as if all the files were piled together in a messy bucket. If that were the case, regardless of how powerful your computer would be, you could have the assurance that searching would be anything but a seemingly instantaneous experience.

The core reason why search queries can work so well is not (just) the mind-blowing speeds of our current processors, but the way in which these algorithms preprocess, arrange, and sort the data that they are going to be searching on. And the best way of illustrating this might be through a simple real-world example.

Imagine that you wake up in the middle of the night to a text from your friend that reads "the craic was mighty!!". Confused and concerned as you may be, you choose to look up that strange "craic" term in a dictionary, but—in an attempt to be old-school cool—you give up the chance of taking out your phone and you decide to dust off that hardbound dictionary that is lying on your shelf. The question is the following: how would you search for the word "craic" in that big tome? Would you begin at the word "a" and evaluate, one by one, all the words in sequential order until you reached "craic"? In principle, you would be guaranteed to find that word eventually (provided that it is included in the dictionary!), but we can all probably agree that it would take you quite a lot of time to complete this task. And this is the deal: no matter how fast you read, this would inevitably take a huge amount of time—it would likely take more than an hour even for the fastest human reader ever to have lived on the face of the Earth!

And where does the issue lie? In the fact that this search procedure could take *as many queries* as words there are in the dictionary: as many queries as items there are in the search space. Using the tools of computational complexity, if N is the size of the search space, this linear search takes $O(N)$ queries to complete, which is not ideal as N grows large. Recall, incidentally, that a function $f(N)$ (in this case, the number of queries) is $O(g(N))$ if there exists a constant M and an N_0 such that, for all $N \geq N_0$, $f(N) \leq M \cdot g(N)$; essentially, this is telling us that, when N grows large, $f(N)$ does not grow worse than $g(N)$.

To learn more...

For a full introduction to computational complexity and the big O notation, you can head to *Appendix C*.

Nevertheless, as we all probably know, there is a better way of searching in a dictionary, as the words are *sorted*. Thus, we can open the dictionary in half and, if the words on the page go after "craic" alphabetically, we can flip a few pages back at around a quarter of the dictionary; conversely, if they go before it, we can flip a few pages forward toward three quarters of the dictionary. Repeating this process iteratively, we will reduce the search space by half in each step, and we will eventually reach the page where "craic" is. If we do everything correctly, the number of iterations needed will be only $O(\log(N))$, which is a dramatic improvement! Assuming an English dictionary has two hundred thousand entries, we would only need $\log_2 2 \cdot 10^5 \approx 18$ queries at most. Now think about that: with a simple linear search—that made no assumptions about the way in which the words were sorted—we could have needed 200.000 iterations. That is quite a huge difference!

To learn more...

In a computer system, when we are given an array of (sorted) data, we can perform a search procedure analogous to the one that we have discussed: this is known as a **binary search** [62].

The situation with a dictionary search is perfectly analogous to the one that we find in any other computational search task. It doesn't matter how powerful and fast a processor may be: the key behind a fast search rests on the efficiency of the search algorithm being used, and in how the search data is structured, sorted, and organized. Maybe at some stage, perhaps after a software update, you have tried to perform a search but were met with a message claiming that your data was being "indexed". That is a fancy way of saying that it was being organized so that your future searches could be as fast as you expect them to be!

Thus, we have seen how the backbone of fast searches is data sorting. There is nevertheless one problem, which is that, in many real-world problems, we may have to perform searches over immense *unsorted* datasets. And that's where classical algorithms face a challenge. If you are given a dataset with N elements that are not sorted, and you are tasked with finding a specific entry within it, no search algorithm can yield a better efficiency than $O(N)$. To see why, notice that, in a classical algorithm, you are bound to querying the elements in the search space one by one, and, if the elements are unsorted, nothing forbids the element you are looking for from being the last one you would query. So, in the worst-case scenario, you will have to query all the elements in the search space!

This naturally begs the question: why not just sort the elements in a dataset and leave it there? Why would we be interested in an algorithm that would allow us to search over an unsorted search space? There are quite a few scenarios in which such an algorithm would come in handy:

- Sorting a list of elements is computationally more costly than looking for one of them. For example, sorting an array of elements in which you are able to compare pairs of elements (deciding which element should go first in the array), would take a number of comparisons of order at least $N \log N$ (for a proof, we refer you to the book *Introduction to algorithms* [62]). Thus, if you are going to perform a single search, it might just not be worthwhile sorting the elements.

  For example, if you were asked on a single occasion to look for a word in the dictionary that ends in "xz", you would be better off doing a linear search than resorting the whole dictionary.

- In some problems, there may not be a way to sort data at all! Imagine that you are given an abstract mathematical problem in the form of a function f (which you know how to compute), for which you are asked to find an entry x such that $f(x) = 1$. In this case, there is no way of sorting the elements that would enable you to perform a binary search.

The last scenario is the one that Grover will be best at tackling. You see, right now, quantum computers may have a problem or two accessing and storing databases with classical data, but for mathematical problems… they can achieve very surprising things!

Thus, for this kind of search problem, quantum computing will now come to the rescue with Grover's algorithm. As we are about to find out, Grover's algorithm will enable us to solve a search problem with just $O(\sqrt{N})$ queries to… you guessed it! An oracle! Thus, Grover's algorithm will provide a quadratic speed-up over any other possible classical alternative. For instance, if you are searching through one million elements for a particular one, a classical algorithm might require up to one million steps. But Grover's search will only need about one thousand queries. Impressive, right? There will be a few caveats in the application of the method (as with anything quantum), but we will take care of them at their own time.

Important note

In a search problem, Grover's algorithm will provide a quadratic speed-up when compared to any classical algorithm. In particular, if the size of the search space is N and the data in it is not sorted or organized in any way, any classical algorithm would require $O(N)$ queries to the search space, while Grover's algorithm would only require $O(\sqrt{N})$ calls to an oracle.

We should remark that $O(\sqrt{N})$ is still less efficient than $O(\log N)$, but, unlike binary search, Grover's algorithm does not need the data to be sorted.

And now that we know what Grover's algorithm promises to deliver, let's dive into its details.

12.2 Grover's algorithm

We now have a general picture of what Grover's algorithm is capable of doing, but, before we can discuss all of its details, we should fix some notation and make our hypotheses and conditions clear.

In the search problem that we will be working with through this section, we will consider an n-bit Boolean function $f(x) = f(x_1, \ldots, x_n)$, and we will assume that $f(x) = 0$ for all possible inputs x except for a single n-bit **marked** string, s, for which $f(s) = 1$. Our goal is to find that n-bit string s. Later in this section, we will see what can be done if the number of marked strings (strings that make f return 1) is any fixed number k, but, for now, we will keep things simple and assume that only one input verifies this condition.

This framework might not seem too powerful at first, but it can actually encode any search problem. Assume, for example, that you have an unsorted database with a billion entries laid out in an array-like form, where each entry has its own index number. And let's say that you are looking for one entry in particular. Of course, you can refer to each of the entries by its index in the database, so you can define f to be the function that takes the binary representation of the index of an item and returns 1 if it is the item you are looking for, and 0 otherwise.

The aim of abstracting and encoding the search problem into a binary function is two-fold. First, it will allow us to unambiguously quantify the number of times the search space is accessed, and, second, using a binary input will make it easier to work with qubits, as we are so well accustomed to.

So, we have an n-bit Boolean function that returns zero for all inputs except for a special "marked" entry s. This works beautifully in the World of Thought, but how are we to access such a function in an algorithm? Well, in a classical algorithm, we would just have a subroutine that would compute f: in the database example that we discussed previously, we would have some SQL queries or something like that. In the quantum world, however, things won't be quite as direct. As we anticipated before, we will have to resort to our old good friends, the oracles.

12.2.1 Oracles, with a twist

We have already worked with oracles on plenty of other quantum algorithms, so you may be puzzled as to why we are devoting a whole subsection to them. As it turns out, we will use oracles in a slightly different way when analyzing the behavior of Grover's algorithm.

In *Chapter 6* and *Chapter 9*, we introduced the following general definition. For any n-bit Boolean function f, an oracle O_f for f is the $(n+1)$-qubit gate that takes any state $\left|x\right\rangle\left|b\right\rangle$

in the computational basis (where $|x\rangle$ is an n-qubit state and $|b\rangle$ is a one-qubit state) to the state $|x\rangle\,|b \oplus f(x)\rangle$. In a circuit, we can depict this as follows:

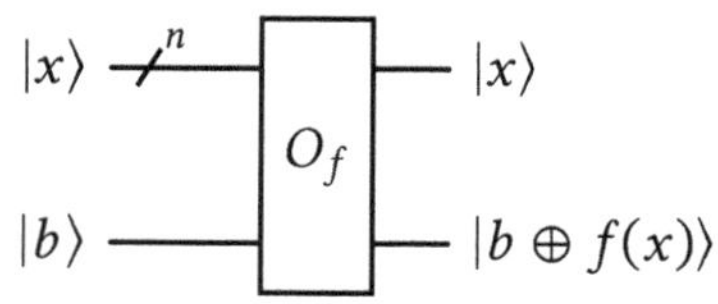

These are also the types of oracles that are used in Grover's algorithm and, when we implement it in *Chapter 13*, that is the way we will go. But for the mathematical analysis of the algorithm, it will be convenient for us to (momentarily) replace them with an equivalent version. Let's see how.

Assume that we are given an oracle O_f (for a function f) like the one we've considered previously. When using O_f in Grover's algorithm, we will set the last qubit to the state $|-\rangle = H\,|1\rangle$. Then,

$$\begin{aligned}
O_f\,|x\rangle\,|-\rangle &= O_f\left(|x\rangle \otimes \frac{1}{\sqrt{2}}\left(|0\rangle - |1\rangle\right)\right) = \frac{1}{\sqrt{2}}\left(O_f\,|x\rangle\,|0\rangle - O_f\,|x\rangle\,|1\rangle\right) \\
&= \frac{1}{\sqrt{2}}\left(|x\rangle\,|0 \oplus f(x)\rangle - |x\rangle\,|1 \oplus f(x)\rangle\right) = \frac{1}{\sqrt{2}}\,|x\rangle\left(|0 \oplus f(x)\rangle - |1 \oplus f(x)\rangle\right) \\
&= \frac{1}{\sqrt{2}}\,|x\rangle\,(-1)^{f(x)}\left(|0\rangle - |1\rangle\right) = (-1)^{f(x)} \cdot \frac{1}{\sqrt{2}}\,|x\rangle\left(|0\rangle - |1\rangle\right) \\
&= (-1)^{f(x)}\,|x\rangle\,|-\rangle\,,
\end{aligned}$$

where we have used the fact that, as we proved in *Section 6.3*, for any bit b,

$$|0 \oplus b\rangle - |1 \oplus b\rangle = (-1)^b(|0\rangle - |1\rangle)\,.$$

Thus, by setting the last qubit to $|-\rangle$ and disregarding it as an ancillary qubit, we can effectively assume that we are applying a gate $\hat{O}_f$ on the first n qubits that will take any state $|x\rangle$ from the computational basis to the state $(-1)^{f(x)}\,|x\rangle$:

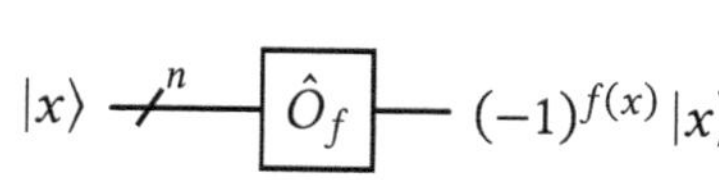

These gates $\hat{O}_f$ are known as **phase oracles**, and in its construction, we have seen a new instance of a phase kickback, as the $(-1)^b$ phase that appears to affect only the $|-\rangle$ state ends up affecting the whole state.

Important note

For convenience reasons, in our analysis of Grover's algorithm, we will assume that we have access to n-bit Boolean functions f through **phase oracles**. These oracles are the n-qubit gates that take any computational basis state $|x\rangle$ to $(-1)^{f(x)}|x\rangle$. They can be constructed from the ones that we employed in *Chapter 6* and *Chapter 9* by setting the last qubit to $|-\rangle$ and disregarding it as an ancillary qubit.

However, in the actual implementation of Grover's algorithm, regular oracles for Boolean functions are usually applied. The distinction is mostly unimportant, because their effect is exactly the same and, if we ignore the ancillary qubit (that remains unchanged throughout the whole process), their behavior is completely identical.

This fully explains how we will be using oracles in our mathematical analysis of Grover's algorithm, but you might be having some reservations about the idea of using oracles altogether. Back in *Section 9.2*, we explored in detail how oracles could be constructed and, as we have just seen, phase oracles can be directly constructed from them. Nevertheless, this might make you wonder: by using oracles, wouldn't we be cheating? If we have access to an oracle in the first place, wouldn't that mean that the search problem should be trivial? As it turns out, that is not the case. An oracle is given to us as a black box. We cannot know what's inside, we can only evaluate it on inputs to obtain outputs. All oracles look exactly the same until you evaluate them, so you gain nothing from receiving them if you don't use them!

To learn more...

Even if you had access to the inner workings of an oracle for a Boolean function f, that might not help you much in finding an element s such that $f(s) = 1$. To justify why, we can refer ourselves to a related problem in computational complexity theory. This problem assumes that you are given a classical circuit (that you can inspect to your heart's desire, unlike an oracle) for a Boolean function, and asks you whether any input makes the function output 1. This is known as the **CIRCUIT-SAT** problem and it is widely believed that it cannot be solved efficiently.

To learn more about computational complexity and the reasons why CIRCUIT-SAT (and other related problems) are believed to be intractable in practice, we recommend that you check Chapter 5 in the wonderful book *The nature of computation* [77].

Also, if you would like to explore the nature of oracles in more detail, we invite you to read Section 6.1.1 from the book *Quantum Computation and Quantum Information: 10th Anniversary Edition* by Nielsen and Chuang [13].

Having discussed oracles in full depth, now it's time for us to dive into the mechanics of Grover's algorithm and see for ourselves how they can be used to deliver the amazing speed-ups that we have been talking about for so long.

12.2.2 The initial setup

Grover's algorithm for an n-bit Boolean function needs n qubits and it starts as most quantum algorithms do: with all qubits initialized to $|0\rangle$ and the subsequent application of a Hadamard gate on each qubit:

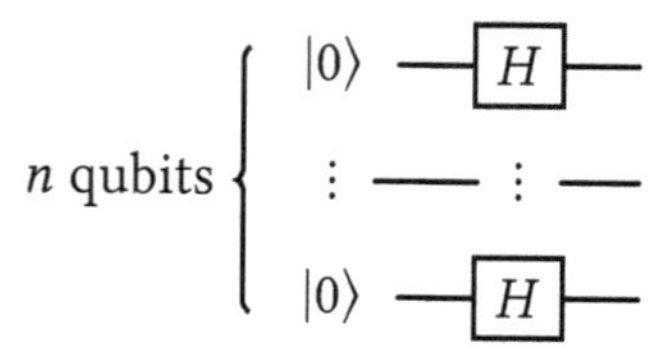

As you proved in *Exercise 8.2*, the resulting state $|\alpha\rangle$ will be a balanced superposition of all the computational basis states, which we can write as

$$
\begin{aligned}
|\alpha\rangle := \frac{1}{\sqrt{2^n}} \sum_{k=0}^{2^n-1} |k\rangle &= \frac{1}{\sqrt{2^n}} \left(|s\rangle + \sum_{k=0,k\neq s}^{2^n-1} |k\rangle \right) \\
&= \frac{1}{\sqrt{2^n}} |s\rangle + \frac{\sqrt{2^n - 1}}{\sqrt{2^n}} \left(\frac{1}{\sqrt{2^n - 1}} \sum_{k=0,k\neq s}^{2^n-1} |k\rangle \right) \\
&= \frac{1}{\sqrt{2^n}} |s\rangle + \sqrt{\frac{2^n - 1}{2^n}} \, |\text{other}\rangle ,
\end{aligned}
$$

where we recall that s is the marked element we are searching for (and $|s\rangle$ is its corresponding computational basis state). Also, in the preceding expression, we have implicitly defined $|\text{other}\rangle$ to be the uniform superposition of all the $2^n - 1$ computational basis states that are not $|s\rangle$.

Notice that if, at this point, we decided to measure the state, the probability of obtaining s (the element that we want to find) is a tiny $1/2^n$ (because the amplitude of $|s\rangle$ is $1/\sqrt{2^n}$). The goal of Grover's algorithm is to iteratively increase the amplitude of $|s\rangle$ so that the probability of finding the marked element when measuring will be close to 1.

From this point, there is an intuitive and geometrical way of understanding Grover's algorithm. As we are about to find out, the state of the system throughout the algorithm can be represented as a unit vector in the two-dimensional Euclidean plane, in such a way that the unit vector $\vec{u}_X = (1, 0)$ represents $|\text{other}\rangle$, which is a state with probability 1 of measuring a non-marked element, and the unit vector $\vec{u}_Y = (0, 1)$ represents $|s\rangle$, a state with probability 1 of measuring the marked element s (see *Figure 12.1*). The initial state of the algorithm $|\alpha\rangle$ will, under this representation, be very close to $\vec{u}_X$, but we will iteratively perform a sequence of counter-clockwise rotations until we bring it just close enough to $\vec{u}_Y$.

Let us now make things precise and discuss how this representation can be set up. If we consider a state in a superposition of $|\text{other}\rangle$ and $|s\rangle$ with real amplitudes (as will always be the case in the application of Grover's algorithm), we can write it as $\cos\theta\,|\text{other}\rangle + \sin\theta\,|s\rangle$

(for some angle θ) because of the normalization condition of quantum states. Indeed, if a and b are two real numbers such that $a^2 + b^2 = 1$, there must exist some angle θ such that $a = \cos\theta$ and $b = \sin\theta$. Thus, we will represent any superposition

$$|\psi\rangle = \cos\theta\,|\text{other}\rangle + \sin\theta\,|s\rangle\,,$$

with the unit vector $(\cos\theta, \sin\theta)$, which makes a counter-clockwise angle θ with $\vec{u}_X$ (and with the positive horizontal axis). It can be readily checked how, indeed, $|\text{other}\rangle$ is represented as $\vec{u}_X$ and $|s\rangle$ as $\vec{u}_Y$, as intended (just think what are the sine and cosine of 0 and $\pi/2$, respectively).

In the case of $|\alpha\rangle$, which is depicted (not to scale!) in *Figure 12.1*, we let ω be this angle θ, and we have

$$\sin\omega = \frac{1}{\sqrt{2^n}},$$

which is extremely small for large n, as is ω itself.

To learn more...

The reason why this representation is natural and well-defined is that $|s\rangle$ and $|\text{other}\rangle$ are orthogonal and they span a two-dimensional vector space. If you are not familiar with these terms, you may wish to review *Appendix A*.

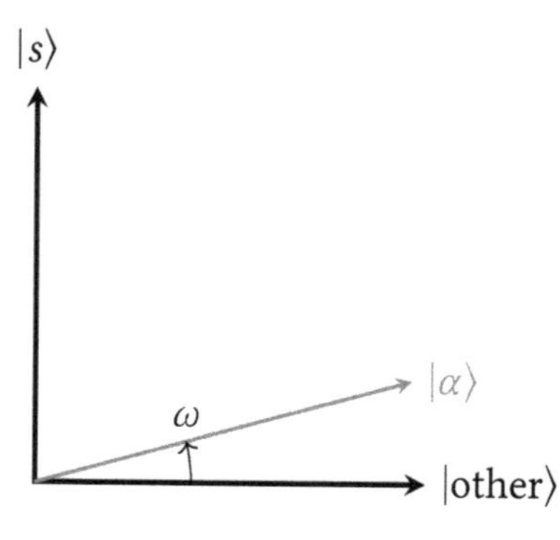

Figure 12.1: Representation (not to scale) on the two-dimensional plane of the states $|s\rangle$ and $|\text{other}\rangle$, and of the state $|\alpha\rangle$, which is a balanced superposition of all the computational basis states. If this were to scale, ω should be much smaller

The probability of measuring s in a state of the form $\cos\theta\,|\text{other}\rangle + \sin\theta\,|s\rangle$ is $|\sin\theta|^2$, as $|\text{other}\rangle$ is a superposition of all the computational basis states different from $|s\rangle$. As we mentioned previously, in the case of $|\alpha\rangle$, this probability is extremely small, $1/2^n$, and of course, we would like to change this. We will do so by transforming the state of the system through rotations in a way that will progressively increase the angle that its state makes with respect to $|\text{other}\rangle$ in our representation, bringing it closer to $\pi/2$ (thus bringing $|\sin\theta|^2$ closer to 1) and increasing our odds of retrieving s as a measurement outcome. We shall now study how these rotations can be implemented and understood.

12.2.3 Amplitude amplification

The circuit that implements the rotation that we have been talking about for so long is depicted in *Figure 12.2*, and it composes two different operations. The first hides no mysteries for us: it's applying the phase oracle $\hat{O}_f$ for the function f that we are considering. The second is slightly more intricate, and it involves a combination of gates known as **Grover's diffusion operator**.

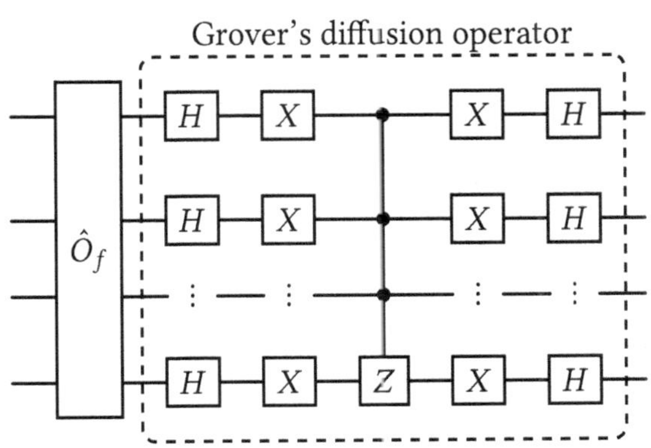

Figure 12.2: The phase oracle and Grover's diffusion operator, used in the amplitude amplification process in Grover's algorithm

As shown in *Figure 12.2*, applying Grover's diffusion operator is equivalent to performing the following sequence of steps:

1. Apply a Hadamard gate on every qubit.

2. Apply a NOT gate on every qubit.
3. Apply a multi-controlled Z gate with the last qubit as a target and all other qubits as controls.
4. Apply a NOT gate on every qubit.
5. Apply a Hadamard gate on every qubit.

Incidentally, this "multi controlled Z" gate works as follows: given any $(n-1)$-qubit state $|x_1 \cdots x_{n-1}\rangle$ from the computational basis and a one-qubit state $|b\rangle$, the gate will leave $|x_1 \cdots x_{n-1}\rangle |b\rangle$ untouched unless $|x_1 \cdots x_{n-1}\rangle = |1 \cdots 1\rangle$, in which case it will apply a Z gate on the last qubit, effectively transforming the state of the whole system into $-|x\rangle |b\rangle$. This pretty much works like a CCNOT gate, but using a Z gate instead of an X, and with a few more control qubits.

Exercise 12.1

Prove that, in the multi-controlled Z gate used in Grover's diffusion operator, we could have picked any qubit as a target, having the remaining qubits as controls.

Let us now try to analyze what the combination of the phase oracle and the diffusion operator is doing and, to that end, consider any state $|\beta\rangle = \cos\theta\,|\text{other}\rangle + \sin\theta\,|s\rangle$ with real amplitudes making an angle θ with the $|\text{other}\rangle$ state in the plane representation that we have been considering.

We recall that, by definition, for any computational basis $|x\rangle$, $\hat{O}_f\,|x\rangle = (-1)^{f(x)}\,|x\rangle$. Therefore, since $f(s) = 1$ and f evaluates to 0 for any other entry, $\hat{O}_f$ will take $|s\rangle$ to $-|s\rangle$ and it will leave all the other computational basis states untouched. We may then deduce that the action of $\hat{O}_f$ on $|\beta\rangle$ will transform the state into

$$
\begin{aligned}
|\beta_1\rangle := \hat{O}_f\,|\beta\rangle &= \cos\theta \cdot \hat{O}_f\,|\text{other}\rangle + \sin\theta \cdot \hat{O}_f\,|s\rangle \\
&= \cos\theta \left(\frac{1}{\sqrt{2^n - 1}} \sum_{k=0,k\neq s}^{2^n-1} \hat{O}_f\,|k\rangle \right) + \hat{O}_f \sin\theta\,|s\rangle
\end{aligned}
$$

$$= \cos\theta\left(\frac{1}{\sqrt{2^n-1}}\sum_{k=0,k\neq s}^{2^n-1}|k\rangle\right) - \sin\theta\,|s\rangle = \cos\theta\,|\text{other}\rangle - \sin\theta\,|s\rangle$$

$$= \cos(-\theta)\,|\text{other}\rangle + \sin(-\theta)\,|s\rangle\,,$$

where we have used the fact that, for any real number x, we have $-\sin(x) = \sin(-x)$ and $\cos(x) = \cos(-x)$. This means that, in our plane representation, the state $|\beta_1\rangle$ will be a reflection of $|\beta\rangle$ over the $|\text{other}\rangle$ axis, and hence it will make an angle $-\theta$ with respect to this axis. This is depicted in *Figure 12.3*.

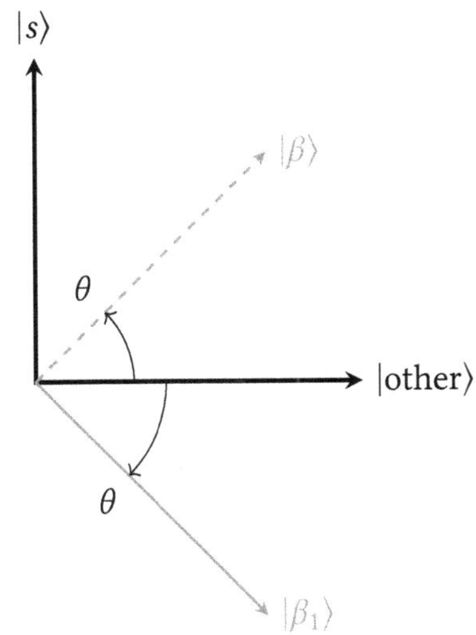

Figure 12.3: The action of the oracle $\hat{O}_f$ on a state $|\beta\rangle$, transforming it into $|\beta_1\rangle$

That was fairly straightforward. Now, studying the behavior of Grover's diffusion operator is a much more difficult task, but, for our purposes, it suffices to know that, from a geometric point of view, Grover's diffusion operator is just changing the sign of the component of $|\beta_1\rangle$ that is perpendicular to $|\alpha\rangle$. In terms of our plane representation, this will lead to $|\beta_1\rangle$ being reflected about $|\alpha\rangle$ into a state $|\beta_2\rangle$, as depicted in *Figure 12.4*. Notice how the counter-clockwise angle that $|\beta_1\rangle$ made from $|\alpha\rangle$ was $-(\theta + \omega)$. Therefore, since $|\beta_2\rangle$ is a reflection of $|\beta_1\rangle$ about $|\alpha\rangle$, the counter-clockwise angle that $|\beta_2\rangle$ will make from $|\alpha\rangle$ must be $(\theta + \omega)$ (see *Figure 12.4* for visual reference). Moreover, the counter-clockwise angle that β_2 will make from $|\text{other}\rangle$ will be $\omega + (\omega + \theta) = \theta + 2\omega$.

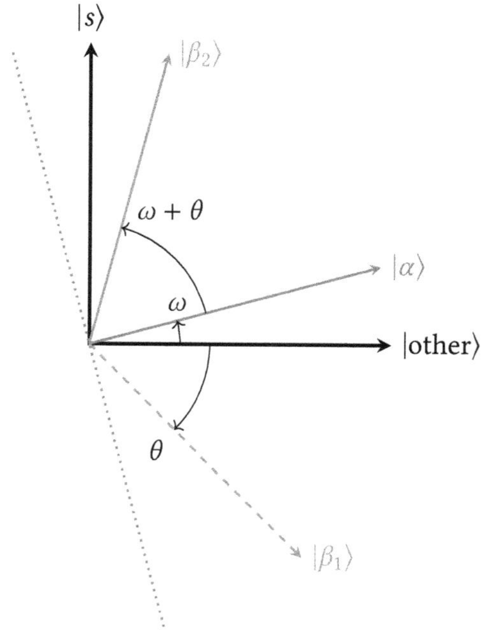

Figure 12.4: The action of Grover's diffusion operator on the state $|\beta_1\rangle$, reflecting it about $|\alpha\rangle$ and transforming it into $|\beta_2\rangle$. The direction perpendicular to $|\alpha\rangle$ is represented with a dotted line; the state $|\beta_2\rangle$ can be obtained from $|\beta_1\rangle$ by changing the sign of its component on this line

We can now try to justify why the circuit that we provided in *Figure 12.2* for Grover's diffusion operator does indeed behave like the geometric operation that we've described: changing the sign of the component of a state that is perpendicular to $|\alpha\rangle$. To that end, we will briefly discuss the effects of the steps it performs. Before we do that, though, there is one simple fact that we should highlight, which is that quantum gates preserve orthogonality. This means that, given any quantum gate U, if two states $|x\rangle$ and $|y\rangle$ are orthogonal, then so will be $U|x\rangle$ and $U|y\rangle$. We cover this in *Appendix A*, but, if you wish, you can just take our word for it (or even try to prove it yourself; you just need to to take into account that $UU^\dagger$ is the identity matrix). With that being said, let's get to analyze what our circuit is doing:

1. The first set of Hadamard gates brings
   - the state $|\alpha\rangle$ to the state $|0\cdots 0\rangle$
   - any state $|\psi\rangle$ perpendicular to $|\alpha\rangle$ to a state orthogonal to $|0\cdots 0\rangle$
2. The NOT gates then take
   - $|0\cdots 0\rangle$ to $|1\cdots 1\rangle$
   - any state orthogonal to $|0\cdots 0\rangle$ to a state orthogonal to $|1\cdots 1\rangle$
3. At this point, the multi-controlled Z gate
   - multiplies the state $|1\cdots 1\rangle$ by a phase -1
   - leaves any state orthogonal to $|1\cdots 1\rangle$ untouched
4. Then, the NOT gates take
   - $-|1\cdots 1\rangle$ to $-|0\cdots 0\rangle$
   - any state orthogonal to $|1\cdots 1\rangle$ to a state orthogonal to $|0\cdots 0\rangle$
5. Lastly, the final set of Hadamard gates brings
   - the state $-|0\cdots 0\rangle$ to $-|\alpha\rangle$
   - any state orthogonal to $|0\cdots 0\rangle$ to a state orthogonal to $|\alpha\rangle$, which will be the original state $|\psi\rangle$ as the gates X and H are their own inverses, and the multi-controlled Z did nothing in this case

We can then see how this circuit multiplies $|\alpha\rangle$ by -1 and does nothing to any state that is perpendicular to it (exactly the opposite of what our geometrical interpretation of Grover's diffusion operator does!). Thus, up to a global phase of -1, which we can safely ignore, the circuit implements Grover's diffusion operator as we have described it.

To learn more...

In order to better and more compactly describe Grover's diffusion operator, we need to use **projection operators**. Given any state $|\psi\rangle$, its projection operator $P_{|\psi\rangle}$ takes any vector $|\beta\rangle$ to

$$P_{|\psi\rangle} |\beta\rangle = \langle\psi|\beta\rangle |\psi\rangle .$$

Here, $\langle\psi|\beta\rangle$ denotes the scalar product of $|\psi\rangle$ and $|\beta\rangle$. This is discussed more extensively in *Appendix A* and *Appendix B*.

It can then be verified that Grover's diffusion operator can be equivalently represented as $2P_{|\alpha\rangle}-I$, where I is the identity matrix. From this definition, one can deduce that the action of this operator transforms a state of the form $|\beta_1\rangle = a|\alpha\rangle + b|\psi\rangle$ (where $|\psi\rangle$ is assumed to be orthogonal to $|\alpha\rangle$) into

$$\begin{aligned}
|\beta_2\rangle := (2P_{|\alpha\rangle} - I)|\beta_1\rangle &= (2P_{|\alpha\rangle} - I)(a|\alpha\rangle + b|\psi\rangle) \\
&= 2aP_{|\alpha\rangle}|\alpha\rangle - a|\alpha\rangle + 2b\cancel{P_{|\alpha\rangle}|\psi\rangle} - b|\psi\rangle \\
&= 2a|\alpha\rangle - a|\alpha\rangle - b|\psi\rangle \\
&= a|\alpha\rangle - b|\psi\rangle ,
\end{aligned}$$

where $P_{|\alpha\rangle}|\psi\rangle = 0$ as $|\alpha\rangle$ and $|\psi\rangle$ are orthogonal and, therefore, $\langle\alpha|\psi\rangle = 0$; and $P_{|\alpha\rangle}|\alpha\rangle = |\alpha\rangle$ because $\langle\alpha|\alpha\rangle = 1$.

This shows how, as we claimed, this operation changes the sign of the component of $|\beta_1\rangle$ that is perpendicular to $|\alpha\rangle$.

The action of Grover's diffusion operator is known as an **inversion about the mean**, since it effectively reflects or "inverts" the amplitudes of a state about the mean of all the amplitudes. This effect can be readily verified using the representation of the operator that we have just introduced.

Thus, the net effect of the composition of the phase oracle and Grover's diffusion operator is to increase by 2ω the counter-clockwise angle that any input state makes with respect to $|\text{other}\rangle$ (in the plane representation).

Important note

Given an n-bit Boolean function f with a unique entry s such that $f(s) = 1$, the composition of its phase oracle $\hat{O}_f$ with Grover's diffusion operator is an n-qubit gate that can be realized by the following circuit:

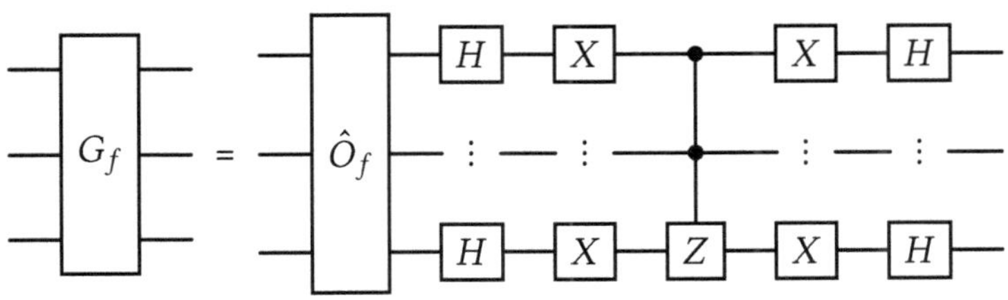

Defining $|\text{other}\rangle := \frac{1}{\sqrt{2^n-1}} \sum_{k=0,k\neq s}^{2^n-1} |k\rangle$, this operator transforms any state of the form $\cos\theta\,|\text{other}\rangle + \sin\theta\,|s\rangle$ (for a real angle θ) into the state

$$\cos(\theta + 2\omega)\,|\text{other}\rangle + \sin(\theta + 2\omega)\,|s\rangle,$$

where $0 < \omega < \pi/2$ is the only angle such that $\sin\omega = 1/\sqrt{2^n}$.

Let's zoom out and recap for a moment. We mentioned earlier that the initial step of Grover's algorithm was the application of a Hadamard gate on each of the n qubits it used, thus setting the state of the system to

$$|\alpha\rangle = \cos\omega\,|\text{other}\rangle + \sin\omega\,|s\rangle,$$

where $\sin\omega = 1/\sqrt{2^n}$ is, for large n, extremely small. However, we would naturally like the amplitude of $|s\rangle$ to be as close to 1 as possible in order to maximize the probability of getting s upon measuring all the qubits. And this would be equivalent to modifying the state in such a way that the amplitude $\sin\theta$ of $|s\rangle$ were as close to $\sin(\pi/2) = 1$ as possible. In this regard, we have just introduced a composition of gates that transforms an input state $\cos\theta\left|\text{other}\right\rangle + \sin\theta\left|s\right\rangle$ into $\cos(\theta + 2\omega)\left|\text{other}\right\rangle + \sin(\theta + 2\omega)\left|\text{other}\right\rangle$. Thus, we have all the ingredients needed to realize our goals: all we have to do is iteratively apply the composition of the phase oracle of f with Grover's diffusion operator until we make that angle θ close to $\pi/2$. This technique is an instance of **amplitude amplification**, since, as the name suggest, we are amplifying one of the amplitudes in a state, namely that of $\left|s\right\rangle$. It is worth highlighting that amplitude amplification can be applied to a variety of computational tasks [78].

The question that we now must address is the following: how many iterations will we need in order to achieve our goals?

12.2.4 Let's play with the odds

From our previous discussion, after k iterations in the algorithm, we will have a state of the form $\cos(\theta_k)\left|\text{other}\right\rangle + \sin(\theta_k)\left|s\right\rangle$, where

$$\theta_k = \omega + 2k\omega = (2k+1)\omega.$$

This means that, after k iterations, the probability of finding our marked element will be

$$\left|\sin(\theta_k)\right|^2 = \left(\sin\left((2k+1)\omega\right)\right)^2;$$

in *Figure 12.5*, you can find a sample depiction of how this probability would evolve over time.

Our task is thus to determine at which index k_0 we will have $\theta_{k_0} \approx \pi/2$, since that is the smallest k for which the preceding probability will be close to 1. Keep in mind that, in principle, we could also aim to find a k_1 such that $\theta_{k_1} \approx 3\pi/2$, but this would go against our interests, as $k_0 < k_1$, $\omega < \pi/2$, and we seek to minimize the number of iterations of the algorithm. Of course, we do this with the goal of making the search procedure as short as possible; in terms of complexity metrics, we do it in order to minimize the number of calls to the oracle of f (remember that each iteration makes an individual call to the oracle).

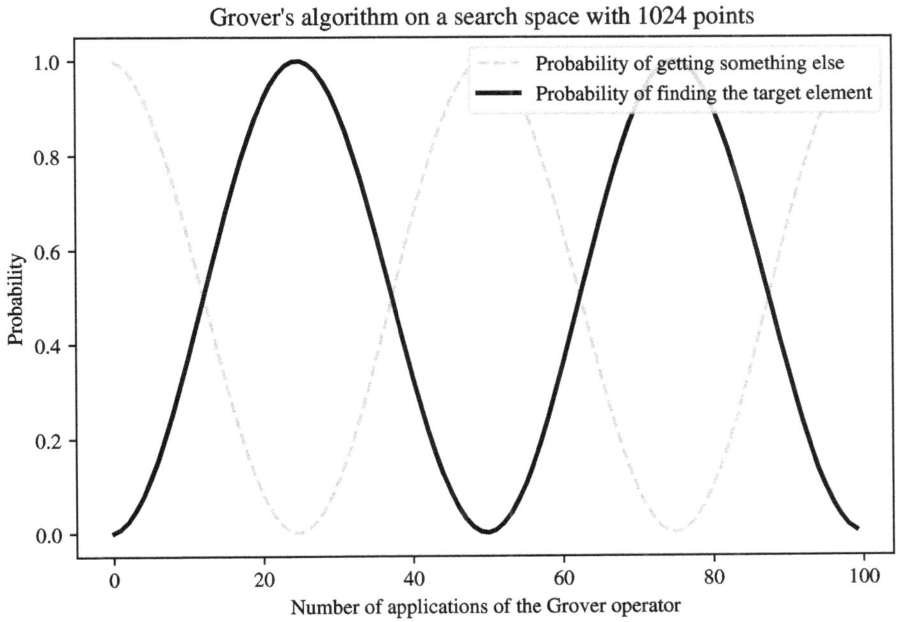

Figure 12.5: Evolution of the probability of finding a marked element among 1024 *elements using Grover's algorithm*

The iteration k_0 at which $\theta_{k_0} = \pi/2$ will be the one at which

$$(2k_0 + 1)\omega = (2k_0 + 1)\arcsin\left(\frac{1}{\sqrt{2^n}}\right) = \frac{\pi}{2},$$

where we have used the fact that $\sin\omega = 1/\sqrt{2^n}$ and, therefore, $\omega = \arcsin\left(1/\sqrt{2^n}\right)$. Of course, there may not exist an integer k_0 for which this equation will be satisfied, but we can round the solution to the nearest integer with the assurance that $\theta_{k_0} \approx \pi/2$. Now, when we look at the preceding expression… we might be mildly overwhelmed by having to deal with a trigonometric function, which is sure to complicate our computations. Nevertheless, since we are assuming 2^n to be huge, and, thus, $1/\sqrt{2^n}$ to be extremely small, we can approximate $\arcsin\left(1/\sqrt{2^n}\right) = 1/\sqrt{2^n}$. That's because $\sin x \approx x$ for small x, as you can see in *Figure 12.6*. If you'd like a more formal justification, we invite you to complete the following exercise:

Exercise 12.2

In order to prove that, for values of $h > 0$ that are close to zero, $\sin h \approx h$, show that

$$\lim_{x \to 0} \frac{\sin x}{x} = 1.$$

You may use L'Hôpital's rule.

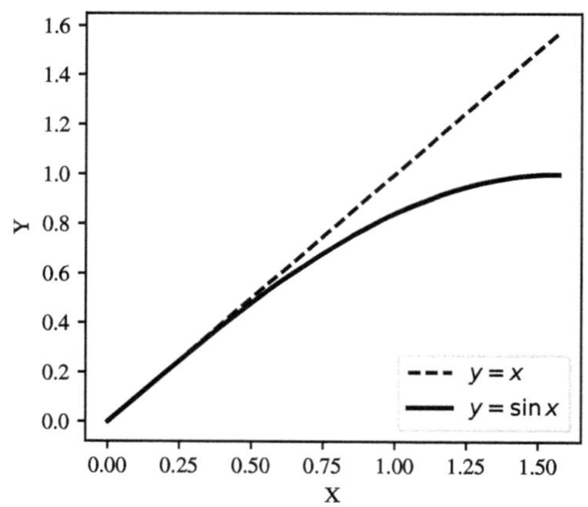

Figure 12.6: It can be seen in the graph how $\sin x \approx x$ when $x > 0$ is small

This approximation can greatly simplify our equation for finding k_0, leading us to

$$(2k_0 + 1)\frac{1}{\sqrt{2^n}} = \frac{\pi}{2} \implies 2k_0 + 1 = \frac{\pi\sqrt{2^n}}{2} \implies k_0 = \frac{1}{2}\left(\frac{\pi\sqrt{2^n}}{2} - 1\right).$$

Assuming 2^n to be sufficiently large, and letting N denote the size of our search space (that is, $N := 2^n$), we can further approximate this as

$$\boxed{k_0 \approx \left\lfloor \frac{\pi}{4}\sqrt{N} \right\rfloor.}$$

Here, we are using $\lfloor \quad \rfloor$ to denote the **floor** of a real number: the greatest integer less than or equal to that number.

Let us now discuss a simple example. If we consider a search problem with $N = 2^{10} = 1024$ elements in an unsorted search space and a single marked element, we would only need to perform $\lfloor(\pi/4)\sqrt{1024}\rfloor = 25$ iterations of Grover's algorithm, and hence we would only need 25 queries to the oracle of the problem. Indeed, with 25 queries, the probability of finding s would be

$$|\sin\theta_{25}|^2 = |\sin((2 \cdot 25 + 1)\omega)| = \left|\sin\left(51 \arcsin\frac{1}{\sqrt{2^{10}}}\right)\right|^2 \approx 99.94\%.$$

This contrasts with the potential 1024 queries that a classical algorithm would need! Moreover, if we only performed 20 calls to the oracle, we would still find s with probability

$$|\sin\theta_{20}|^2 = |\sin((2 \cdot 20 + 1)\omega)| = \left|\sin\left(41 \arcsin\frac{1}{\sqrt{2^{10}}}\right)\right|^2 \approx 92\%.$$

Exercise 12.3

Consider a search problem with $N = 2^{13} = 8192$ unsorted elements and a single marked element.

(a) How many iterations would be necessary in order for the probability of finding the marked element with Grover's algorithm to be close to 1?

(b) If only 50 iterations were performed, what would be the probability of finding the marked element?

Exercise 12.4

Our estimation for k_0 returns a very close approximation of the number of applications of the Grover operation that will bring the probability of measuring the marked element very close to 1. Nevertheless, if we are willing to be flexible and relax our need for determinism, we can accept as well other values for k. Show that, if

$$\frac{\pi}{8}\sqrt{N} < k < \frac{3\pi}{8}\sqrt{N},$$

the probability of measuring our marked element is still bigger than or equal to $1/2$. Of course, along the way, you should feel free to use the same kind of approximations that we have employed in our computation of k_0.

Before moving on, there is a very significant remark that we should make. The value of k_0 is the optimal value of k that will bring the probability of getting s in a measurement as close to 1 as possible (and, actually, extremely close to 1!). If we exceed that number of iterations, the probability of getting s in a measurement will actually decrease! This behavior becomes evident in *Figure 12.5*, and it can be explained by the fact that, as k increases, the state will keep rotating (in our plane representation) and θ_k will get further and further away from $\pi/2$. In this process, we won't be close to measuring s with high probability until we approach $k_1 \approx 3k_0$ (at which point, $\theta_{k_1} \approx 3\pi/2$, as you can easily verify by yourself). This contrasts with most other iterative algorithms, in which the accuracy is guaranteed to grow with the number of iterations.

Important note

For a large n, let f be an n-bit Boolean function with a unique marked element s such that $f(s) = 1$. Given an oracle for f, Grover's algorithm is an n-qubit search algorithm for the search of s among the $N = 2^n$ possible inputs of f. In its execution, as an initial step, all qubits must be initialized to $|0\rangle$ and then a Hadamard gate must be applied on each of them. Subsequently, $\lfloor(\pi/4)\sqrt{N}\rfloor$ iterative applications of the phase oracle of f composed with Grover's diffusion operator must be applied. At this stage, a measurement of the n qubits will return s with a probability practically equal to 1.

We recall that the phase oracle of f could be implemented by applying the ordinary oracle with an ancillary qubit set to $|-\rangle$. The state of this ancillary qubit will not change by the action of the oracle and can thus be reused.

The procedure is summarized in the following circuit, where D denotes Grover's diffusion operator:

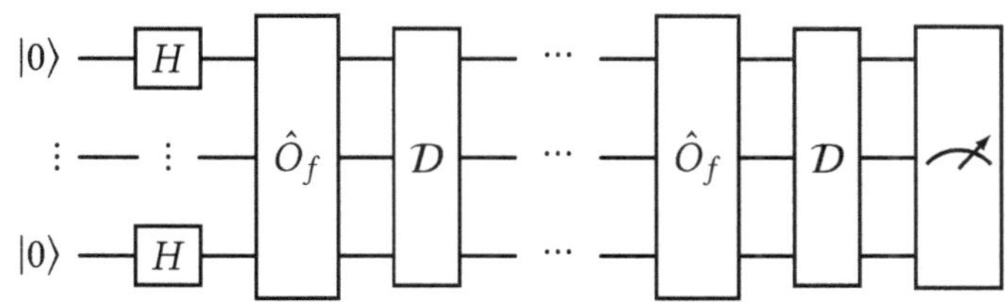

Since every iteration makes a single call to $\hat{O}_f$, the oracle for f is only evaluated $\lfloor(\pi/4)\sqrt{N}\rfloor$ times.

This is all good and nice, but so far, this version of Grover's algorithm only works when we are looking for a single marked element in a search space, that is, when $f(s) = 1$ for a unique entry s. But what if there were more marked elements? What if f were equal to 1 for more than one value? That's what we will be taking care of next.

12.2.5 Not one, but many

Assume now that we have an n-bit Boolean function f such that, for a collection $M = \{s_1, \dots, s_m\}$ of n-bit string, $f(s_j) = 1$ and $f(x) = 0$ for any x not in that collection. In this case, we have m possible inputs for which f takes the value 1. So, now, how do we go about applying Grover's algorithm?

As it turns out, in this scenario, the procedure is identical: the only difference is in the number of iterations that are needed. Let's see why and how many iterations will be needed in this case.

After applying a Hadamard gate on each qubit, we reach the state $|\alpha\rangle$, which we can decompose as

$$
\begin{aligned}
|\alpha\rangle &= \frac{1}{\sqrt{2^n}} \sum_{k=0}^{2n-1} |k\rangle = \frac{1}{\sqrt{2^n}} \left(\sum_{k=1}^{m} |s_k\rangle + \sum_{k=0, k\notin M}^{2^n-1} |k\rangle \right) \\
&= \sqrt{\frac{m}{2^n}} \left(\frac{1}{\sqrt{m}} \sum_{k=1}^{m} |s_k\rangle \right) + \sqrt{\frac{2^n - m}{2^n}} \left(\frac{1}{\sqrt{2^n - m}} \sum_{k=0, k\notin M}^{2^n-1} |k\rangle \right)
\end{aligned}
$$

$$= \sqrt{\frac{m}{2^n}}\,|S\rangle + \sqrt{\frac{2^n - m}{2^n}}\,|\text{other}\rangle = \sin\omega\,|S\rangle + \cos\omega\,|\text{other}\rangle,$$

where we have implicitly defined the angle ω and the states $|S\rangle$ and $|\text{other}\rangle$.

Following a reasoning analogous to the one we considered before, it can be readily checked that the composition of the phase oracle with Grover's diffusion operator will transform any state of the form $\cos\theta\,|S\rangle + \sin\theta\,|\text{other}\rangle$ into the state $\cos(\theta + 2\omega)\,|S\rangle + \sin(\theta + 2\omega)\,|\text{other}\rangle$, thus implementing the same kind of rotation that we had in the plane representation that we used when $m = 1$. The only difference is that, in this case, such a representation would have $|S\rangle$ in the Y axis and our new $|\text{other}\rangle$ in the X axis.

As before, our goal would be to bring the amplitude of $|S\rangle$ as close to 1 as possible, in such a way that the state of the system would be close to being a superposition of computational basis states $|s_t\rangle$ for which $f(s_t) = 1$. In full analogy with the $m = 1$ case, and resorting to the approximation that $\sin\omega = \sin\left(\sqrt{m/2^n}\right) \approx \sqrt{m/2^n}$, the smallest number k_0 of iterations that will make this possible is the one that will satisfy

$$(2k_0 + 1)\sqrt{\frac{m}{2^n}} = \frac{\pi}{2} \implies 2k_0 + 1 = \frac{\pi\sqrt{2^n}}{2\sqrt{m}}$$
$$\implies k = \frac{1}{2}\left(\frac{\pi\sqrt{2^n}}{2\sqrt{m}} - 1\right).$$

Approximating and simplifying, this yields the value

$$\boxed{k_0 \approx \left\lfloor \frac{\pi}{4}\sqrt{\frac{N}{m}} \right\rfloor.}$$

After k_0 iterations, we are pretty much guaranteed to get a value s such that $f(s) = 1$ when measuring the n qubits of the algorithm. The particular marked string that we get will be random, and all values s_t have equal probability of being returned as an outcome because of the way in which $|S\rangle$ is constructed.

To illustrate this, let us discuss an example. If we consider a search problem with $N = 2^{11} = 2048$ unsorted elements and 7 marked elements, we would only need to perform $\lfloor (\pi/4)\sqrt{2048/7} \rfloor = 13$ iterations of Grover's algorithm, with their 13 queries to the oracle of the problem. Moreover, if we only performed 8 calls to the oracle, we would still find a marked element with probability

$$|\sin \theta_8|^2 = |\sin((2 \cdot 8 + 1)\omega)| = \left| \sin \left(17 \arcsin \frac{1}{\sqrt{2^{11}/7}} \right) \right|^2 \approx 70\%.$$

Important note

For a large $N = 2^n$, let f be an n-bit Boolean function for which there only exist m inputs s such that $f(s) = 1$. If m/N is close to zero, Grover's algorithm will work as intended, provided that we run $(\pi/4)\sqrt{N/m}$ iterations instead of the $(\pi/4)\sqrt{N}$ that we used when $m = 1$.

Notice, by the way, how we have not only assumed that 2^n is large but also that $m/2^n$ is very small. This shouldn't be a problem, because if the proportion of elements for which f is 1 turns out to be large… we probably don't need to apply quantum computing in the first place: picking values at random might just get the job done after a few iterations! Indeed, if elements are picked at random, the probability of finding a marked element after k iterations will be

$$1 - \left(1 - \frac{m}{2^n}\right)^k.$$

This can be deduced rather easily. At each iteration, the probability of finding a marked element is $m/2^n$, hence that of not finding it is $1 - m/2^n$; therefore, the probability of not finding a marked element in k iterations must be $(1 - m/2^n)^k$, so the probability of having found at least one element must be 1 minus that. For example, if $m/2^n = 1/2$, this means that after $k = 3$ iterations, the probability of having found a marked element will be $7/8 = 87.5\%$.

Exercise 12.5

Assume you are given an oracle for a function f on 1024 elements (and, therefore, the phase oracle works on 10 qubits).

(a) If we wanted to perform a search with 512 marked elements (i.e., $f(x) = 1$ for 512 entries x), would it make sense to use Grover's algorithm? Why? If we were to pick elements at random, how many elements would we have to pick in order to find a marked element with a probability bigger than 90%?

(b) If there were only 5 marked elements, how many iterations would we need to perform in Grover's algorithm? What will be the probability of finding a marked element by (classically) picking that number of elements at random? What is Grover's probability? Is it worth using Grover's algorithm in this case?

As a final example to illustrate the power of Grover's algorithm, if a quarter of all the elements in a search space are marked (that is, if $m = N/4$), then, using Grover's algorithm, just a single iteration (and, thus, a single call in the oracle) will be sufficient in order to retrieve a marked element with certainty. Indeed, in this case, the probability of finding a marked element after a single iteration is

$$|\sin\theta_1|^2 = |\sin((2\cdot 1+1)\omega)| = \left|\sin\left(3\arcsin\sqrt{\frac{1}{4}}\right)\right|^2 = \left|\sin\left(3\cdot\frac{\pi}{6}\right)\right|^2 = \left|\sin\left(\frac{\pi}{2}\right)\right|^2 = 1.$$

To learn more...

Grover's algorithm is fairly impressive on its own, but maybe you are ambitious and you have hopes of beating it with a better algorithm that may use less calls to an oracle asymptotically. Sadly, that can't be possible. The BBBV theorem (named after Charles H. Bennett, Ethan Bernstein, Gilles Brassard, and Umesh Vazirani) shows that Grover's algorithm is (asymptotically) an optimal oracle-based quantum algorithm for search [79].

And this pretty much sums it up for our discussion of Grover's algorithm.

To learn more...

All of our work so far has been resting on the assumption that we know, in advance, the number of marked elements m that we expect in our search problem. In the following section, we will introduce a procedure that will enable us to compute that number m. Nevertheless, there is a way for quantum computers to still yield a quadratic speedup in search algorithms without having to know m beforehand—and it is based on a clever application of Grover's algorithm! If you would like to learn more about it, we suggest reading "Tight Bounds on Quantum Searching" [80] by Boyer et al.

To conclude this chapter, we will introduce a procedure that will make use of some of the key ingredients that we have employed over the last couple of chapters: the operators used in Grover's algorithm and the quantum Fourier transform.

12.3 Counting with the quantum Fourier transform

If we are given an oracle for an n-bit Boolean function f, is there any way for us to (quantumly) approximate the number m of inputs s for which we have $f(s) = 1$? Yes, there is, and in fact, it has advantages over classical algorithms! In order to unveil it, we will first need to analyze the composition of the phase oracle $\hat{O}_f$ with Grover's diffusion operator, which we will denote as G_f.

As we have mentioned on countless occasions already, G_f maps states $\cos(\theta)\left|\text{other}\right\rangle + \sin(\theta)\left|S\right\rangle$ to states $\cos(\theta + 2\omega)\left|\text{other}\right\rangle + \sin(\theta + 2\omega)\left|S\right\rangle$. This means that we can pretend that G_f only acts on the space of states of the form $x\left|\text{other}\right\rangle + y\left|S\right\rangle$, and, on this restricted space, G_f simply behaves like a counter-clockwise rotation by 2ω. Therefore, the coordinate matrix of G_f on this space will be the following one:

$$R(2\omega) = \begin{pmatrix} \cos 2\omega & -\sin 2\omega \\ \sin 2\omega & \cos 2\omega \end{pmatrix},$$

Remember that, as we discuss in *Appendix A*, this coordinate matrix is constructed by putting, in its first column, the coordinates of $G_f \left|\text{other}\right\rangle$ (as $\left|\text{other}\right\rangle$ is the unit vector of the positive horizontal axis) and, in its second column, those of $G_f \left|S\right\rangle$ (as $\left|S\right\rangle$ is the unit vector of the positive vertical axis).

You may recall that the eigenvalues of a matrix are the numbers λ for which there exists a vector v such that $Av = \lambda v$. Finding the eigenvalues of $R(2\omega)$ is as easy as it is crucial for our next steps, so we entrust this process to you. Incidentally, for a quick review of what eigenvalues are and how they are computed, you may take a look at *Appendix A*.

Exercise 12.6

Prove that $R(2\omega)$ is diagonalizable and its eigenvalues are $e^{\pm i(2\omega)}$.

Since $R(2\omega)$ is diagonalizable, every vector in the space on which it is defined must be a linear combination of its eigenvectors (we discuss this in *Appendix A*). In particular, this means that any state of the form $\cos(\theta)\left|\text{other}\right\rangle + \sin(\theta)\left|S\right\rangle$ must be a superposition (i.e., a linear combination) of the eigenvectors of G_f corresponding to the eigenvalues $e^{\pm i2\omega}$.

Now, what do these eigenvalues have to do with the estimation of m? Simple. As we saw in the previous section,

$$\sin\omega = \sqrt{\frac{m}{2^n}}.$$

Thus, since the value of n is given, if we have access to an estimation of ω, we can also get an estimation of m. This means that, in order to estimate m, we just need to find an estimate of ω.

Luckily for us, back in *Chapter 11*, we introduced an algorithm that enabled us to do just that: finding estimations of the eigenvalue associated to a state. This algorithm was the quantum phase estimation algorithm, which made use of the quantum Fourier transform, one of the most important quantum operations out there.

You may recall that, given a quantum gate with an eigenvector $|\psi\rangle$, running the quantum phase estimation algorithm enabled us to estimate the phase of the eigenvalue associated to $|\psi\rangle$; remember how this means that, if the eigenvalue is $e^{i2\pi x}$, with $0 \leq x \leq 1$, the quantum phase estimation algorithm will yield an estimate of x.

In our case, in order to estimate ω, we can simply provide the state $|\alpha\rangle$ to the quantum phase estimation algorithm for G_f. Since $|\alpha\rangle$ is, as we discussed before, a superposition of eigenvectors associated to $e^{i2\omega}$ or $e^{-i2\omega} = e^{i2\pi(1-\omega)}$, the quantum phase estimation algorithm will return ω/π or $1 - \omega/\pi$ (with a probability determined by the amplitudes of the corresponding eigenvectors in $|\alpha\rangle$). This follows trivially from the linearity of quantum phase estimation, as we will now discuss in detail.

Assume that $|\omega_+\rangle$ and $|\omega_-\rangle$ are the normalized eigenvectors associated to the eigenvalues $e^{i2\omega}$ and $e^{-i2\omega}$ respectively. Also, let $|\alpha\rangle = a\,|\omega_+\rangle + b\,|\omega_-\rangle$ (for some complex numbers a and b). If E_f is the gate that encapsulates the phase estimation algorithm for f, then, using linearity,

$$E_f(|\alpha\rangle) = E_f(a\,|\omega_+\rangle + b\,|\omega_-\rangle) = aE_f(|\omega_-\rangle) + bE_f(|\omega_+\rangle).$$

Hence, at the time of measurement, quantum phase estimation will work as if estimating ω/π with probability $|a|^2$, and as if estimating $1 - \omega/\pi$ with probability $|b|^2$. Nevertheless, this is not such as big issue. Indeed, as $\omega < \pi/2$ by construction (and, therefore, $\omega/\pi < 1/2$), if an estimate ξ is bigger than $1/2$, we can just replace it by $1 - \xi$ and thus pretend we are always estimating ω/π.

And that's how the quantum Fourier transform can save the day and—through the use of quantum phase estimation—yield an estimation of m.

Now, we mentioned at the beginning of this section that this method yields an advantage over classical methods, so let's quantify this advantage. When trying to approximate the number of marked elements m in an unsorted space with N elements, it can be shown that, assuming $m < N/2$ and with the tools that we have discussed, it is possible to estimate m with $O(\sqrt{m})$ accuracy just using $O(\sqrt{N})$ queries to the oracle. On the other hand, pretty much as in the case of search algorithms, any classical algorithm aiming would need to perform $O(N)$ queries. Thus, quantum computing can provide again in this case a quadratic speed-up! And, if you want to use this counting method to estimate m and then run Grover's algorithm, your total number of oracle calls will still be $O(\sqrt{N})$. Neat, huh?

To learn more...

For a complete discussion on the complexity of approximating m using quantum phase estimation, you can read Section 6.3 of *Quantum Computation and Quantum Information: 10th Anniversary Edition* [13].

This concludes the content of our chapter. We've done quite a lot over the last few pages, so let's wrap up!

Summary

In this chapter, we have introduced Grover's algorithm, which tackles the problem of searching through an unsorted list and provides a quadratic speed-up when compared to any classical analogue.

We began by exploring what searching looks like in classical computers, and we saw how the reason behind the speed of classical search procedures lies not just in the speed of classical computers but in the sorting and structuring of data. We thus understood how, when data is not sorted or structured, classical computing is bounded to algorithms with $O(N)$ complexity on the number of queries. On the other hand, quantum computing can use Grover's algorithm to solve these search problems with $O(\sqrt{N})$ complexity, thus obtaining a quadratic speed-up.

Once the scene was set, we began to explore Grover's algorithm and we discussed how—after creating a perfectly balanced superposition between all the computational basis states—it consisted in the iterative application of the composition of an oracle for a Boolean function with Grover's diffusion operator. We explored the action of this composition in detail and saw how it just behaves like a rotation in a two-dimensional space. Taking advantage of this geometrical understanding, we were able to look for the number of iterations that would bring the probability of retrieving our marked element at measurement close to 1.

We finished by exploring how the quantum Fourier transform can help us approximate the number of marked elements through quantum phase estimation, again quadratically faster than with classical algorithms.

In the following chapter, we will discuss how to implement all the methods that we have introduced so far using Qiskit. Get ready to have the quantum Fourier transform and Grover's diffusion operator running in front of your eyes!

13

Coding Shor and Grover's Algorithms in Qiskit

Knowing is not enough; we must apply.
Willing is not enough; we must do.

— Johann Wolfgang von Goethe

Now that we have ample experience working with Qiskit, it's time for us to implement two very important algorithms: Shor's and Grover's. We covered all the theoretical details of both methods in the two previous chapters, so now we can just dive into the code and see these algorithms shine through some practical examples.

We will start by implementing the quantum Fourier transform (QFT), which is central to Shor's algorithm. Then, we will implement Shor's factoring method. Lastly, we will implement Grover's search in all its glory.

We have a busy agenda ahead of us!

The topics that we will cover in this chapter are the following:

- The QFT in Qiskit
- Shor's algorithm
- Grover's algorithm

After reading this chapter, you will know how to implement the quantum Fourier transform on any number of qubits using Qiskit. You will also know how to apply it to factor numbers with Shor's algorithm. Last, but certainly not least, you will be able to implement Grover's algorithm to search for the elements marked by any Boolean oracle.

Throughout the book, we've been building up to this moment. Let's get started!

13.1 The QFT in Qiskit

As you surely remember from *Section 11.4.1*, the quantum Fourier transform on m qubits is the linear transformation that acts on computational basis states by taking $|j\rangle$ to

$$\frac{1}{\sqrt{2^m}}\sum_{k=0}^{2^m-1} e^{\frac{2\pi i jk}{2^m}}\left|k\right\rangle.$$

The quantum circuits for the QFT on one and two qubits are, as we discussed in *Section 11.4.3*, pretty straightforward. In fact, for a single qubit, it consists of just a Hadamard gate. For two qubits, it is as follows:

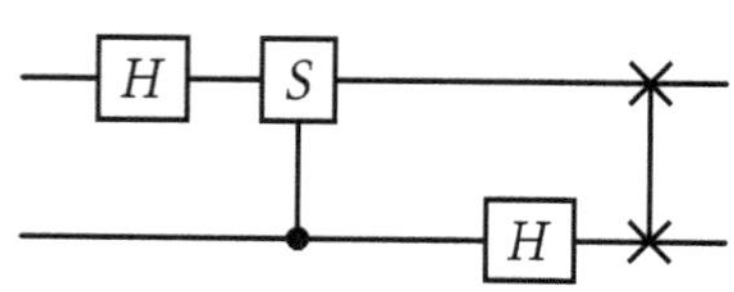

We can implement this circuit in Qiskit with the following piece of code:

```
from qiskit import QuantumCircuit

qft2 = QuantumCircuit(2)
```

```
qft2.h(0)
qft2.cs(1,0)
qft2.h(1)
qft2.swap(0,1)
qft2.draw("mpl")
```

There are a couple of new things here, so let's stop for a moment to clarify them. The first one is that we've used the `cs` method to implement the controlled-S gate. In its call, the first parameter (`1`) is the control qubit and the second one (`0`) is the target one. In addition, we have used the `swap` method to apply the SWAP gate. This method receives two parameters, which are exactly the qubits we want to swap.

Upon running this code, you will see a figure like the following one, which shows that our code in fact implements the QFT on two qubits:

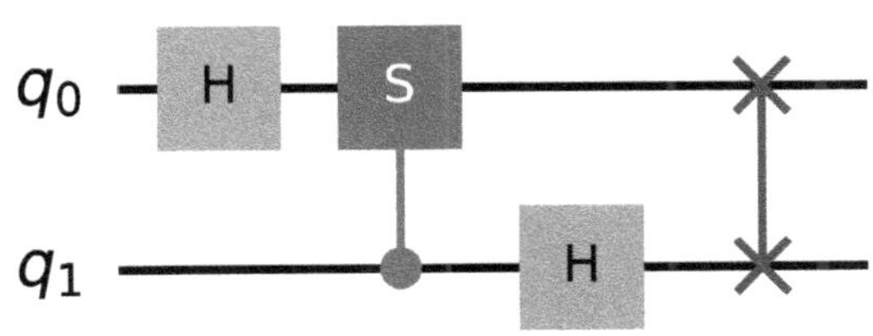

Perfect! Let's then move on to the QFT on three qubits. Its circuit, as you may remember, is as follows:

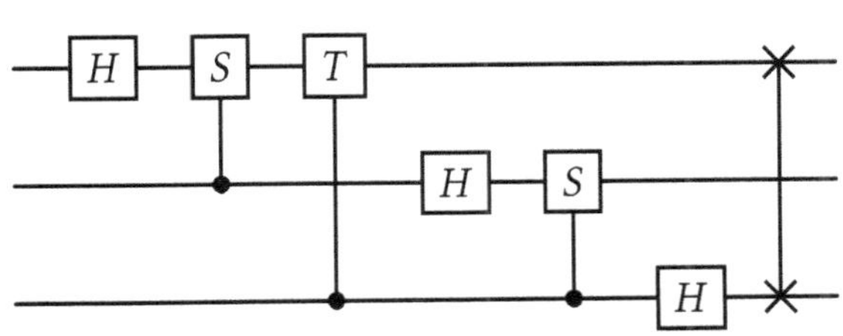

The only new thing here is that we need a controlled-T gate. Unfortunately, there is no specific method that implements this gate in Qiskit, but we can resort to the `cp` method, which implements a controlled version of the phase gate, which we introduced in *Section 2.3.1*. Recall that the coordinate matrix of the phase gate $P(\theta)$ is

$$P(\theta) = \begin{pmatrix} 1 & 0 \\ 0 & e^{i\theta} \end{pmatrix},$$

where θ is a real value. Thus, we can recover the T gate by setting $\theta = \pi/4$. The `cp` method receives three parameters: the first one is the rotation angle θ, the next one is the qubit that acts as a control, and the third one is the target qubit. Then, to implement a controlled-T gate from qubit 2 to qubit 0, we can use `cp(np.pi/4,2,0)`, provided that we have previously imported the NumPy package to have access to π via `np.pi`. The rest of the circuit is quite straightforward, so we will leave it to you to write the code and check it.

Exercise 13.1

Write Qiskit code to implement the QFT on three qubits. Check that your code is correct by drawing the circuit and comparing it to the one above.

Obviously, we can go on and implement a circuit for the QFT on four qubits, another one for the QFT on five, and so on and so forth. But we can do much better than that: we can define a function that, when given an integer m, creates a circuit for the QFT on m qubits.

To that end, let's take a closer look at the general quantum circuit for the QFT:

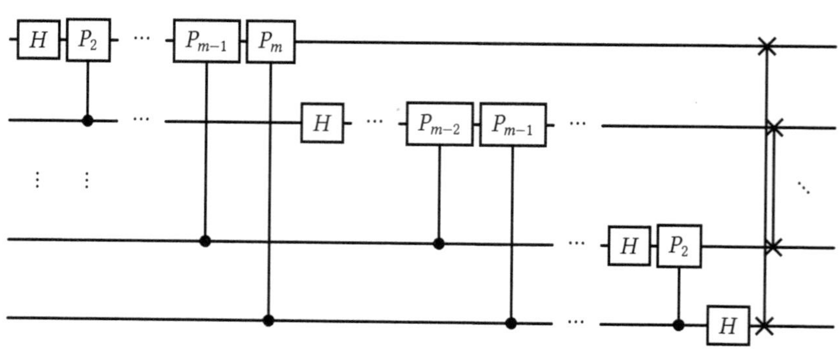

Remember that we have adopted the convention that $P_k = P(\pi/2^k)$. Then, the following piece of code implements the QFT on a given number of qubits:

```
import numpy as np

def qft(m):
    circ_qft = QuantumCircuit(m)
    for i in range(m):
        circ_qft.h(i) # Hadamard gate
        for j in range(i+1,m):
            circ_qft.cp(np.pi/2**(j-i),j,i) # Controlled-P gates
        circ_qft.barrier()
    # Final swaps
    for i in range(m//2):
        circ_qft.swap(i,m-i-1)
    return circ_qft
```

Here, we have added barriers just for sake of clarity, but you can remove them if you'd like.

We can now use this function, for instance, to create a circuit for the QFT on four qubits, as follows:

```
qft4 = qft(4)
qft4.draw("mpl")
```

This will produce the following beautiful figure:

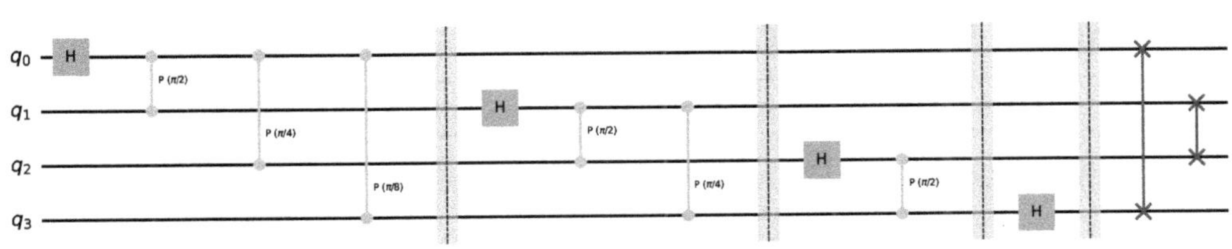

We only have one last thing to do. For our implementation of Shor's algorithm, we need the **inverse** QFT. But with what we already have, that is pretty straightforward. Remember that, in order to invert a circuit, we only need to read it from right to left and invert each gate. The SWAP and Hadamard gates are their own inverses, and the controlled-phase gates can be inverted just by taking the angle in the opposite direction. Thus, the function that we need is the following:

```
def iqft(m):
    circ_iqft = QuantumCircuit(m)
    # Initial swaps
    for i in range(m//2-1,-1,-1):
        circ_iqft.swap(i,m-i-1)
    for i in range(m-1,-1,-1):
        for j in range(m-1,i,-1):
            circ_iqft.cp(-np.pi/2**(j-i),j,i) # Controlled-P gates
        circ_iqft.h(i) # Hadamard gate
        circ_iqft.barrier()

    return circ_iqft
```

And that's it! We now have the main ingredient that we need to implement Shor's algorithm, so let's get to it.

13.2 Shor's algorithm

In the previous section, we implemented the inverse quantum Fourier transform, which is central to Shor's algorithm. However, we still need an additional element if we want to be able to build and run quantum circuits to factor integers: quantum gates for modular multiplication.

Remember that, in order to factor an integer N using Shor's algorithm, we first choose at random another integer a with no common factors with N. Then, we need to create the following circuit:

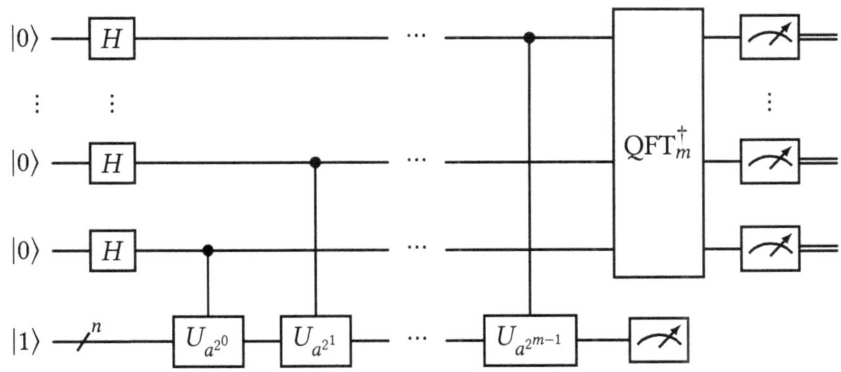

Here, the $U_{a^{2^j}}$ gates implement the operation of multiplying an integer times a^{2^j} and taking the remainder of the division by N. These are the gates that we need to implement now before we can run Shor's algorithm.

How to implement these quantum gates is a problem that has been extensively studied in the literature, and several very efficient options exist (see, for instance, [72]–[76]). However, describing them in detail is beyond the scope of this book, so we will adopt a more straightforward approach.

In order to implement the modular multiplication gates, we will use a Qiskit feature that we have not highlighted yet. With the `UnitaryGate` class, you can create a custom quantum gate from any (unitary) matrix. Thus, if you have a matrix object, `M`, that represents a unitary operator, you can obtain a quantum gate, `U`, that you can later use in your circuits by simply executing `U = UnitaryGate(M)`.

Imagine, for instance, that we are working with $N = 4$ and $a = 3$, and we want to implement a gate U for multiplication by 3 mod 4. Numbers mod 4 can be represented with just two bits because we only need to consider 0, 1, 2, and 3, whose binary expansions are 00, 01, 10, and 11, respectively. Thus, we will need a 4×4 matrix, because we will be working with 2 qubits. Moreover, we know that it should hold that

$$U|00\rangle = |00\rangle, \quad U|01\rangle = |11\rangle, \quad U|10\rangle = |10\rangle, \quad U|11\rangle = |10\rangle,$$

because

$$3 \cdot 0 = 0 \mod 4, \quad 3 \cdot 1 = 3 \mod 4, \quad 3 \cdot 2 = 2 \mod 4, \quad 3 \cdot 3 = 1 \mod 4.$$

Thus, the matrix for U must be

$$\begin{pmatrix} 1 & 0 & 0 & 0 \\ 0 & 0 & 0 & 1 \\ 0 & 0 & 1 & 0 \\ 0 & 1 & 0 & 0 \end{pmatrix}.$$

This expression follows from the fact that, for a non-negative integer j, the ket $|j\rangle$ is a column vector whose components are all zeroes with the exception of the $(j+1)$-th position, which is 1. Thus, the vector $U|0\rangle = U|00\rangle$ is the first column of U, the vector $U|1\rangle = U|01\rangle$ is its second column, and so on.

However, when using `UnitaryGate`, we need to take into account an additional detail: Qiskit expects binary strings representing integers to be sorted out in reverse. Thus, the actual matrix that we would need to use is

$$\begin{pmatrix} 1 & 0 & 0 & 0 \\ 0 & 1 & 0 & 0 \\ 0 & 0 & 0 & 1 \\ 0 & 0 & 1 & 0 \end{pmatrix},$$

where we have interchanged the roles of $|01\rangle$ and $|10\rangle$ to account for Qiskit's convention. You can check this by simply swapping the second and third columns, and then the second and third rows of the original matrix.

For this reason, in order to implement the modular multiplication gates in Qiskit, we will need the following function:

```
def invert(j,n): # Inverts integer j when represented with n bits
    str_j = bin(j)[2:] # Binary string for j without the leading "0b"
```

```
    str_j = (n-len(str_j))*'0' + str_j # Pad with initial zeroes up to n bits
    str_j = str_j[::-1] # Reverse the string
    return int(str_j,2) # Convert the string to a decimal integer
```

Here, we have used the `bin` function to obtain a string with the binary representation of `j`. This function always adds `0b` at the beginning of the binary strings, so we have removed it with the slice `[2:]`. Then, we have added initial zeroes as needed to get the string to length `n` and reversed the string. Finally, with `int(str_j,2)`, we have converted the binary string stored in `str_j` to a decimal value.

With this auxiliary function in our toolkit, we can now implement another function that will take a and N as parameters and return a quantum gate that performs multiplication by a mod N. For that, we can use the following piece of code:

```
from qiskit.circuit.library import UnitaryGate

def gate_mult(a,N):
    m = len(bin(N-1))-2 # Number of qubits needed to store numbers mod N
    M = 2**m

    matrix = []
    for i in range(M):
        matrix.append(M*[0]) # Initialize matrix to all zeroes

    for j in range(N):
        i = (a*j)%N # Position that we need to set to 1
        ii = invert(i,m) # Index of row for i according to Qiskit
        ij = invert(j,m) # Index of column for j according to Qiskit
        matrix[ii][ij] = 1

    for j in range(N,M):
```

```
        # Numbers bigger than N-1 are not considered, so leave them fixed
        ij = invert(j,m)
        matrix[ij][ij] = 1

    return UnitaryGate(matrix)
```

In this function, we first iterate over the numbers j from 0 to $N-1$, computing $j \times a \bmod N$, and we set the corresponding matrix entry to 1, as we did with our example matrix. Numbers greater than or equal to N are not part of the operation, so we just leave them fixed.

We are now ready to give our implementation of the quantum circuit for Shor's algorithm. We just need to use the function that we have just defined and the implementation of the inverse quantum Fourier transform from the previous section. The final result is the following:

```
def circuit_shor(a,N,m):
    # a is the number we want to obtain the period of
    # N is the number to factor
    # m is the number of qubits for the upper register

    n = len(bin(N-1))-2 # Number of qubits needed to store numbers mod N

    shor = QuantumCircuit(m+n,m)
    # Upper register is of size m
    # Lower register is of size n
    # We only measure the upper register, hence m classical bits

    shor.x(m+n-1) # Set the bottom register to |1>

    for i in range(m):
```

```
        shor.h(i) # Column of Hadamard's in the upper register

    for i in range(m):
        gate = gate_mult(a**(2**i),N).control(1) # Mult by a^(2^j) mod N
        shor.append(gate,[m-i-1]+list(range(m,m+n))) # Add controlled mult

    shor.barrier()

    shor.append(iqft(m),range(m)) # Inverse QFT

    shor.barrier()

    for i in range(m):
        shor.measure(i,i) # Measure the upper register

    return shor
```

Notice that, in addition to a and N, we need to provide m, the number of qubits for the register that will contain the information about the order of a. The bigger this number, the higher the precision; nevertheless, as we mentioned in *Chapter 11*, $m = 2n$, where n is the size of N in bits, is usually a good choice. In the function, we create a quantum circuit of size $m + n$, and only the upper m qubits will be measured. The bottom n qubits will store the numbers mod N. After setting the lower register to $|1\rangle$ and adding Hadamard gates to the qubits in the upper register, we add gates for modular multiplication for $a^{2^0}, a^{2^1}, \ldots, a^{2^{m-1}}$ controlled by the corresponding qubits. Notice how we use the `control` method to easily create a controlled version of a quantum gate. Also notice that we have used the `append` method, since our gate is custom-made and there is no specific instruction for it in Qiskit. Then, we append the inverse quantum Fourier transform circuit and measure the upper m qubits. Et voilà, it's done!

Let's try our new, shiny toy by creating and drawing a circuit for a simple case. Let's choose $a = 3$, $N = 4$ and $m = 4$, just for illustration purposes. We can do this with the following instructions:

```
shor = circuit_shor(3,4,4)
shor.draw("mpl")
```

This will give you something like the following:

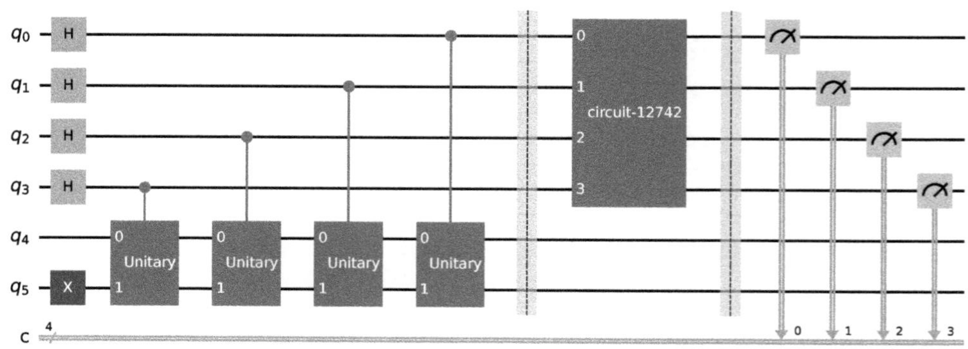

This looks exactly like what we wanted! But wait a minute... What are those mysterious boxes where the modular multiplication gates and the IQFT should be? Well, it turns out that Qiskit uses these blocks with generic names such as "unitary" and "circuit" to encapsulate operations that may involve a number of simpler gates. You can try to see what's inside them by running `shor.decompose().draw("mpl")`, `shor.decompose().decompose().draw("mpl")`, and so on, but let us warn you that what you will see will not be pretty... or easy to understand. It turns out that using classes such as `UnitaryGate` does not usually result in efficient implementations. But, as we mentioned before, we are not aiming for efficiency with this demonstration. Our goal is to simply illustrate how to use the quantum circuits that make up Shor's algorithm. Another small detail is that you may have noticed that we have not measured the bottom qubits, whereas we did in *Chapter 11*. In fact, that measurement is not strictly needed and was only used back there to simplify our discussion, so we have decided to omit it here.

OK, so now it's time to run our circuit and see what we get. We will start with a simple example, trying to factor 15 (spoiler: its factors are 3 and 5) and choosing $a = 2$. Since the numbers from 0 to 15 can be written with $n = 4$ bits, we will select $m = 2 \cdot n = 8$, as we discussed previously. We can build the quantum circuit and run it with the following instructions:

```
shor = circuit_shor(2,15,8)

from qiskit_aer.primitives import SamplerV2 as Sampler
from qiskit_aer import AerSimulator
from qiskit.transpiler.preset_passmanagers import generate_preset_pass_manager

backend = AerSimulator()
sampler = Sampler(seed = 1234)
pm = generate_preset_pass_manager(backend=backend, optimization_level=1)
t_shor = pm.run(shor)
job = sampler.run([t_shor], shots = 100)
```

Since we are using custom gate operations through the `UnitaryGate` class, we need to transpile these gates in order to be able to run them on the simulator. That is what the call to `generate_preset_pass_manager` is for.

Now, we need to process the results. In this case, $a^4 = 1 \mod 15$; so, as we discussed in *Section 11.4.2*, we should obtain values of the form $k2^m/4 = 64k$, with k an integer. But remember that Qiskit reverses the binary strings when giving the results, so we need to massage the output a little bit. We will use the following piece of code for that:

```
results = job.result() # Access the results
d = results[0].data.c
res = d.get_counts()
for k in res:
    c = int(k[::-1],2) # Reverse the string and convert it to an integer
```

```
    print("Value:",c,"Frequency:",res[k])
```

Here, we are just accessing all the strings measured in the execution of the circuit, reversing them to convert them to the integer value that we need, and showing the number of times they were measured. The output is as follows:

```
Value: 64 Frequency: 28
Value: 128 Frequency: 34
Value: 192 Frequency: 23
Value: 0 Frequency: 15
```

As you can see, they are all multiples of 64, as expected. Now, if we divide the non-zero values by $2^8 = 256$, we will obtain the fractions $\frac{1}{4}$, $\frac{1}{2} = \frac{2}{4}$, and $\frac{3}{4}$, and, from their denominators, we can recover the order of a, which is $r = 4$. Following the process described in *Section 11.2*, we compute $x = a^{\frac{r}{2}} + 1 \bmod N$, $y = a^{\frac{r}{2}} - 1 \bmod N$, $gcd(x, N)$, and $gcd(y, N)$ to discover that the factors of 15 are 3 and 5. Bingo!

Let's now try to factor 21. For that, we will first choose $a = 16$. Notice that 21 can be written using 5 binary digits, but we will choose $m = 8$, as previously, to make the circuits run a little bit faster (as you will see, this will not affect our ability to factor 21). Then, we just need to run the following piece of code:

```
shor = circuit_shor(16,21,8)

pm = generate_preset_pass_manager(backend=backend, optimization_level=1)
t_shor = pm.run(shor)
job = sampler.run([t_shor], shots = 100)

results = job.result() # Access the results
d = results[0].data.c
res = d.get_counts()
for k in res:
```

```
c = int(k[::-1],2) # Reverse the string a convert it to integer
print("Value:",c,"Frequency:",res[k])
```

With this, we will obtain the following output:

```
Value: 87 Frequency: 3
Value: 0 Frequency: 36
Value: 85 Frequency: 21
Value: 170 Frequency: 6
Value: 171 Frequency: 18
Value: 55 Frequency: 1
Value: 84 Frequency: 5
Value: 51 Frequency: 1
Value: 86 Frequency: 7
Value: 169 Frequency: 1
Value: 172 Frequency: 1
```

Hmmm, what are these values telling us? Remember that we should be getting integers that are close to $k2^m/r$, where k is an integer and r is the order of a. So, let's take 85, which is the most frequent non-zero value, and divide it by 2^m to see whether we can get k/r for some integer and recover r. In this case, $85/256 = 0.33203125$, which is quite close to $1/3$, so it seems that $r = 3$. Indeed, $16^3 \equiv 1 \mod 21$, so we got our order!

To learn more...

Notice how, when running our circuit, we have obtained results around 171 and 85. This is because these are the integers that, when divided by 256, are closest to $1/3$ and $2/3$, respectively.

The problem here is that you will never obtain an integer that, when divided by 2^m, will give you exactly $1/3$ or $2/3$, because these fractions cannot be represented with a finite binary expansion. However, if you increase m, you will obtain better and better approximations. For instance, if you use $m = 10$ with $a = 16$ and $N = 21$, you will likely obtain values around 341 and 683 when running your circuit. These will give you 341/1024 and 683/1024, which are better approximations to 1/3 and 2/3 than 85/256 and 171/256.

In any case, once you have run your circuit and obtained the measurements, the standard way of recovering r is by using a method based on continued fractions (you can check out the book by Nielsen and Chuang [13] for all the details). If you want to use this method in Python, it is very easy. For instance, to obtain r from the results of running the circuit with $m = 8$, $a = 16$, $N = 21$, and the measurement 85, you can use the following instructions:

```
from fractions import Fraction
m=8
N=21
Fraction(85,2**m).limit_denominator(N)
```

The output will be `Fraction(2,3)`. The denominator will be the order r, which is 3 in this case.

We have found out that $16^3 \equiv 1 \mod 21$. But wait a minute. We need r to be even to be able to compute $a^{\frac{r}{2}}$. D'oh, we are back to square one! Let's try now with $a = 10$. We can create our circuit with `shor = circuit_shor(10,21,8)` and run it just as the previous one to obtain the following results:

```
Value: 171 Frequency: 10
Value: 42 Frequency: 3
Value: 0 Frequency: 15
Value: 213 Frequency: 11
Value: 44 Frequency: 1
Value: 128 Frequency: 19
Value: 43 Frequency: 9
```

```
Value: 167 Frequency: 1
Value: 85 Frequency: 12
Value: 212 Frequency: 2
Value: 170 Frequency: 5
Value: 166 Frequency: 2
Value: 83 Frequency: 1
Value: 86 Frequency: 3
Value: 39 Frequency: 1
Value: 214 Frequency: 2
Value: 211 Frequency: 1
Value: 87 Frequency: 1
Value: 41 Frequency: 1
```

The most frequent non-zero values this time are 43, 85, 128, 171, and 213. If we divide them by 256, we obtain good approximations of $1/6, 2/6, 3/6, 4/6$, and $5/6$, so it seems that the order of $a = 10$ should be $r = 6$. Indeed, if you compute 10^6 and divide it by 21, you will find that the remainder is 1, as needed. Moreover, 6 is even, so we are on the right track! If we now compute $x = a^{\frac{r}{2}} + 1 \bmod N$, $y = a^{\frac{r}{2}} - 1 \bmod N$, $gcd(x, N)$, and $gcd(y, N)$, we will be able to factor 21 as $3 \cdot 7$.

Exercise 13.2

Repeat this process to factor 15, but using $a = 7$. Then, do the same with $N = 21$ and $a = 11$. Finally, repeat it with $a = 2$ and $N = 35$.

Now that we have implemented the quantum part of Shor's algorithm successfully, it is time to do some soul-searching… with the help of Grover's method.

13.3 Grover's algorithm

In *Chapter 12*, we studied Grover's algorithm in detail, and, as you surely remember, there are two main ingredients that it relies on: a Boolean function oracle that encodes the elements that we want to find and Grover's diffusion operator. In *Section 10.3.1*, we gave a general (but not necessarily efficient) way of implementing oracles, so we will just focus on coding Grover's diffusion operator.

Grover's diffusion operator is in charge of performing the inversion about the mean operation. As we discussed in *Section 12.2.3*, this can be implemented with the following circuit:

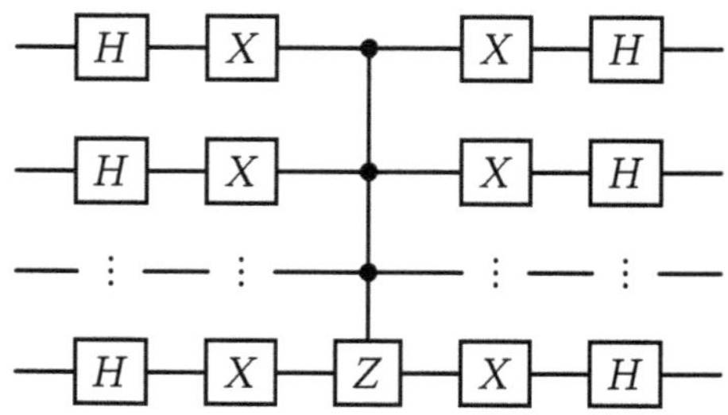

We already know how to write Qiskit code for every gate in the circuit with the only exception being the multi-controlled Z gate. In order to implement it, we will just use a `ZGate` object (which will instantiate a Z gate) and add to it a bunch of controls with the `ZGate().control(n-1)` instruction. Thus, we can use the following function to create Grover's diffusion circuits at will:

```
from qiskit import QuantumCircuit
from qiskit.circuit.library import ZGate

def diffusion_circuit(n):
    dc = QuantumCircuit(n)
    for i in range(n):
        dc.h(i)
        dc.x(i)
    mcz = ZGate().control(n-1)
    dc.append(mcz, range(n))
    for i in range(n):
        dc.x(i)
```

```
        dc.h(i)
    return dc
```

We can see an example of the circuits that this function generates by using the following instructions:

```
dc = diffusion_circuit(4)
dc.draw("mpl")
```

This will yield the following representation:

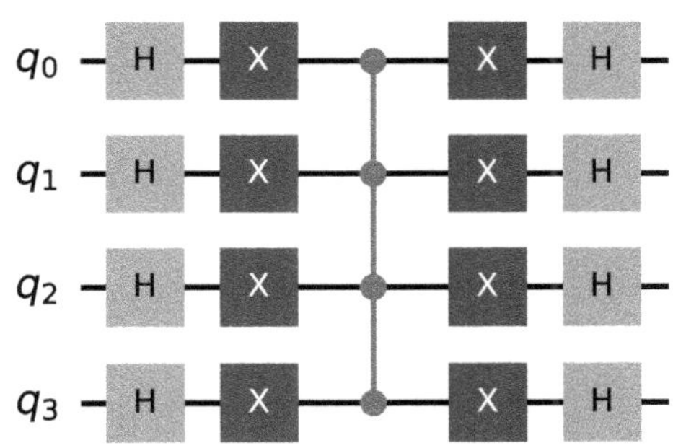

We are now ready to implement Grover's algorithm in all its glory. For that, we will construct a function receiving a quantum oracle for a Boolean function and a number of iterations, and returning the circuit for Grover's search with those parameters. For this, we will rely on the `build_oracle` function that we developed in *Section 10.3.1* and that we have reproduced here for convenience:

```
def build_oracle(strings_one):
    # If the function is never 1, the oracle is the identity.
    # Hence, we return an empty circuit.
    if len(strings_one) == 0:
        return QuantumCircuit()
```

```
    # Number of bits that the function takes as input:
    n = len(strings_one[0])

    qc = QuantumCircuit(n+1)
    for x in strings_one:
        # Find the positions in the string x where the bit is 0.
        # For this, we find the list of indices i such that x[i]=='0'.
        bits_zero = []
        for i in range(len(x)):
            val = x[i]
            if val == '0':
                bits_zero.append(i)

        # Step 1 in our construction.
        for bit in bits_zero:
            qc.x(bit)

        # Step 2.
        qc.mcx(list(range(n)), n)

        # Step 3.
        for bit in bits_zero:
            qc.x(bit)

    return qc
```

Notice that this function creates a quantum Boolean oracle that uses an ancillary qubit to store its output. As you surely recall, in *Chapter 12*, our analysis relied on phase oracles instead. But don't worry. We can use an ancillary qubit set to $|-\rangle$ to induce a phase kickback (as discussed in *Section 12.2.1*) and transform the Boolean oracle into a phase oracle. Easy peasy!

Then, the piece of code that we need to implement Grover's algorithm is as follows:

```
def grover_circuit(oracle, k, measure = False):
    # Number of qubits in the circuit (same as the oracle).
    # If we are working with an n-bit function, nqubits = n + 1.
    # We only measure the top qubits (hence nqubits-1 classical bits).
    nqubits = oracle.num_qubits
    qc = QuantumCircuit(nqubits,nqubits-1)

    # Set bottom qubit to |1>
    qc.x(nqubits-1) # Ancillary qubit for phase kickback
    # Apply Hadamard gates to all qubits.
    for i in range(nqubits):
        qc.h(i)

    # Apply k iterations of oracle + diffusion operator
    dc = diffusion_circuit(nqubits-1)
    qc.barrier()
    for i in range(k):
        qc.append(oracle, range(nqubits))
        qc.barrier()
        qc.append(dc, range(nqubits-1))
        qc.barrier()

    # Measure the top qubits if so instructed
    if measure:
        qc.measure(range(nqubits-1), range(nqubits-1))
```

```
    return qc
```

In this implementation, we first set all the top qubits to $|+\rangle$ with Hadamard gates, and the bottom one to $|-\rangle$ with an X gate followed by a Hadamard gate. Then, we apply the oracle and the diffusion operator k times. Finally, we measure the top qubits (we have made this optional, depending on the `measure` parameter, to more easily compute the probability of finding the element that we are looking for).

Let's try this with an example. We will work with a Boolean oracle that marks a single element, let's say `111`, out of 8 possibilities. From *Section 12.2.4*, we know that we should set a number of iterations $k = \left\lfloor \frac{\pi}{4}\sqrt{8} \right\rfloor = 2$. Then, we can use the following instructions:

```
oracle = build_oracle(["111"])
grover = grover_circuit(oracle, 2, measure = True)
grover.decompose().draw("mpl")
```

This will create and draw the circuit that we need in order to run Grover's algorithm, which looks as follows:

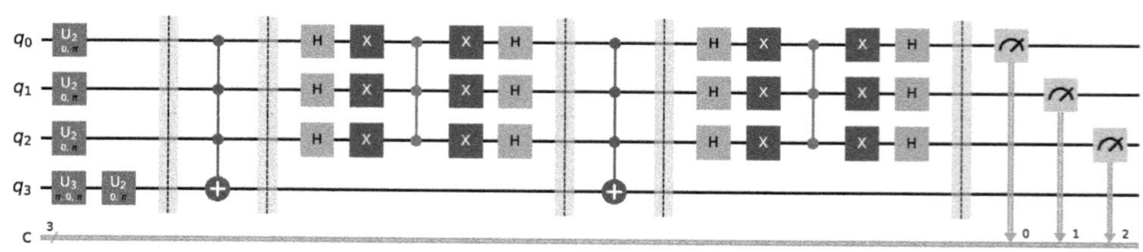

In order to run it and get some results, we can use the following instructions:

```
from qiskit_aer.primitives import SamplerV2 as Sampler
from qiskit_aer import AerSimulator

backend = AerSimulator()
sampler = Sampler(seed = 1234)
job = sampler.run([grover.decompose()], shots = 100)
results = job.result()
```

```
d = results[0].data.c
print(d.get_counts())
```

Notice that we need to decompose the circuit in order to run it on `AerSimulator`. An alternative way of achieving this could be to use `generate_preset_pass_manager`, as we did with Shor's circuit in the previous section. In any case, the output that you'll get when running the instructions above is the following:

```
{'111': 95, '010': 1, '001': 1, '100': 1, '011': 1, '101': 1}
```

This means that, 95 out of 100 times, we have indeed found the marked element!

We can also explicitly compute the probability of obtaining `'111'` as an output by using `Statevector` after removing the measurements from the circuit, as follows:

```
from qiskit.quantum_info import Statevector

grover = grover_circuit(oracle, 2)
sv = Statevector(grover)
print("The probability of 111 is", abs(sv['0111'])**2+abs(sv['1111'])**2)
```

Notice that our circuit has an ancillary qubit (the bottom one) and that Qiskit reverses the order of the strings in the statevector amplitudes. That's why we have added the probabilities of both `'0111'` and `'1111'`, which are the two results compatible with measuring `'111'` on the upper register. Notice also that we have used `abs` to compute the absolute value of the amplitudes and, then, we have squared them with `**2`. The result that we obtained is the following:

```
The probability of 111 is 0.9453124999999958
```

This nicely fits our experimental results when running and measuring the circuit!

Exercise 13.3

What is the probability of measuring `'111'` if we use the same oracle but a number of iterations in the Grover algorithm that varies from 0 to 8?

This concludes this section and our implementation of Shor's and Grover's algorithms. In the next chapter, we will move on to discuss what might be in store for quantum computing in the near future.

Summary

In this chapter, we have implemented two of the most important quantum algorithms out there: Shor's and Grover's. We started by writing code to create circuits for the all-important quantum Fourier transform. From this point, we easily derived an implementation for the inverse quantum Fourier transform too.

Then, we focused on implementing modular multiplication with quantum gates, although we did this with a brute-force approach through the computation of their unitary matrices. This allowed us to construct circuits to demonstrate the use of Shor's algorithm with some small integers.

After that, we shifted our attention toward Grover's search algorithm. We started by constructing circuits for Grover's diffusing operator and, then, we revisited the function that we introduced in *Chapter 10* to create quantum Boolean oracles. Putting everything together, we wrote a function to build the circuit for Grover's algorithm; with it, we were able to search for marked elements and even to compute our probability of success.

In the next chapter, we will move on to explore what the future of quantum computing may look like in the next few years, including discussions on quantum error correction and quantum supremacy.

Part 5

Ad Astra: The Road to Quantum Utility and Advantage

Over the last few parts of the book, we have been able to explore numerous quantum algorithms, which have enabled us to inductively build our understanding on quantum computing and its potential.

Nevertheless, as beautiful and surprising as the algorithms that we have introduced may be, they do need to run son some quantum hardware… and we haven't said much about that yet. Thus, in this part of the book, we will go outside of the realm of theory and address some of the questions that concern physical quantum computers.

This part includes the following chapters:

- *Chapter 14*, Quantum Error Correction and Fault Tolerance
- *Chapter 15*, Experiments for Quantum Advantage

14

Quantum Error Correction and Fault Tolerance

Don't find fault; find a remedy.

— Henry Ford

Throughout this book, we have explored some problems that quantum computers can tackle, and we have gained a lot of perspective on the potential of quantum computing. Analyzing whole functions in a single shot, breaking the limits of classical information, factoring huge prime numbers, navigating through unsorted sets… quantum computers can do all that, and this is just the beginning. The future of quantum computing is promising, to say the least, but we should take some time to address the elephant in the room. All these promises for the future are great, but… can we build the hardware that will make them a reality?

Ideal quantum computers are nowhere to be seen. If you'd like to run a quantum algorithm on real quantum hardware without any errors, you are simply out of luck. At the moment, we *do* have some quantum computers with seemingly impressive numbers of qubits, but they are **noisy**. This means, to put it succinctly, that errors may arise during the execution of circuits: maybe the state of a qubit is corrupted by some environmental phenomena, or maybe the implementation of a quantum gate or a measurement operation is imperfect and introduces errors of its own—all sorts of things could happen!

Most of the algorithms that we have introduced in this book need to run on close-to-ideal quantum hardware. So, we need to find some solutions. We should nevertheless mention that some interesting quantum algorithms can work on noisy hardware; you can find some of these in our book *A Practical Guide to Quantum Machine Learning and Quantum Optimization: Hands-on Approach to Modern Quantum Algorithms* [16].

Given these constraints, you could just sit back for a few decades waiting for physicists and engineers to figure quantum hardware out and come up with better quantum computers; that's an option. Nevertheless, we will take a different path. You know what they say: if life gives you lemons, make lemonade.

In this chapter, we are going to introduce some techniques that will show us how we can use noisy quantum hardware and overcome the errors that it may introduce in our quantum circuits. The topics covered in this chapter are the following:

- The need for error correction
- Quantum error correction: The Shor code
- Implementing the Shor code in Qiskit
- Fault-tolerant quantum computing

Through these sections, you will learn about the importance of error correction techniques, and you will see how they're already very much present in the world of classical computing. You will then see how we can bring these techniques into one-qubit systems through the **Shor code**, and we will show you how to implement it on Qiskit. Lastly, we will dedicate some time to discussing **fault-tolerant quantum computing**, that is, what the path ahead is for getting quantum algorithms to run without errors on noisy quantum hardware.

Alpha	**E**cho	**I**ndia	**M**ike	**Q**uebec	**U**niform	**Y**ankee
Bravo	**F**oxtrot	**J**uliett	**N**ovember	**R**omeo	**V**ictor	**Z**ulu
Charlie	**G**olf	**K**ilo	**O**scar	**S**ierra	**W**hiskey	
Delta	**H**otel	**L**ima	**P**apa	**T**ango	**X**ray	

Table 14.1: The ICAO phonetic alphabet

That's the plan, and now we should get started. Before diving into all the quantum business, let us first get a global perspective on (classical) error correction. This will lay the foundation for us to better understand quantum error correction later on.

14.1 The need for error correction

If an air traffic controller wants to instruct a pilot to go through some taxiways identified with the letters R and J, they will tell them to "taxi via *Romeo* and *Juliet*". Much to the disappointment of romantics and literature scholars, this is not an act of appreciation for the work of William Shakespeare, but one of practicality. You see, when communicating over radio, speaking out letters can easily lead to misunderstandings. The letter M could easily be mistaken for an N or even an L, just as a P could be misheard to be a T, or a C to be an E.

Given how noisy radio communications can be, there are way too many opportunities for communication errors to take place when identifying letters verbally. For this reason, instead of speaking out letters with their usual names, people communicating over radio refer to each letter by a fixed codeword. In particular, in *Table 14.1*, you can find the set of codewords that are used in aviation and military communications; keep in mind that each of them begins with the letter that it represents.

By using these codewords, communication errors are much less likely to take place. The word "em" (for M) can be easily misheard as the word "en" (for N), but it is much more difficult to mistake the words "Mike" and "November". Not only does this strategy prevent misunderstandings, but it also makes the communication more robust against noise. If, say, there were a lot of interference in the channel and, when saying "Mike", the first letter "m" were lost, the receiver would have no trouble in retrieving the message, as "Mike" is the only code word that ends in "ike". All of these advantages, however, come at the cost of slowing down communication, as these codewords are longer than the usual names that we give to letters.

This simple example that we have considered perfectly illustrates what **error correction codes** are and what they aim to achieve. In general, an error correction code is a mechanism that encodes a message creating redundancy in such a way that it is then possible to identify certain communication errors and correct them. In our preceding example, we encoded letters into codewords and this created redundancy: the codewords were much longer than the names of the original letters, and then it was possible to retrieve full codewords from parts of them (we illustrated this using the codeword "Mike" as an example). This then made it possible to easily identify communication mistakes and reverse them.

Important note

An **error correction code** encodes messages in a way that creates redundancy and provides ways of identifying certain communication errors and correcting them.

We should also mention that, in certain scenarios, it may suffice to be able to identify a set of communication errors (without needing to provide a means for correcting them). For this, you can resort to **error detection** mechanisms, which work in a similar way to error correction codes. As a sample use case, error detection mechanisms are built into bank account and credit card numbers, where certain digits are used as **control digits** to protect the integrity of financial transactions against typing errors.

To learn more...

We only have space to discuss the most trivial aspects of classical error detection and correction mechanisms. If you would like to learn more about classical coding theory, we invite you to read *Coding Theory: A First Course* [81] by Ling and Xing.

So far, all of our work has been heuristic, and not particularly formal. We will now consider our first (formal) error correction code on the simplest system possible: a one-bit system.

This will lay the foundation for us to then introduce quantum error correction for one-qubit systems.

14.1.1 Our first error correction code

Consider a channel of communication that allows the transmission of one-bit messages. We will assume that, with a certain probability p, the bits in this channel are "flipped". This means that, with probability p, if a 0 was sent through the channel, a 1 arrived at the receiving end (and vice versa). In this situation, and assuming p to be reasonably small, is it possible to send bits reliably?

The easiest solution to this problem would be to just send each bit three times. That is, instead of just 0, we will send 000, and instead of 1 we will send 111. In this way, every time we receive a set of three bits, if the three bits aren't the same, we can be certain that at least one bit was flipped. Moreover, we can assume with a high probability that the original message is the bit that appears twice. For example, if we get the bits 001, we could infer that the original message was 000. This follows from the fact that the original message was either 000 or 111 and—as we are assuming the probability of bit flips to be small—it's more likely for one bit to have flipped ($000 \to 001$) than for two ($111 \to 001$) to have gone through errors. This way of correcting errors is known as **majority voting**.

Now let's try to make things a little bit quantitative. First, let's assess the probability of identifying whether an error (a bit flip) has taken place in the transmission of data. As we have mentioned, if the three bits that we receive aren't the same, we can be sure that a bit flip has taken place, but we seek to address the converse question: if an error does take place, what is the probability that the three bits won't be equal (and that we will be able to detect the presence of an error)? The only way for errors to go undetected is if all the bits are flipped, and that would happen with probability p^3. Therefore, the probability of correctly identifying whether or not an error has taken place is

$$P_{\text{identify}} = 1 - p^3,$$

which is large for small p. For instance, if p is 10^{-2} (a probability of 1 in 100 of a bit flipping), then the probability of having an undetected transmission error goes down to just 10^{-6}, or one in a million.

Now, let's consider a different question. What is the probability that we will be able to correctly recover the original message? Well, our method will only work if no bits are flipped or if just one of them is. If two or more bits suffer transmission errors, we will get the wrong message. So, let's compute the probability that no bits or just one bit gets flipped. The probability that no bits are flipped is $(1-p)^3$. The probability that only one fixed bit is flipped is $p(1-p)^2$, but this can happen at three different places. Therefore, the probability that our method will work is

$$P_{\text{correct}} = (1-p)^3 + 3p(1-p)^2 = 1 - 3p^2 + 2p^3,$$

where we have added the factor 3 to account for the fact that there are three bits.

In Figure 14.1, we can see the evolution of the probability of detecting errors and that of recovering the correct message as p changes. From this plot, we can deduce an obvious fact, which is that the probability of properly correcting an error is bounded above by the probability of detecting them. Moreover, we can see how, as p grows, our method becomes less and less efficient, but it should work fine for small values of p, as you are about to see in an exercise.

Exercise 14.1

For the method that we have introduced, compute the probability of correctly identifying whether or not an error occurred and of properly correcting any potential bit-flip errors for:

(a) $p = 5\%$,

(b) $p = 10\%$,

(c) $p = 30\%$,

(d) $p = 50\%$.

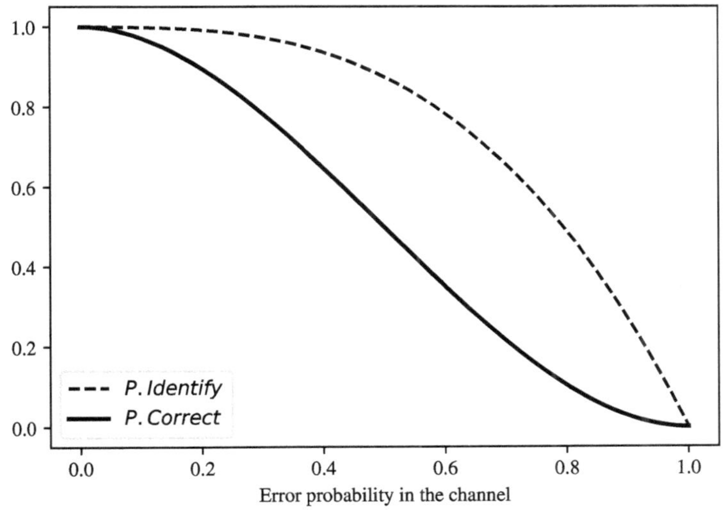

Figure 14.1: The probability of correctly identifying whether or not bit flips have taken place and correctly recovering the original bits in a naive error correction code, as a function of the probability of error p

As you have been able to see for yourself, our method works well enough when p is reasonably small, but not for big values of p. In general, no error correction method can protect against any kind of error occurring with any probability. After all, this is mathematics, not magic!

Simple as it may appear to be, this has been our first proper error correction code. With it, we have encoded one bit into three bits (creating redundancy) and, through majority voting, we have built a method that is robust against a bit-flip in the transmission of the three bits.

This is all we have to say about classical error correction, and we can now begin to explore how quantum error correction actually works. You may be surprised to learn that the codes that we are about to introduce will have some analogies with the innocent classical code that we have been discussing! So, without further ado, let's break free from bits and try to dominate some quantum noise.

14.2 Quantum error correction

Just as in the last section we introduced a simple error correction code for one-bit systems, in this section we will introduce some error correction codes for one-qubit systems. This might seem like a humble and easy task, but—as you are about to find out—the jump from classical error correction to quantum error correction is not exactly trivial.

For starters, there are a few obstacles that we need to take into account before we set out to define quantum error correction codes. Firstly, when we designed our naive classical code for a one-bit system, we simply copied its state into three bits. In the quantum world, however, the no-cloning theorem would make that impossible straight away. But that's not all.

In a quantum system, there are far more possible errors than in a classical one. When you have a single bit, your catalog of errors is rather small: either it gets flipped or it gets lost, but that's it. On the other hand, when you have a qubit, lots of things could happen. These are some examples:

- As in the classical case, your qubit may suffer a **bit flip**, in such a way that $\left|0\right\rangle$ becomes $\left|1\right\rangle$ and vice versa. This would be the same as having the state go through an unwanted X gate.
- In the same way that the bit could go through an unwanted X gate, it could also go through an unwanted Z gate, thus introducing a **phase flip**, which would take the state $\left|+\right\rangle$ to $\left|-\right\rangle$, and $\left|-\right\rangle$ to $\left|+\right\rangle$.

In real hardware, any unwanted quantum gate U could be mysteriously applied to a qubit (or several!) as a consequence of some physical process that hasn't been accounted for. You see, in our crude and finite reality, no physical system can be *perfectly* isolated from its surrounding environment, and physical qubits are no exception. Moreover, these qubits are extremely sensitive to any outside **noise** that could originate from the interactions of the qubits with their environment or from imperfections in the quantum hardware itself. However, this isn't at all the full story; in a quantum computer, lots of others kinds of errors may arise!

To learn more...

If you'd like to learn more about the current implementations of quantum hardware and the kinds of errors that they are vulnerable to, you might want to read *Hardware for Quantum Computing* [82] by Easttom.

If you learn more about the physical implementation of quantum computers, you will see how some errors can't even be addressed from an algorithmic point of view. Thus, from the theoretical side of things, our only choice is to work within some fixed **error models**: assuming that a fixed set of errors may occur and finding ways of dealing with them. Following this approach, we will introduce error correction codes for one-qubit systems step by step: first discussing a way of correcting bit-flip errors, then tackling phase-flip errors, and finally combining both approaches into a single error correction code. As you will see, this will end up allowing us to correct a significant set of one-qubit errors. Let's get to it!

14.2.1 Bit flips

Let us consider a simple scenario: assume that we know that a qubit is going to be exposed to some **noise**, in such a way that it may suffer a bit-flip. Maybe this is a consequence of our performing some operations to apply a quantum gate, or maybe of our transmitting the qubit through some media—those details don't quite matter for now. All we know is that, with a small probability, our qubit will suffer a bit-flip.

This scenario is somewhat analogous the one we considered in our classical error correction code, and the approach that we will take will be surprisingly similar.

Assume that the one-qubit state we want to preserve against bit-flip errors is

$$\left|\alpha\right\rangle = \alpha_0 \left|0\right\rangle + \alpha_1 \left|1\right\rangle.$$

Back in our study of classical codes, we got an idea of how error correction codes are based on the creation of some redundancy. For the classical code that we considered, we achieved this redundancy by creating two new copies of the bit state that we wanted to protect, but that's not something that we can do directly with a qubit state. Nevertheless, we can take an approach that is similar in spirit. Indeed, if we consider two ancillary qubits, we can perform two CNOT operations, transforming our state as follows:

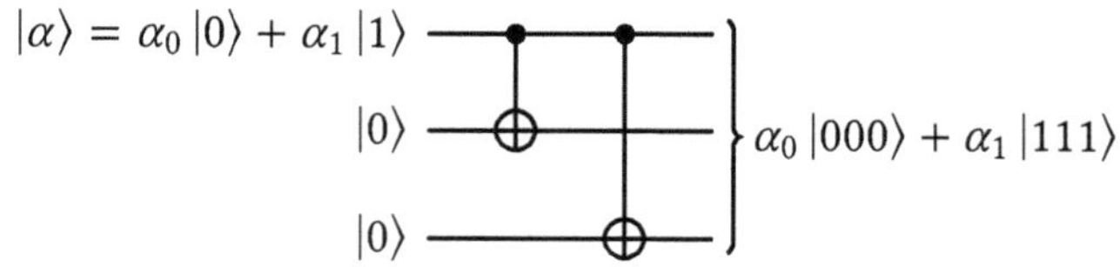

Needless to say, this is not a state where we have three copies of $|\alpha\rangle$ (this is not $|\alpha\rangle \otimes |\alpha\rangle \otimes |\alpha\rangle$, as that would go against the no-cloning theorem), but you can see how, in some way, we are exploiting a similar notion of redundancy to the one on which we built our classical code. The fact that this circuit transforms the state as we have claimed holds no mystery; you can verify it yourself using the definition of the CNOT gate.

Exercise 14.2

Verify that, indeed, the circuit above transforms the state $(\alpha_0 |0\rangle + \alpha_1 |1\rangle) |0\rangle |0\rangle$ into the state $\alpha_0 |000\rangle + \alpha_1 |111\rangle$.

The circuit that we have introduced is called the **encoding circuit**, and it encompasses all the operations that need to be performed in order to prepare our system for error correction. At this point, some bit flips may occur in some of the qubits, and we need to find some ways to correct them. Our solution will be to apply an **error correction circuit**, which will perform a sequence of two CNOT operations controlled by the first qubit followed by a Toffoli gate targeting it, as shown in the following representation:

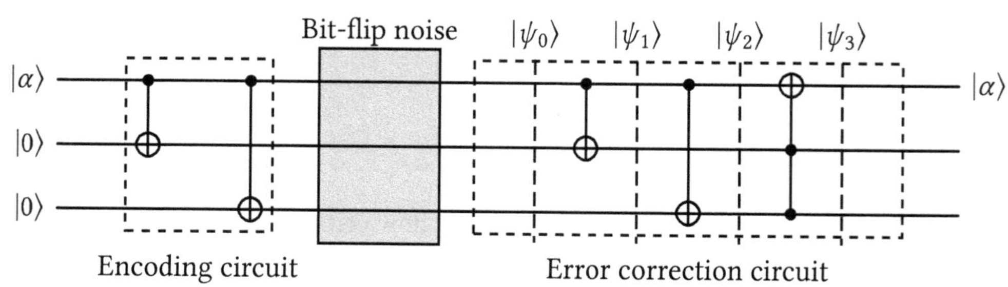

After this procedure, if no more than one qubit has suffered a bit flip, we can be certain that the state of the first qubit will be $|\alpha\rangle$. We will later provide a general proof of how and why this works, but, for now, let us analyze each possible scenario in which at most a single qubit suffers a bit flip.

- If no qubit has suffered a bit flip, then $|\psi_0\rangle = \alpha_0 |000\rangle + \alpha_1 |111\rangle$, hence

$$|\psi_1\rangle = \alpha_0 |000\rangle + \alpha_1 |101\rangle ,$$

  as the action of the first CNOT flips the state of the second qubit when the first qubit takes the value 1. Consequently,

$$|\psi_2\rangle = \alpha_0 |000\rangle + \alpha_1 |100\rangle = |\alpha\rangle |00\rangle ,$$

  and the last Toffoli gate will have no effect on the system as the last two qubits are in state $|0\rangle$. Hence, $|\psi_3\rangle = |\psi_2\rangle$ and the first qubit will be left in the state $|\alpha\rangle$, as desired.

- If the first qubit suffered a bit flip, then we have

$$|\psi_0\rangle = \alpha_0 |100\rangle + \alpha_1 |011\rangle ,$$

  which is transformed under the action of the CNOT gates as

$$\begin{aligned} |\psi_1\rangle &= \alpha_0 |110\rangle + \alpha_1 |011\rangle , \\ |\psi_2\rangle &= \alpha_0 |111\rangle + \alpha_1 |011\rangle \\ &= (\alpha_0 |1\rangle + \alpha_1 |0\rangle) |11\rangle , \end{aligned}$$

  hence—as the last two qubits are both in state $|1\rangle$, the action of the Toffoli gate will apply a negation on the first qubit, thus bringing the system to the state $|\alpha\rangle |11\rangle$, and restoring $|\alpha\rangle$ in the first qubit, just as we wanted.

- If the second qubit suffered a bit flip, we have

$$|\psi_0\rangle = \alpha_0 |010\rangle + \alpha_1 |101\rangle,$$

so the action of the CNOT gates will transform the state as

$$\begin{aligned} |\psi_1\rangle &= \alpha_0 |010\rangle + \alpha_1 |111\rangle, \\ |\psi_2\rangle &= \alpha_0 |010\rangle + \alpha_2 |110\rangle \\ &= |\alpha\rangle |10\rangle, \end{aligned}$$

and the Toffoli gate will have no effect as the third qubit is in state $|0\rangle$. Hence, $|\psi_3\rangle = |\psi_2\rangle$ and the first qubit will be in the state $|\alpha\rangle$, as claimed.

- Lastly, if there were a bit flip on the third qubit, we would have

$$|\psi_0\rangle = \alpha_0 |001\rangle + \alpha_1 |110\rangle,$$

and, as you can check for yourself, the action of the CNOT and Toffoli gates would bring the system to the final state

$$|\psi_3\rangle = |\alpha\rangle |01\rangle,$$

and the first qubit ends up in state $|\alpha\rangle$.

Thus, we have seen how, if there is at most one bit flip, the first qubit always ends up in state $|\alpha\rangle$. What is more, you may have noticed that the last two qubits end up in a state that allows us to identify whether there was a bit flip and which qubit suffered it. As it turns out, all of this can be deduced from a general analysis of the action of our error correction circuit, which we will now present. Now that you are well acquainted with the way these gates operate, this will be easy.

If we feed the error correction circuit with a computational basis state of the form $|\psi_0\rangle = |x\rangle |y\rangle |z\rangle$, the action of the CNOT gates will transform it into $|\psi_2\rangle = |x\rangle |y \oplus x\rangle |z \oplus x\rangle$, and, therefore, the last Toffoli gate will bring this to

$$|\psi_3\rangle = |(x \oplus (y \oplus x) \cdot (z \oplus x))\rangle \, |y \oplus x\rangle \, |z \oplus x\rangle$$

Now the states of the last two qubits are very easy to interpret: they will be 1 if and only if their original state (y or z) was different from the one of the first qubit (x). That's why these two bits reveal whether or not a bit flip may have taken place. If the two qubits are 0, this means that all the states matched. If the two are 1, it means that the two last qubits disagreed with the first. If only one of the last two qubits takes the state 1, it means that it just had a different value compared to the first qubit. Measuring these two qubits thus helps us identify whether an error occurred and where it occurred; in the jargon of quantum error correction, we refer to this kind of measurement as **syndrome measurement**.

The state of the first qubit is slightly more tricky to analyze, but not too much. The expressions $(y \oplus x)$ and $(z \oplus x)$ are 1 if and only if y and z differ from x, respectively. Therefore, $(y \oplus x)(z \oplus x)$ will be 1 if and only if both y and z are different from x. Consequently, $x \oplus ((y \oplus x) \cdot (z \oplus x))$ will be $x \oplus 1$ (that is, x flipped) if both y and z differ from x, or else it will be x. In short, this expression is simply implementing the majority voting that we used in our classical error correction code: it is equal to the bit that appears at least twice among x, y, z.

For those of you who enjoy algebra and are hungry for a more symmetric expression, we can take a slightly longer route. Using the properties of Boolean algebra,

$$\begin{aligned} x \oplus (y \oplus x)(z \oplus x) &= x \oplus \left(yz \oplus xy \oplus xz \oplus x^2\right) \\ &= x \oplus x \oplus yz \oplus xy \oplus xz \\ &= yz \oplus xy \oplus xz. \end{aligned}$$

where we have used the fact that, for any bit x, we have $x^2 = x$ and $x \oplus x = 0$. We have also used the commutativity, associativity, and distributivity of the operations of Boolean algebra. Now, as expected, the value that this will take is, precisely, the bit that appears at least twice in x, y, z, as you can now see in an exercise.

Exercise 14.3

Show that $yz \oplus xy \oplus xz$ implements majority voting among the bits x, y, z. That is, show that the expression evaluates to 1 or 0 if and only if at least two of x, y, and z are 1 or 0, respectively.

In summary, this analysis reveals that our method ends up behaving in the same way as the classical error correction code that we introduced back in *Section 14.1.1*. It creates redundancy by encoding a one-qubit state into three entangled qubits; it can detect a bit-flip error if the values of the three qubits don't agree; and it can correct it if no more than one qubit has suffered a bit flip. This analogy means that our performance assessments for our bit-flip classical code also apply to our bit-flip quantum code. Thus, if the noise induces a bit flip with probability p, our quantum error correction code will correctly identify if quantum bit flips have occurred with probability $1 - p^3$, and it will return the original qubit with probability $1 - 3p^2 + 2p^3$.

To learn more...

We have presented a simple formulation of the bit-flip error correction code, but there is an alternative one. In Chapter 10 of *Quantum Computation and Quantum Information: 10th Anniversary Edition* [13], you can find a different approach to this code, in which some qubits are measured a priori to determine whether bit flips have taken place and—based on the outcomes of those measurements—the necessary corrections are then performed. In that book, you can find similar alternative formulations for all the codes that we will discuss in this chapter.

Our quantum error correction code needs three qubits in order to encode a single qubit. In the lingo of quantum computing, we say that this code uses three **physical qubits** to implement one **logical qubit**.

Important note

Consider some noise that induces bit flips with a certain probability. If a one-qubit state $|\alpha\rangle$ is encoded, exposed to the noise, and processed as

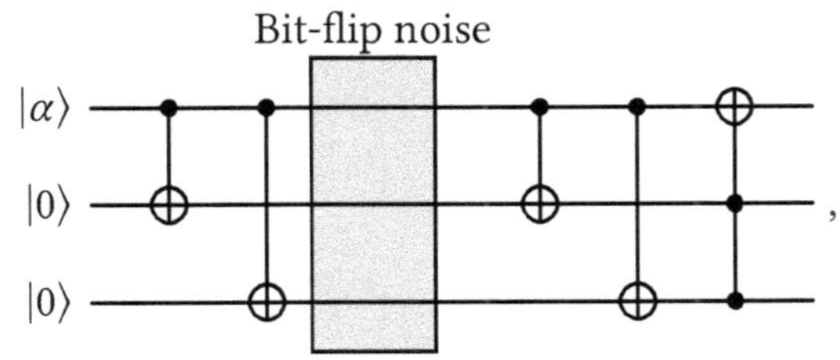

then the state in the first qubit will still be $|\alpha\rangle$ if at most one qubit suffers a bit flip. Moreover, the state of the last two qubits will be in the computational basis and, if one or two qubits have suffered a bit flip, this state will be different from $|00\rangle$.

That's our quantum error correction code for bit flips. As you can see, in spite of the complexities that quantum computing introduces, there are clear and direct analogies connecting this code to the classical error correction code that we discussed in *Section 14.1.1*.

Now let's try to go bigger than this. Imagine that instead of a bit-flip error (an unexpected X gate), we have a phase-flip error (an unexpected Z gate). In this case, will our setup work?

- If the original state were $|0\rangle$, then the encoded state would be $|000\rangle$, and the Z gate would leave it as it is, so the final state on the first qubit would be $|0\rangle = Z|0\rangle$.
- If the original state were $|1\rangle$, the encoded state would be $|111\rangle$, which, after the application of a Z gate, would become $-|111\rangle$. The error correction circuit would then yield the final state $-|1\rangle = Z|1\rangle$ on the first qubit.

By linearity, we can conclude that, for any state $|\alpha\rangle$, if a phase-flip takes place, the final state (on the first qubit) would be $Z|\alpha\rangle$, so phase-flip errors are just ignored in this setting. In a similar fashion, you can check for yourself that if the error is ZX or XZ, the final state will be, respectively, $Z|\alpha\rangle$ and $-Z|\alpha\rangle$ (which is $Z|\alpha\rangle$ up to an irrelevant global phase). This means that—while our current code is hopeless against phase-flip errors—it, in a vague way, "commutes" with them, meaning that it ignores them and simply passes any phase flips to the corrected state. This will be relevant later on in the chapter.

In light of the limitations of our error correction code, we will now discuss a quantum error correction code for phase flips. This may seem more difficult to tackle, but don't let anything intimidate you—as you are about to find out, we have already done most of the heavy lifting!

14.2.2 Phase flips

Bit flips, which we have just taken care of, occur when an unwanted X gate is applied to a qubit. Phase flips, on the other hand, take place when an unwanted Z gate is applied. Thus, bit flips transform a state $\alpha_0 |0\rangle + \alpha_1 |1\rangle$ into $\alpha_0 |1\rangle + \alpha_1 |0\rangle$, whereas phase flips transform it into $\alpha_0 |0\rangle - \alpha_1 |1\rangle$.

Bit flips and phase flips may seem like completely different phenomena, but they are actually related in a very easy way: through Hadamard gates! Indeed, as it turns out, $X = HZH$. In terms of circuits:

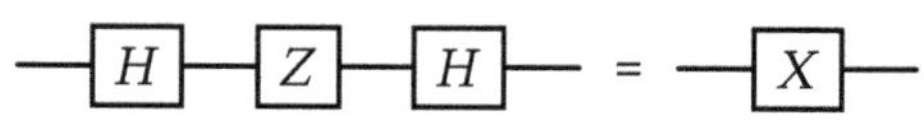

Thus, if we apply a Hadamard gate to a qubit before a phase flip, and we apply a Hadamard gate afterward, the phase flip will just act like a bit flip! What is more, if no bit flip takes place, adding those Hadamard gates will have no effect, as H is its own inverse.

Exercise 14.4

Prove that, indeed, $X = HZH$. Deduce that also $Z = HXH$.

This has a remarkable implication, which is that, if we want to construct an error correction code for phase flips, we can just reuse our error correction code for bit flips! Indeed, all we have to do is apply a Hadamard gate to every qubit right at the end of the encoding circuit and at the beginning of the error correction circuit; in this way, every potential phase flip will become a bit flip (and if there are no phase flips, this will have no net effect on the

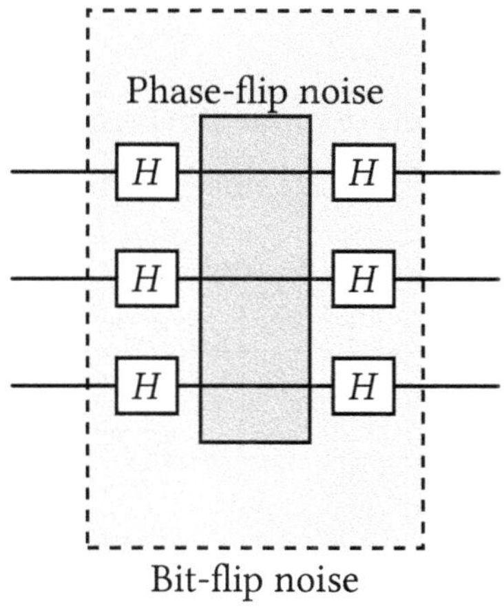

Figure 14.2: Hadamard gates can be used to transform a phase-flip error into a bit-flip error

state), and we can correct that bit flip with the very same techniques that we have already developed. That's all it takes! This is illustrated in *Figure 14.2*.

Exercise 14.5

To get a better feel for what we are doing, consider a general state of the form $a\,|000\rangle + b\,|111\rangle$ and compute, step by step, its evolution under the application of (1) a Hadamard gate on each qubit, (2) a Z gate on the first qubit, and (3) a Hadamard gate on each qubit.

Conclude that, indeed, the application of Hadamard gates before and after a potential phase-flip error transforms it into a bit-flip error.

That is how a phase-flip error correction code can be constructed: by transforming phase flips into bit flips with Hadamard gates and reusing our bit-flip error correction code. Needless to say, the performance of this code will be equivalent to that of the bit-flip code: it will be able to detect up to two phase flips, and it will properly correct the state if at most one phase flip takes place. The success probabilities will also be the same.

Important note

Consider some noise that may induce phase flip errors with some probability. If a one-qubit state $|\alpha\rangle$ is encoded, exposed to noise, and processed as

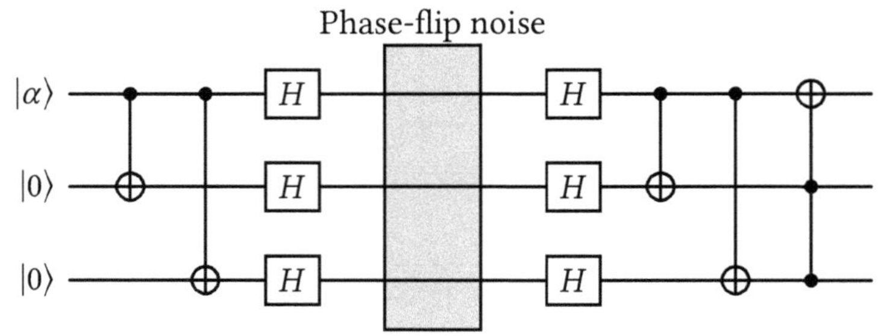

then the state in the first qubit will still be $|\alpha\rangle$ if at most one qubit suffers a phase flip. Moreover, the state of the last two qubits will be in the computational basis and, if one or two qubits have suffered a phase flip, this state will be different from $|00\rangle$.

With this, we have tools to detect and correct bit-flip and phase-flip errors. Now we will introduce a quantum error correction code that combines these tools in a surprisingly powerful way.

14.2.3 The Shor code

The **Shor code** is a quantum error correction code that uses nine physical qubits to encode a single logical qubit. That's three times as many physical qubits as all of our previous codes! But the cost is justified, for this code will allow us to correct any error induced by the application of an undesired gate on a single qubit. How? Let's take it step by step. First, we will introduce the code, then we will discuss why it protects against both bit-flip and phase-flip errors simultaneously, and then we will be able to conclude that it can protect against any single-qubit error.

The Shor code simply combines the phase-flip and bit-flip error correction codes that we have just introduced, and it does it in a rather straightforward way, which is represented in *Figure 14.3*. Now, this circuit might seem overwhelming, but it actually brings nothing new to the table. In the Shor code, we are just applying our usual phase-flip error correction

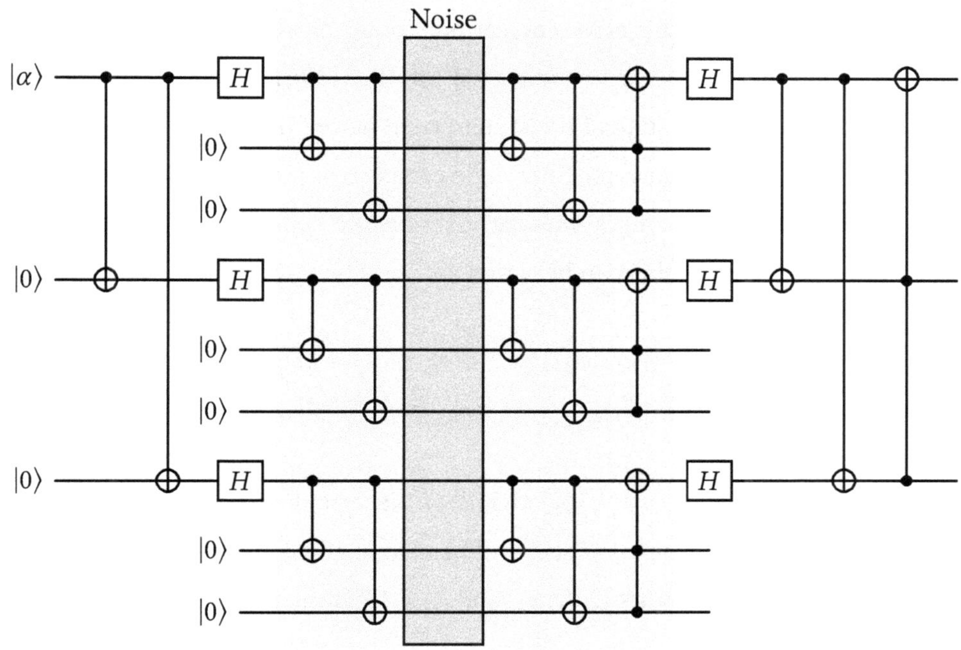

Figure 14.3: The Shor code

code, with a small tweak: right after the encoding circuit and before the error correction circuit, we apply the bit-flip error correction code to each of the three qubits, leading to the use of a total of nine qubits; this is what is represented in the shaded area of the circuit. In this way, we make sure that each of the three qubits that leaves the shaded area is free of any bit-flip errors, and only phase-flip errors are left to be corrected. Keep in mind that this works since, as we showed earlier on, the bit-flip error correction code ignores any phase-flip errors and simply passes them on to the output state. That's why this code can handle both phase-flip and basis-flip errors simultaneously!

So, we have a code that can handle bit-flip and phase-flip errors simultaneously. That's impressive enough, isn't it? Well, there's one more thing to discuss here.

Let $k \leq 9$ be a natural number and let X_k and Z_k be the X and Z gates on the k-th qubit; also, let E denote the nine-qubit error correction circuit, placed right after the noise in *Figure 14.3*. Given any input state $|\alpha\rangle$ being fed into the Shor code, we will use $|\psi_\alpha\rangle$ to denote the encoded state—the state of the system right before the nine qubits are exposed to noise. Based on what we know, the Shor code can correct phase-flip and bit-flip errors (either on their own or occurring simultaneously on the same qubit). In terms of our recently introduced notation, this can be stated as

$$E(I\,|\psi_\alpha\rangle) = |\alpha\rangle\,|\varphi_0\rangle\,, \qquad E(X_k\,|\psi_\alpha\rangle) = |\alpha\rangle\,|\varphi_x\rangle\,,$$
$$E(Z_k\,|\psi_\alpha\rangle) = |\alpha\rangle\,|\varphi_z\rangle\,, \qquad E(X_k Z_k\,|\psi_\alpha\rangle) = |\alpha\rangle\,|\varphi_{xz}\rangle\,,$$

for some eight-qubit states $|\varphi_0\rangle$, $|\varphi_x\rangle$, $|\varphi_z\rangle$ and $|\varphi_{xz}\rangle$. Nevertheless, because of the linearity of quantum operations, this means that E will be able to correct errors produced by any linear combination of the operators I, X_k, Z_k, and $X_k Z_k$! Indeed, given any scalars $\lambda_1, \ldots, \lambda_4$,

$$E\left((\lambda_1 I + \lambda_2 X_k + \lambda_3 Z_k + \lambda_4 X_k Z_k)\,|\psi_\alpha\rangle\right) =$$
$$= \lambda_1 E(I\,|\psi_\alpha\rangle) + \lambda_2 E(X_k\,|\psi_\alpha\rangle) + \lambda_3 E(Z_k\,|\psi_\alpha\rangle) + \lambda_4 E(X_k Z_k\,|\psi_\alpha\rangle),$$

which will produce a state of the form $|\alpha\rangle \otimes |\varphi'\rangle$. Here comes the bomb: as it turns out, *every* single-qubit gate acting on the k-th qubit can be written as a linear combination of I, X_k, Z_k, and $X_k Z_k$, as you are about to prove in an exercise.

Exercise 14.6

Prove that any 2×2 matrix can be written as a linear combination of the matrices I, X, Z, and XZ, where, in this case, I denotes the 2×2 identity matrix.

Deduce that in an n-qubit circuit, any gate acting on the k-th qubit can be written as a linear combination of the gates I, X_k, Z_k, and $X_k Z_k$, where I denotes the $2^n \times 2^n$ identity matrix.

Thus it follows that the Shor code can actually correct any error induced by the application of a single-qubit gate on any of the nine encoding qubits.

Important note

The Shor code, which is represented in *Figure 14.3*, encodes a single logical qubit into nine physical qubits, and it makes it possible to correct any single-qubit error affecting those nine qubits, assuming errors to consist in the application of some unwanted quantum gate.

This error correction code was introduced in the article "Scheme for reducing decoherence in quantum computer memory" [83] by Shor, and it is one of the canonical examples of quantum error correction codes.

To learn more...

We have shown how the Shor code protects against errors induced by the introduction of an undesired gate into one of the encoded qubits. However, it can do even more than that. It can also protect against **decoherence**: transformations in which an (encoded) qubit gets entangled with its environment. In "Scheme for reducing decoherence in quantum computer memory" [83], you can explore all the decoherence processes that the Shor code can detect and correct.

With this, we have built a basic foundational knowledge of quantum error correction. Nevertheless, there is much more for you to learn if you want to invest the time and effort.

To learn more...

In Chapter 10 of the book *Quantum Computation and Quantum Information: 10th Anniversary Edition* [13], you can deepen your knowledge of quantum error correction, studying **CSS codes** and the **stabilizer formalism** for quantum codes.

You can also refer to "Topological codes" [84] to learn about other kinds of codes, such as **surface codes** and **color codes**. Color codes, in particular, have been demonstrated by a research team at Google [85] to be a viable means toward fault-tolerant quantum computing, which we will discuss later in the chapter.

Now that we have understood how the Shor code works, it's time for us to implement it, so let's get our hands dirty with Qiskit!

14.3 Implementing the Shor code in Qiskit

In order to get started qiskiting, we first need to import the quantum circuit class, as we will inevitably need it:

```
from qiskit import QuantumCircuit
```

Remember that we are using version 2.1 of Qiskit, and that you can find instructions on how to set up your computing environment in *Appendix D*.

Now that we are all set, let us implement the Shor code. We will do it defining a function, `ShorCode`, which will take as input two arguments:

- a nine-qubit circuit `qc`, which we will assume has its first qubit set to the input state and has the remaining qubits set to $|0\rangle$,
- and a function `noise`, which should take as input a quantum circuit and append to it a one-qubit gate introducing an error.

Our function will add the necessary gates to `qc` as well as the noise defined by `noise`. Since all of this will be stored in `qc`, there is no need for the function to return any values (recall that, in Python, objects are passed by reference, not by value).

Here is our implementation:

```
def ShorCode(qc, noise):
    qc.barrier()
```

```
# Encoding
qc.cx(0, 3)
qc.cx(0, 6)
for qubit in [0,3,6]:
    qc.h(qubit)
    qc.cx(qubit, qubit + 1)
    qc.cx(qubit, qubit + 2)

# Noise
qc.barrier()
noise(qc)
qc.barrier()

# Error correction
for qubit in [0,3,6]:
    qc.cx(qubit, qubit + 1)
    qc.cx(qubit, qubit + 2)
    qc.ccx(qubit + 2, qubit + 1, qubit)
    qc.h(qubit)
qc.cx(0, 3)
qc.cx(0, 6)
qc.ccx(6, 3, 0)
```

Notice how we have added some barriers along the circuit. We included them for two reasons: first, because they mimic the real implementation of the Shor code, where the noise is unwanted and therefore cannot be "optimized" together with the rest of the circuit. Second, and perhaps most importantly, because these barriers will make visual representations of circuits cleaner.

Having defined the Shor code implementation, let's test it out. We will prepare a nine-qubit circuit and we will take $|+\rangle = H|0\rangle$ to be our input state. For the noise, we will apply a Z gate on the third qubit.

```
qc = QuantumCircuit(9)                # Initialize the circuit.
qc.h(0)                               # Prepare the input state.
ShorCode(qc, lambda qc: qc.z(2))      # Apply the code with noise.
```

Notice that, in order to specify the noise, we have used **lambda notation**, which provides a very convenient way of defining functions without the need to name them. For example, the object `lambda x: x + 1` is a function that takes an argument `x` and returns `x + 1`.

We can now use the `draw` function to obtain a visual representation of the circuit.

```
qc.draw(output="mpl")
```

The resulting image is shown in *Figure 14.4*, and it allows us to verify that our implementation matches the Shor code that we introduced in *Figure 14.3*. We can also see how the input state has been properly initialized and how the error is correctly added.

That's all we have to do in terms of implementation. Now we should probably test whether our code actually works. To this end, we will retrieve the final state of our circuit and save it in a `statevector` variable.

```
from qiskit.quantum_info import Statevector
statevector = Statevector(qc)
```

If you try to print `statevector`, you will likely be overwhelmed. It's a list with $2^9 = 512$ amplitudes, which doesn't seem too appealing. However, since most of its entries are zero, we can run a few lines of code to find the computational basis states with non-zero amplitudes and the value of those amplitudes.

For example, we can run the following instructions:

```
from numpy import isclose
```

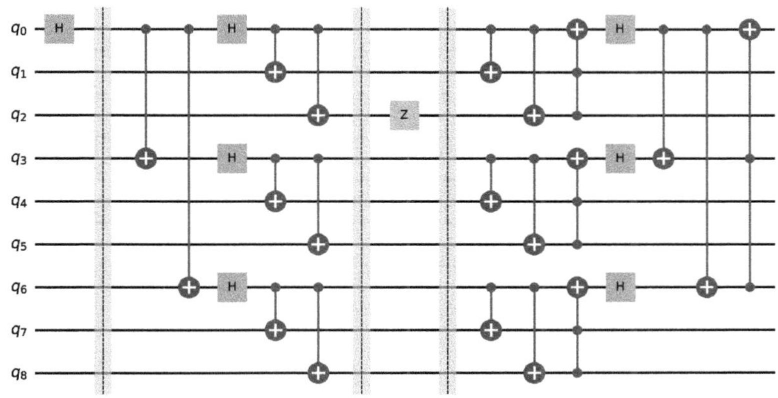

Figure 14.4: Our implementation of the Shor code with Qiskit

```
for index, value in enumerate(statevector):
    if not isclose(value, 0):
        # Use bin to get a binary string for the index.
        # Remove the first two characters (which are '0b').
        index_string = bin(index)[2:]

        # Add leading zeros.
        index_string = index_string.zfill(9)

        # Flip the string (to account for Qiskit's conventions).
        index_string = index_string[::-1]

        print(f"{value} |{index_string}>")
```

Here, we have used the **bin** function, which is a built-in function that, when given a number, returns a string with its binary representation; however, since this representation is preceded by the characters '0b', we also added [2:] to discard them. We also used the zfill method, which adds leading zeros to a string until it has at least the number of characters provided as an argument. Lastly, we reversed the string in order to make the results consistent with our usual notation, accounting for the "peculiar" way in which Qiskit labels qubits—assigning the topmost qubit to the rightmost qubit in a tensor product. We already addressed this at length in *Chapter 7* and *Chapter 10*.

As a side note, notice how we have used the isclose function to check whether value was zero. We did this in order to account for any inaccuracies that may arise from the use of floating-point arithmetic in the simulator.

Upon running the preceding code, we get this result:

```
(0.7071067811865471+0j) |000100100>
(0.7071067811865471+0j) |100100100>
```

This is simply the state $|+\rangle\,|00100100\rangle$, which agrees with the expected behavior of the Shor code. The first qubit is in our original state (free of errors) and the last qubits are telling us that a phase error was corrected in the first block of three encoded qubits.

Let's now consider a more sophisticated example, using as noise a $R_Y(4)$ gate on the third qubit (remember that we introduced Y-rotation gates in *Chapter 3*). To run this, we can execute the following lines of code (they are analogous to what we've just done):

```
qc = QuantumCircuit(9)
qc.h(0)
# Here we use a different error:
ShorCode(qc, lambda qc: qc.ry(4,3))
statevector = Statevector(qc)

for index, value in enumerate(statevector):
    if not isclose(value, 0):
        index_binary = bin(index)[2:]
```

```
    index_string = index_binary.zfill(9)
    index_string = index_string[::-1]
    print(f"{value} |{index_string}>")
```

With the preceding code, we get the following output:

```
(-0.294260250091814+0j) |000000000>
(-0.294260250091814+0j) |100000000>
(0.6429703766239176+0j) |000111000>
(0.6429703766239176+0j) |100111000>
```

This is telling us that, if we let $|x\rangle = |00000000\rangle$, and $y = |00011100\rangle$, then, for $a \approx -0.294$ and $b \approx 0.643$, the final state is equal to

$$|\psi\rangle = a\,|0\rangle\,|x\rangle + a\,|1\rangle\,|x\rangle + b\,|0\rangle\,|y\rangle + b\,|1\rangle\,|y\rangle\,.$$

From this expression, we can factor out the state of the first qubit as

$$\begin{aligned} |\psi\rangle &= a(|0\rangle + |1\rangle)\,|x\rangle + b(|0\rangle + |1\rangle)\,|y\rangle \\ &= \frac{|0\rangle + |1\rangle}{\sqrt{2}} \otimes \sqrt{2}(a\,|x\rangle + b\,|y\rangle), \end{aligned}$$

thus showing that the topmost qubit is in state $|+\rangle$, just as we wanted. Notice that, in this case, the remaining qubits are not in a computational basis state, but in a superposition. That's a consequence of the fact that the error that was corrected was neither X, Z, nor XZ.

With this, we have been able to implement and test the Shor code with Qiskit. To finish the chapter, we would like to share a few words about fault tolerance and the near future of quantum hardware.

14.4 Fault-tolerant quantum computing

Right now, quantum computers—having more than a handful of qubits—do exist, and some are even available for public use at no cost. Our issue is that, as we discussed in *Chapter 1*, at the moment, we live in the noisy intermediate-scale quantum era, also known as the NISQ era. This means that our quantum hardware still has a limited number of qubits and that there is no way for us to make it work like ideal hardware would.

This is problematic because, as we mentioned in the introduction, most of the algorithms that we have discussed in this book are designed to run on ideal error-free quantum computers, and they have no chance of being effective in current NISQ hardware. Other quantum algorithms can be more or less resilient against noise and can run on NISQ hardware, but for Shor and Grover… that's not going to work.

Nevertheless, where there is a problem, there are people willing to solve it, and a lot of effort is being invested in making quantum computers **fault-tolerant**: making them usable (almost) as if they were ideal, in spite of their imperfections. Right now, our path toward fault tolerance can be found in quantum error correction codes, which we have briefly introduced in this chapter.

It is important to remark that what we have discussed in this chapter is just a drop in the ocean. The world of quantum error correction is vast, and we hope to have given you at least a small glimpse of how some of its most basic techniques can work. If you would like to learn more, you can consult the references that we have dropped throughout the chapter. Incidentally, if you were looking for a theoretical result that would give you reasons to believe in the potential of quantum error correction, the **threshold theorem** establishes that, as long as error rates are below a certain threshold, arbitrarily long quantum computing processes are feasible using techniques from quantum error correction; you can find this theorem in Section 10.6.1 of *Quantum Computation and Quantum Information: 10th Anniversary Edition* [13].

We should also mention that there are other techniques that—in conjunction with error correction codes—will help pave the way toward the implementation of fault-tolerant architectures. A notable example is the quantum gate teleportation mechanism, introduced by Gottesman and Chuang [86]. From a bird's-eye view, quantum gate teleportation leverages entanglement to enable the fault-tolerant implementation of quantum gates using some special quantum states known as **magic states**.

As our tools to achieve fault tolerance increase and are further refined, the future where quantum computers will be fully usable gets closer and closer. Based on the progress that we have seen in recent times, we have many reasons to be optimistic about the short- or medium-term future. Across the industry, very significant advances are being made toward making quantum hardware fault-tolerant and, therefore, fully useful [87], [88]. Chief among all of these achievements is the breakthrough that a team at Google made in 2024, introducing a quantum chip with an error rate low enough for effective quantum error correction, thus paving the way for scalable and practical quantum hardware [85].

Who would've thought that some simple bit flips could have kept us so entertained? Indeed, as the short story "Single-Bit Error" [89] by Ken Liu shows, a single bit flip can have consequences far beyond what we could even imagine. In any case, and in spite of those sneaky quantum errors, the future of quantum computing appears to be more than bright. On that optimistic note, let's wrap things up!

Summary

In this chapter, we have discussed the general limitations of current quantum hardware and we have taken a look at the techniques that may help us overcome them.

We began our journey by introducing classical error correction and constructing our first classical error correction code, which encoded a single bit into three bits. We also performed a performance assessment of this code, computing the probability that it would properly identify errors and safeguard the transmission of data.

Building on this work, we then introduced a couple of simple quantum error correction codes that worked against bit-flip and phase-flip errors; we showed how in many ways they were analogous to the classical code that we considered, and we were even able to reuse our computations for the assessment of their performance.

Then, we combined our quantum error correction codes into the Shor code, which encodes a qubit state into nine qubits and enables us to correct any single-qubit error induced by the introduction of an unwanted gate. We also learned how to implement this code in Qiskit and we were able to test it in a simple case.

Finally, we concluded the chapter with some general remarks on fault tolerance and the prospects for near-term quantum hardware.

To put in succinctly, in this chapter, we've had a glimpse at some of the techniques that, in the future, will most certainly allow us to overcome the limitations and flaws of quantum hardware. In the following chapter, we will explore the different ways in which we can assess whether or not quantum hardware is capable of producing results that are beyond the possibilities of classical hardware. Get ready for our grand finale!

15

Experiments for Quantum Advantage

There's a million things I haven't done. But just you wait, just you wait.

— Alexander Hamilton

Congratulations! You have reached the last chapter of this book. This has been quite a ride, hasn't it? In this journey together, we've studied quantum algorithms and protocols that have been introduced in the last forty years and we've explored how to use Qiskit to implement them. We've also discovered how quantum phenomena such as superposition, entanglement, interference, and the uncertainty principle can enable us to do things that would be impossible to achieve with classical computers.

Now it is time to take a look ahead and ponder what quantum computing holds in store for the short or medium term. To that end, in this chapter, we are going to discuss different ways in which researchers from all over the world have been pushing the limits of quantum hardware in an attempt to achieve the holy grail of quantum computing: quantum advantage.

We will especially focus on random circuit sampling, a type of task that has been (and still is!) central in quantum advantage experiments. We will expose its theoretical foundations and show how it can be used to benchmark the performance of quantum computers. We will also go through some practical demonstrations with Qiskit. Finally, we will briefly discuss how quantum computing might evolve in the next few years.

The contents of this chapter are the following:

- The race for quantum advantage
- Random circuit sampling
- The best is yet to come

After reading this chapter, you will know about random circuit sampling and its prominent role in current experiments for quantum advantage. You will also understand how random circuit sampling can be used to benchmark the performance of quantum computers and how to compute an important metric, called the cross-entropy benchmark fidelity, to estimate that performance. You will also acquire some tools to think critically and correctly assess the evolution of quantum computing in the next few years.

Let's get to it!

15.1 The race for quantum advantage

Throughout this book, we have studied different protocols and algorithms with which quantum computers offer some kind of advantage over classical computers. For instance, in *Chapter 11*, we discussed Shor's algorithm, which shows a (quasi) exponential speed-up over known classical algorithms for factoring integers. Similarly, in *Chapter 12*, we showed that any classical algorithm uses quadratically more queries than Grover's algorithm when searching on an unstructured database.

With these tools, it would be natural to think that demonstrating in practice that quantum computers can offer an advantage should be straightforward. You may think, for instance, of taking a big integer that you cannot feasibly factor with a classical computer and use Shor's algorithm to find its prime decomposition. Since you can always use a classical computer to multiply back the factors and recover the original number, it should be easy to check the results and show that quantum computers surpass classical ones, shouldn't it? Well, not so fast. This approach works neatly on paper but, unfortunately, there are practical limitations that prevent us from carrying it out in practice with current quantum hardware.

As we discussed in detail in *Chapter 14*, the kind of quantum computers that are available today suffer from errors and noise and are limited in size. Procedures such as Shor's factoring algorithm require quantum error correction schemes that, despite recent advances in the field, are still beyond what is possible with current quantum hardware. Were this not enough, the execution of the algorithm would require a number of qubits that is way higher than the size of the biggest quantum computer existing today.

These limitations have not deterred researchers from all over the world from trying to demonstrate that, in practice, quantum computers can solve some tasks way faster than classical ones. In fact, as you may know, in 2019 a team from Google announced that they had achieved something called **quantum supremacy** [18]. This is such an important feat that it merits a detailed discussion.

So what is quantum supremacy[1]? This expression was coined by John Preskill to describe the moment in which a quantum computer carries out a task in much less time than it would take to do it even with the most powerful classical supercomputer on Earth. It is important to remark that the concrete computational task need not be of practical use. It is only required that it can be solved much faster with the help of quantum computers than with classical devices alone. In the case of the Google team experiments, the particular task

[1]The term "quantum supremacy" has been shunned by part of the quantum computing community because it might evoke the use of "supremacy" by racist groups. We will use it only in this context and for historical reasons. Quantum advantage seems to be the preferred term these days, although some researchers think that it does not completely capture the meaning of quantum supremacy.

was that of **random circuit sampling**, a problem that we will discuss in great detail in the next section, but which has not awakened much interest outside of quantum supremacy experiments (although some possible applications in randomness generation have been proposed by some authors [90]).

Another thing that needs to be taken into account about this kind of experiment is that quantum supremacy is a moving target. Whether quantum computers achieve a significant advantage over classical ones depends, of course, on the capabilities of quantum hardware, but also on the availability of faster classical supercomputers and better classical algorithms. What, today, is beyond the reach of classical computers might become feasible if more efficient algorithms are discovered or if new and better classical hardware is introduced. In fact, the mere idea of quantum supremacy has created some kind of an "arms race" between classical and quantum methods, with new developments on both fronts that sometimes make previous achievements obsolete after a few months.

For instance, in Google's original supremacy paper [18], it was reported that a certain random circuit sampling problem was solved in about 200 seconds, while "the equivalent task for a state-of-the-art classical supercomputer would take approximately 10 000 years". This claim was then disputed by researchers from IBM [91], who suggested a different method that may reduce the classical computing time to a few days. Later, many developments on both the classical [92]–[96] and quantum fronts [97]–[99] have been published, making the race leadership change hands quite a few times. At the time of writing, Google has announced a new breakthrough: with their 105-qubit quantum chip named Willow [100], they've conducted a certain random circuit sampling task in under 5 minutes, but they estimate that it would take 10 000 000 000 000 000 000 000 000 years (much longer than the age of the universe!) to do something similar with the best classical supercomputer currently available. Impressive, right? Moreover, recent results [101] published by researchers from various Chinese institutions seem to keep the competition for the top quantum computer in the world more open than ever.

We hope this has whetted your appetite to learn more about random circuit sampling, because that is exactly what we are about to do in the next section.

15.2 Random circuit sampling

Random circuit sampling (or **RCS** for short) is a task with a very descriptive name, because it consists in sampling from the possible measurement outcomes of a random quantum circuit. But let's make things more explicit and concrete, to fully understand what RCS actually is about and why it can be used to benchmark the performance of quantum chips.

15.2.1 Defining random circuit sampling

In RCS, we are given a quantum circuit C. We know that if we apply C to a set of qubits, all of them in the $|0\rangle$ state, we will obtain a certain quantum state $|\psi_C\rangle$. Therefore, if we measure this state in the computational basis, we will obtain a binary string x with a probability that we will denote p_x. The RCS task consists in generating strings x according to the probability distribution given by p_x.

Notice that we are asked to generate binary strings according to p_x. We are *not* required to run the quantum circuit on an actual quantum computer. So, if we have a shortcut to sample from measurements of $|\psi_C\rangle$, we can use it to carry out the RCS task. This can be done, for instance, if the number n of qubits of C is small enough that you can simulate it classically. However, when n grows, we do not know of any efficient classical method that can address this problem and there is some compelling theoretical evidence [102] that suggests that it is simply impossible to do so.

But imagine that you receive a huge circuit C, you don't have access to a big enough quantum computer, and you want to pretend that you can sample from $|\psi_C\rangle$. Can you cheat? Is it possible, for instance, to just output a bit string x uniformly at random and fool everyone? After all, if C is chosen at random, x will also be random, right? Well, not completely so. There are some noticeable differences between the probability distribution induced on strings x by C and the uniform distribution. In fact, these difference are key in understanding RCS, so let's take a closer look at them.

To make things easier to analyze from a theoretical point of view, we will assume that instead of a random circuit C, we receive a random unitary U that acts on n qubits[2] which is to be applied to the n-qubit state $|0\rangle$ in order to obtain the state $|\psi_U\rangle = U|0\rangle$. This will induce a probability distribution p_U on n-bit strings x. Namely, $p_U(x)$ will be the probability of obtaining x when measuring $|\psi_U\rangle$ in the computational basis.

Now, following the very useful tutorial by M. Sohaib Alam and Will Zeng [104], we consider two different scenarios: we sample strings x_s by measuring $|\psi_U\rangle$ and we sample strings x_r of n bits uniformly at random. The former case corresponds to the situation in which we actually use a quantum computer while the latter is the cheating strategy described previously. It turns out that the probabilities $p_U(x_s)$ and the probabilities $p_U(x_r)$ are different enough that we can distinguish between them, and that will eventually lead us to define a useful benchmark metric for RCS. But before we turn to doing some proper mathematics, it may be helpful to see all this graphically. Let's write some code!

15.2.2 Probability distributions with random unitaries

In Qiskit, we can easily generate random unitaries of n qubits by using the `random_unitary` function. For instance, the following code generates a random unitary matrix of size 2 by 2 and prints it:

```
from qiskit.quantum_info import random_unitary


U = random_unitary(2, seed = 1234).to_matrix()
print(U)
```

Notice that we have set a value for the `seed` parameter to guarantee repeatability of the results. With these settings, the output will be as follows:

```
[[-0.65182701+0.35104045j -0.06872086+0.66870741j]
 [ 0.30111103-0.60101938j  0.3613751 +0.64615469j]]
```

[2]Technically, we say that U is taken uniformly at random according to the Haar measure (see the paper by Mezzadri [103] for more details).

If you wish, you can check that this is indeed a unitary matrix by multiplying it times its conjugate transpose.

Now that we know how to generate random unitary matrices, let's try to study the probability distributions that we were talking about earlier. We will do the following: we will generate 1000 random unitaries acting on 10 qubits and, for each of them, we will generate, uniformly at random, 100 strings of 10 bits and store their probabilities according to p_U. To do that, we will use the following code:

```
import numpy as np

np.random.seed(1234)
n = 10
n_unitaries = 1000 # Number of unitaries we receive
n_samples = 100 # Number of strings to sample

p_r = [] # Probabilities according to U of the sampled strings

for i in range(n_unitaries):

    U = random_unitary(2**n, seed = 1234 + i).to_matrix()
    probs = abs(U[:,0])**2

    for j in range(n_samples):
        r = np.random.randint(0,2**n)
        p_r.append(probs[r])
```

Note that the dimension of the matrix (the first parameter in the call to `random_unitary`) is 2^{10}, because the number of qubits is $n = 10$. We are using 1234 as the seed to generate uniform random strings with the NumPy library. However, since with `random_unitary` we need to specify the seed every time, we add i to 1234 to generate different matrices in each iteration. Then, we obtain the probabilities associated to $|\psi_U\rangle$ by taking the first column of `U` (with `U[:,0]`) and computing the square of its absolute value. This is exactly what we need, because if we apply U to the $|0\rangle$ state, then we are multiplying U times a vector whose components are all 0 but the first one, which is 1. Thus, the amplitudes of the resulting state $|\psi_U\rangle$ are the entries of the first column of U. Finally, we sample a bunch of strings x_r uniformly at random and store their probabilities according to U. Beware that running this code can take several minutes because of the number of unitaries we are generating (you can reduce it if you want it to run faster).

We can now plot the frequencies of the `p_r` probabilities with the following code:

```
%matplotlib inline
import matplotlib.pyplot as plt


plt.hist(p_r, bins = 100)
plt.show()
```

Figure 15.1 shows the resulting graphic. As you can see, this has a rather nice, regular shape and, in fact, it can be shown that this kind of plot approaches what is known as the Porter-Thomas distribution when the number of qubits grows (see the paper by Boixo et al. [105] for more details). Notice, in particular, that most strings concentrate around the smallest probabilities.

Now, we are going to do something similar, but sampling the binary strings according to the probabilities induced by U. To achieve that, we will use the following code:

```
p_s = [] # Probabilities according to U of the sampled strings


for i in range(n_unitaries):


    U = random_unitary(2**n, seed = 1234 + i).to_matrix()
    probs = abs(U[:,0])**2
```

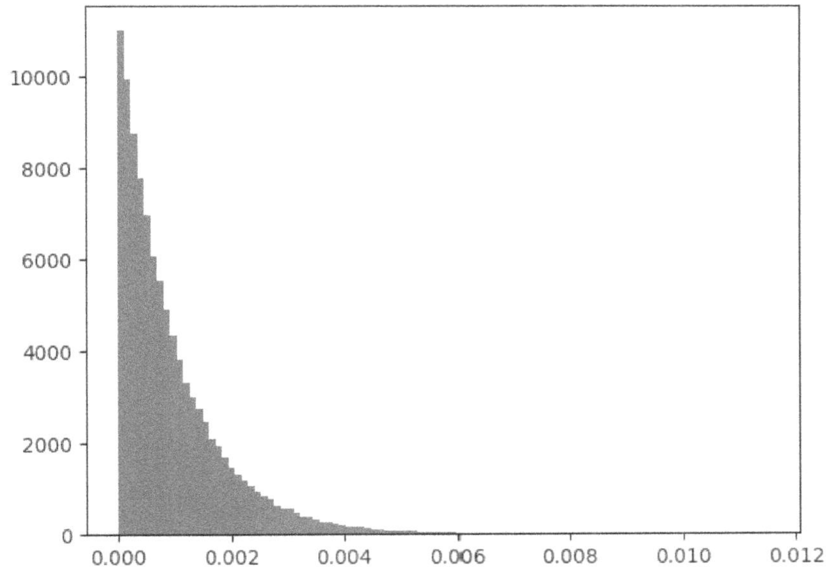

Figure 15.1: Histogram of the probabilities, according to a random unitary U, of binary strings sampled uniformly at random

```
    for j in range(n_samples):
        s = np.random.choice(2**n, p = probs) # Sample s according to probs
        p_s.append(probs[s])

plt.hist(p_s, bins = 100)
plt.show()
```

We have used the `random.choice` function to sample numbers between 0 and $2^n - 1$ according to the probabilities induced by U. The graphic created by executing this code is shown in *Figure 15.2*. Notice the difference in shape compared to *Figure 15.1*. In particular, here the strings tend to have slightly higher probabilities. This makes sense, because now we are sampling x_s according to p_U and then we are considering the probabilities $p_U(x_s)$, so it's natural that strings x_s with higher probabilities will be selected. We will use this to our advantage next, when we define the cross-entropy benchmark fidelity.

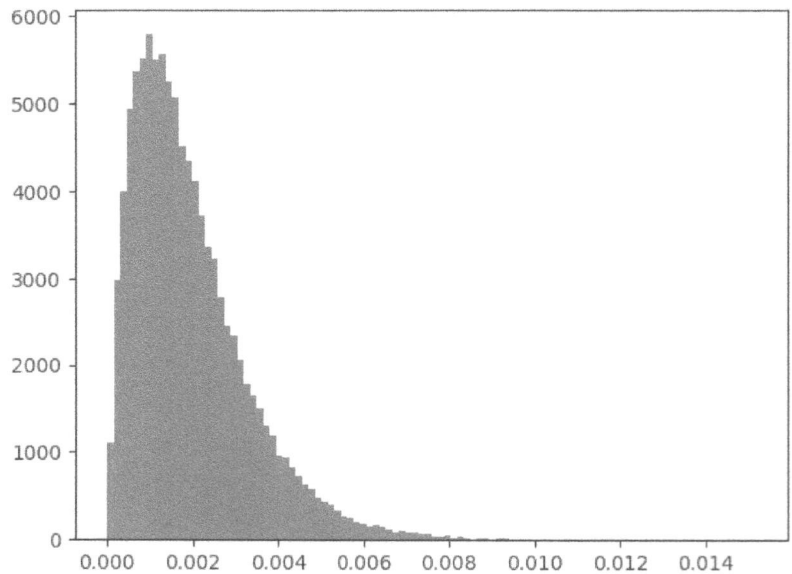

Figure 15.2: Histogram of the probabilities of binary strings sampled according to p_U for random unitaries U

15.2.3 The cross-entropy benchmark fidelity

As we just showed, when given random unitary operators U, there is an apparent difference in the shape of the probability histograms that we obtain when sampling strings uniformly at random and according to p_U. But can we quantify this difference? The answer is that not only can we do it, but it can be achieved in quite an intuitive and elegant way. Let's see how.

When comparing two different distributions, the first thing that comes to mind is to compute their averages. In our case, we can do it with the following code:

```
a_r = round(np.mean(p_r),6)
a_s = round(np.mean(p_s),6)
print("Average probability of uniformly sampled strings:", a_r)
print("Average probability of strings sampled according to U:", a_s)
```

When we run it, we obtain the following output:

```
Average probability of uniformly sampled strings: 0.000976
Average probability of strings sampled according to U: 0.001948
```

As you can see, the average probability a_s (sampling according to U) is about double that of the average probability a_r (sampling uniformly at random). What is more, this latter average is very, very close to $1/2^n$, which is approximately 0.000977. This is no coincidence. It can be shown (see [104] for more details) that a_s is $2/(2^n+1)$, while a_r is approximately $1/2^n$. Obviously, when n grows, a_s is closer and closer to being twice as big as a_r.

This leads to the definition of the **cross-entropy benchmark fidelity**, usually denoted as F_{XEB}. Imagine that we are given a random unitary U of dimension 2^n and we sample some binary strings $x_1, x_2, \ldots, x_k$. This sampling can be done from measuring $|\psi_U\rangle = U|0\rangle$ or with a different method (we don't care). Then, the F_{XEB} of the samples is defined by

$$F_{\text{XEB}}(x_1, \ldots, x_k) := \left(\frac{2^n}{k} \sum_{j=1}^{k} p_U(x_j) \right) - 1,$$

where $p_U(x_j)$ is the probability of obtaining x_j when measuring $|\psi_U\rangle$, as before. Notice that this is just the average of the probabilities of the samples (exactly what we computed earlier) times 2^n and minus 1.

We can compute the F_{XEB} of our samples with the following piece of code:

```
fxeb_r = round(2**n*np.mean(p_r)-1,3)
fxeb_s = round(2**n*np.mean(p_s)-1,3)
print("FXEB of uniformly sampled strings:', fxeb_r)
print("FXEB probability of strings sampled according to U:", fxeb_s)
```

When run, this code gives the following output:

```
FXEB of uniformly sampled strings: -0.001
FXEB probability of strings sampled according to U: 0.995
```

In general, if you sample your strings according to p_U, you will obtain values of F_{XEB} close to 1, while uniform random sampling will achieve values close to 0. What is more, there are theoretical arguments that suggest that obtaining a value of F_{XEB} that is bigger than 0 is computationally hard for classical computers (see the papers by Aaronson and his collaborators [106], [107] for more details). This is the basis for using F_{XEB} as a way of assessing the performance of quantum computers, a process usually called (not very surprisingly) **cross-entropy benchmarking** and abbreviated **XEB**. This is exactly the topic that we will address next.

15.2.4 XEB and random circuit sampling

Obviously, in random circuit sampling, we do not receive random unitary operators but random quantum circuits. However, if these random circuits are constructed in an adequate way, our analysis for random unitaries also applies to random circuits. A typical approach is to create the circuits from a number of consecutive layers of one- and two-qubit gates (see Google's supremacy paper [18] for an example of this).

In Qiskit, the `random_circuit` function allows us to create random circuits in a fashion that is somewhat like that of the RCS experiments. With it, we can create our version of random circuit sampling and study its behavior under different levels of noise. We will start with the case of perfect simulation (no noise), to have a baseline. We will create 1000 random circuits of 10 qubits, from which we will sample 100 strings and compute the resulting F_{XEB}. In order to converge to random unitaries, we will be using 1000 layers. Our code is as follows:

```
from qiskit_aer.primitives import SamplerV2 as Sampler
from qiskit.transpiler.preset_passmanagers import generate_preset_pass_manager
from qiskit_aer import AerSimulator
from qiskit.circuit.random import random_circuit
from qiskit.quantum_info import Statevector

backend = AerSimulator()
```

```
sampler = Sampler(seed = 1234)

p_c = [] # Probabilities according to U of the sampled strings

n_circuits = 1000
n_samples = 100
n = 10
n_layers = 1000

for i in range(n_circuits):

    circ = random_circuit(num_qubits = n, depth = n_layers, seed = 1234 + i,
                          max_operands = 2, measure = True)
    pm = generate_preset_pass_manager(backend=backend, optimization_level=1)
    t_circ = pm.run(circ)
    job = sampler.run([t_circ], shots = n_samples)
    results = job.result()
    d = results[0].data.c
    v_s = [int(s,2) for s in d.get_bitstrings()]
    circ.remove_final_measurements()
    sv = Statevector(circ)
    for s in v_s:
      p_c.append(abs(sv[int(s)])**2)

print("FXEB", round((2**n)*np.mean(p_c)-1,3))
```

Most of the code is self-explanatory, but there are some things that we should highlight here. The `max_operands = 2` parameter in the call to `random_circuit` indicates that we want quantum gates that act on at most 2 qubits. The circuits created by this function may include gates that are not supported by `AerSimulator`, so we need to transpile them with `generate_preset_pass_manager`. Finally, we retrieve all the binary strings measured when running the circuits, we convert them to integers with `int(s,2)`, and compute their actual probabilities using `Statevector`. Notice that, in order to do this, we first need to remove the final measurements from the circuits. With this information, we can then compute the F_{XEB} of the samples. If you run this code (be warned: it can be quite slow), you will obtain an F_{XEB} of 1.04, which, as we already know from our previous analysis, is around what we expect from a random circuit sampling experiment. Notice that this value is higher than 1, which is perfectly consistent with the mathematical definition of F_{XEB} (this is not a probability!).

In actual quantum computers, the execution of circuits is affected by noise and, hence, deviates from the results of perfect simulations. Let's try to replicate that with Qiskit, to study how errors can affect the F_{XEB} that we obtain. In order to achieve that, we will introduce a very simple type of noise: we will consider only errors in measurements, with a certain probability of obtaining 1 instead of 0 and vice versa. The code that we will use is the following:

```
from qiskit_aer.noise import NoiseModel, pauli_error

noise = NoiseModel()
p_error = 0.02
error_meas = pauli_error([("X", p_error), ("I", 1 - p_error)])
noise.add_all_qubit_quantum_error(error_meas , "measure")

noise_options = {"backend_options": {"noise_model": noise}}

sampler = Sampler(seed = 1234, options = noise_options)

p_c = [] # Probabilities according to U of the sampled strings
```

Qiskit Aer includes quite a number of options to run noisy simulations. Here, we are focusing only on the use of `NoiseModel` and a type of error modeled by `pauli_error` (for additional options, you can check the Qiskit Aer documentation at `https://docs.quantum.ibm.com/guides/simulate-with-qiskit-aer`). In this case, we are just adding, for each qubit, a probability of 0.02 of applying an X gate at the moment of measuring (thus changing 1 to 0 and 0 to 1). Then, we can just use the main `for` loop in our previous code block to run 1000 random circuits and compute the resulting F_{XEB}. This can be quite slow (over one hour of total execution time, depending on the specs of your computer), but if you run it, you will obtain an F_{XEB} of 0.808. If you then increase `p_error` to 0.05, you will reduce the F_{XEB} to 0.587. And if you use `p_error = 0.1`, your result will be 0.292. As you can see, the higher the noise, the closer F_{XEB} will be to 0.

In the case of real RCS experiments, of course, there are many other sources of errors and noise: imperfections in the implementation of quantum gates, external interactions, spontaneous decay of qubits from 1 to 0... All this adds up and, in fact, in the original supremacy experiments, the reported F_{XEB} was only 0.00224, which seems pretty low (to put it mildly). But remember that we have reasons to believe [106], [107] that getting values of F_{XEB} that are strictly positive is hard for classical computers when a large number of qubits is involved. And, of course, quantum hardware has improved quite a lot since 2019 and, for example, Quantinuum reported in June 2024 an F_{XEB} of about 0.35 with its 56-qubit trapped-ion quantum computer [98], [108]. That is much better, isn't it?

There is still one question that remains to be answered, though. If simulating these random circuits is hard for classical computers, how can we obtain the exact probabilities that we need in order to compute F_{XEB}? It seems to be an insurmountable problem, because the cases for which we can efficiently compute the cross-entropy benchmark fidelity with classical computers are, by definition, not in the supremacy regime. Nevertheless, certain tricks can be used to *estimate* F_{XEB}. For instance, in the original quantum supremacy problem, the authors explicitly state that "with certain circuit simplifications, we can obtain quantitative fidelity estimates of a fully operating processor running wide and deep quantum circuits". These simplifications include, for instance, removing "a slice of two-qubit gates (a small fraction of the total number of two-qubit gates), splitting the circuit into two spatially isolated, non-interacting patches of qubits". The resulting, smaller circuits *can* be simulated exactly and their F_{XEB} can be computed. Then, the F_{XEB} for the global circuit is computed as the product of the individual F_{XEB} values. Other "tricks" involve the use of modified sequences of gates, resulting in circuits that are much easier to simulate classically, but that still have a high number of layers. The F_{XEB} values for these circuits can be used to verify that the F_{XEB} estimations for the hard cases are accurate. Again, we encourage you to read the original paper [18] for more details.

That is all we wanted to say about random circuit sampling, so it's time to wrap up this chapter... and the whole book!

15.3 The best is yet to come

As we have seen in the previous section, a lot of effort is being put into demonstrating quantum advantage through random circuit sampling experiments, but, of course, that is not the only approach that is being considered. Another prominent technique is that of **boson sampling** [109], which is usually carried out in practice with photonic quantum computers. Some research groups have claimed having achieved quantum supremacy with this technique [110], [111] but, as in the case of RCS, this is a moving target because of advances in classical algorithms for this kind of problem [112], [113].

In order to improve the performance of quantum devices and achieve quantum advantage in these and other tasks, many companies have put forward roadmaps with their plans for developing new, improved quantum chips in the next few years (see [114] for a very complete report). Most efforts are focused on using quantum error correcting codes, along the lines of what we described in *Chapter 14*, to achieve hundreds or even thousands of logical qubits before the end of this decade. This might lead to the possibility of implementing, in a fault-tolerant way, some of the quantum procedures that we have described in this book, also paving the way for quantum advantage in real-world tasks.

At the same time, many researchers in the quantum computing community are exploring practical applications of noisy quantum devices like the ones that are available today.

For instance, quantum machine learning and quantum optimization have emerged in the last decade as promising areas of research both on the theoretical and applied fronts. To learn more about the kind of algorithms that are being used in these fields, including **quantum annealing**, the **quantum approximate optimization algorithm** (**QAOA**), the **variational quantum eigensolver** (**VQE**), **quantum support vector machines**, and **quantum neural networks**, we invite you to check out our book *A Practical Guide to Quantum Machine Learning and Quantum Optimization: Hands-on Approach to Modern Quantum Algorithms* [16]. We believe that it is a natural continuation of what you have already studied here and that it will give you a more complete perspective of the whole field of quantum computing.

In any case, two things seem very clear to us about the future of quantum computing: that it is very difficult to predict with accuracy what new developments we will witness in the next few years, and that they will certainly be exciting. We hope that you have enjoyed this quantum journey and that you will join us for all that lies ahead. As the famous song says… the best is yet to come!

Summary

In this final chapter, we've dived into the race towards quantum advantage and we've discussed why the task of random circuit sampling plays a prominent role in it. We've also discussed about the different probability distributions that arise when considering random unitaries and how they can be used to define the useful cross-entropy benchmark fidelity.

All this was illustrated with Qiskit code that can be used to compute F_{XEB} values for random unitaries and random quantum circuits. In the latter case, also with a simple simulation of the effect of noise.

Finally, we briefly discussed some of the future directions that quantum computing may take in the next few years and pointed at some other possibilities for achieving practical applications of quantum computing—for instance, in the fields of optimization and machine learning.

This brings our journey to its end. We started from very humble beginnings, by studying quantum systems with a single qubit, and we expanded that all the way to describing quantum supremacy experiments. Along the way, we learned about quantum money, quantum key distribution, and quantum teleportation, among many other protocols, and we studied important quantum procedures such as Shor's and Grover's algorithms.

We hope that you have enjoyed reading this book and that you have built a solid understanding of the foundations of quantum computing, learning how quantum computers can be programmed to achieve wonderful feats. Quantum computing is still in its infancy, yet we are sure it will bring us many more surprising and thrilling developments in the future. We hope that you stay around to experience and enjoy them!

Appendices

All those appendices that we've been constantly referencing over the main text have their home here. For those curious, we also share some notes on how this book was produced.

This part includes the following contents:

- *Appendix A*, Complex numbers and basic linear algebra
- *Appendix B*, The bra-ket notation and other foundational notions
- *Appendix C*, Measuring the complexity of algorithms
- *Appendix D*, Installing the tools
- *Appendix E*, Production notes

A

Mathematical Tools

A tensor is something that transforms like a tensor

— Anthony Zee

The goal of this appendix is to provide a short review of all the mathematical tools that we use in this book. In particular, we shall review some basic notions about complex numbers and discuss the foundational concepts of linear algebra. In addition to this, we will briefly touch upon some modular arithmetic.

If you would like to learn more about any of the topics that we discuss here, there are plenty of references throughout the text.

Complex numbers

In this section, we will briefly discuss some general properties of complex numbers. This should be enough to understand the material in the book, but, if you want a deeper dive, feel free to read *Complex analysis* by Bak and Newman [115].

The set of complex numbers is the set of all numbers of the form $a + bi$, where a and b are real numbers and $i^2 = -1$. This might not be the most formal way of presenting them, but it will do for our purposes!

The way you operate with complex numbers is pretty straightforward. Let a, b, x, and y be some real numbers. We add complex numbers as

$$(a + bi) + (x + yi) = (a + x) + (b + y)i.$$

Regarding multiplication, we have

$$(a + bi) \cdot (x + yi) = ax + ayi + bix + byi^2 = (ax - by) + (ay + bx)i.$$

In particular, when $b = 0$, we can deduce that

$$a(x + yi) = ax + (ay)i.$$

Given any complex number $z = a + bi$, its **real part**, which we denote as $\operatorname{Re} z$, is a, and its **imaginary part**, which we denote as $\operatorname{Im} z$, is b. Moreover, any such number z can be represented in the two-dimensional plane as a vector $(\operatorname{Re} z, \operatorname{Im} z) = (a, b)$. The length of the resulting vector is said to be the **module** of z, and it is computed as

$$|z| = \sqrt{a^2 + b^2}.$$

In addition, the counter-clockwise angle that the vector (a, b) makes with the positive X axis is said to be the **argument** of z, denoted as $\arg(z)$.

If $z = a + bi$ is a complex number, its **conjugate** is $z^* = a - bi$. In layman's terms, if you want to get the conjugate of any complex number, all you have to do is flip the sign of its imaginary part. It is easy to check that, given any complex number z,

$$|z|^2 = zz^* = z^*z,$$

which shows us, incidentally, that zz^* is always a non-negative real number.

One of the most well-known formulas involving the use of complex numbers is **Euler's identity**, which reads that, for any real number θ,

$$e^{i\theta} = \cos\theta + i\sin\theta.$$

This formula can be easily derived from the usual series that defines the exponential function. According to Euler's identity and using the usual properties of exponentiation, we must have, for any real numbers a and b,

$$e^{(a+ib)} = e^a e^{ib} = e^a(\cos\theta + i\sin\theta).$$

A complex number is said to be a **phase** if its module is one. Using Euler's identity, we can write any complex number z as a product of its module and a phase, as

$$z = |z| \cdot e^{i\arg(z)}.$$

Consequently, all phases can be written as a complex number of the form $e^{i\theta}$ for a real θ (where θ, to be precise, is the argument of the phase). Conversely, all complex numbers of the form $e^{i\theta}$ are phases if θ is a real number.

For any real number θ, the complex conjugate of $e^{i\theta}$ is

$$(e^{i\theta})^* = \cos\theta - i\sin\theta = \cos(-\theta) + i\sin(-\theta) = e^{-i\theta},$$

where we have used the fact that, for any real x, $\cos(x) = \cos(-x)$ and $\sin(-x) = -\sin(x)$.

Linear algebra

In this section, we will present a very broad overview of linear algebra. More than anything, this is meant to be a refresher. If you would like to learn linear algebra from the basics, we suggest reading Sheldon Axler's wonderful book [116]. If you are all-in with abstract algebra, we can also recommend the great book by Dummit and Foote [117]. With this out of the way, let's do some algebra!

When most people think of vectors, they think of fancy arrows pointing in a direction. But, where others see arrows, we mathematicians—in our tireless pursuit of abstraction—see elements of vector spaces. And what is a vector space? Simple!

Vector spaces

Let $\mathbb{F}$ be the real or the complex numbers. An $\mathbb{F}$-vector space is a set V together with an "addition" function (usually represented by $+$, for obvious reasons) and a "multiplication by scalars" function (denoted like usual multiplication). Addition needs to take any two vectors and return another vector, that is, $+$ needs to be a function $V \times V \longrightarrow V$. Multiplication by scalars, as the name suggests, must take a scalar (an element of $\mathbb{F}$) and a vector, and return a vector, that is, it needs to be a function $\mathbb{F} \times V \longrightarrow V$. Moreover, vector spaces must satisfy, for any arbitrary $\alpha_1, \alpha_2 \in \mathbb{F}$ and $v_1, v_2, v_3 \in V$, the following properties:

- Associativity for addition : $(v_1 + v_2) + v_3 = v_1 + (v_2 + v_3)$
- Commutativity for addition: $v_1 + v_2 = v_2 + v_1$
- Identity element for addition: there must exist a $0 \in V$ such that, for every vector $v \in V, v + 0 = v$
- Opposites for addition: there must exist a $-v_1 \in V$ such that $v_1 + (-v_1) = 0$
- Compatibility of multiplication by scalars with multiplication in $\mathbb{F}$: $(\alpha_1 \cdot \alpha_2) \cdot v_1 = \alpha_1 \cdot (\alpha_2 \cdot v_1)$
- Distributivity with respect to vector addition: $\alpha_1(v_1 + v_2) = \alpha_1 v_1 + \alpha_1 v_2$
- Distributivity with respect to scalar addition: $(\alpha_1 + \alpha_2)v_1 = \alpha_1 v_1 + \alpha_2 v_1$
- Identity for multiplication by scalars: $1 \cdot v_1 = v_1$

To learn more...

If you, like us, love abstraction, you should know that vector spaces are usually defined over an arbitrary **field**—not just over the real or complex numbers! If you want to learn more, we suggest reading the book by Dummit and Foote [117].

These are some examples of vector spaces:

- The set of real numbers with the usual addition and multiplication is a real vector space.
- The set of complex numbers with complex number addition and multiplication is a complex vector space. Moreover, it can be trivially transformed into a real vector space by restricting multiplication by scalars to multiplication of complex numbers by real numbers.
- The set $\mathbb{R}^n$ with the usual component-wise addition and multiplication by scalars (real numbers) is a vector space. If we fix $n = 2, 3$, that's where we can find those fancy arrows everyone is talking about!
- Most importantly for us, the set $\mathbb{C}^n$ with component-wise addition and scalar multiplication by complex numbers is a vector space.
- Just to give a cute example, the set of all smooth functions on a closed finite interval of the real numbers is a vector space. You can try to define addition and multiplication by scalars of functions yourself.

When we refer to a vector space on a set V with addition $+$ and multiplication by scalars $\cdot$, we should denote it as $(V, +, \cdot)$ in order to indicate what function we are considering as the addition function and what function we are taking to be the multiplication by scalars. Nevertheless, in all honesty, $(V, +, \cdot)$ is a pain to write, and we mathematicians—like all human beings—have a natural tendency toward laziness. So we usually just write V and let $+$ and $\cdot$ be inferred from context whenever that is reasonable to do.

Bases and coordinates

Some $\mathbb{F}$-vector spaces V are **finite-dimensional**: this means that there is a finite family of vectors $\{v_1, \dots, v_n\} \subseteq V$ such that, for any vector $v \in V$, there exist some unique scalars $\alpha_1, \dots, \alpha_n \in \mathbb{F}$ for which

$$v = \alpha_1 v_1 + \cdots + \alpha_n v_n.$$

The scalars $\alpha_1, \dots, \alpha_n$ are said to be the **coordinates** of v with respect to the basis $\{v_1, \dots, v_n\}$. Also, any v that can be written in the preceding form is said to be a **linear combination** of $v_1, \dots, v_n$. The natural number n is said to be the dimension of the vector space, and it is a fact of life that any two bases of a vector space need to have the same number of elements, so the dimension is well defined. If you want proof (which you should want!), check your favorite linear algebra textbook; either of the two that we have suggested should do the job.

Two examples of finite-dimensional vector spaces are $\mathbb{R}^n$ and $\mathbb{C}^n$ (with the natural addition and multiplication operations). For example, a basis of $\mathbb{C}^3$ or $\mathbb{R}^3$ would be

$$\{(1, 0, 0), (0, 1, 0), (0, 0, 1)\}.$$

To further illustrate this, if we considered the vector $(i, 3 + 2i, -2)$ in $\mathbb{C}^3$, we would have

$$(i, 3 + 2i, -2) = i \cdot (1, 0, 0) + (3 + 2i) \cdot (0, 1, 0) + (-2) \cdot (0, 0, 1),$$

and this representation in terms of these basis vectors is, clearly, unique. What is more, this basis is so natural and common that it has a name, the **canonical basis**, and its vectors are usually denoted as $\{e_1, e_2, e_3\}$. An analogous basis can be defined on $\mathbb{R}^n$ and $\mathbb{C}^n$ for any n.

Important note

We use the canonical basis extensively in this book, but with a different name and notation. We refer to it as the **computational basis**.

When you have a vector in a finite-dimensional vector space, sometimes it is handy to work with its coordinates with respect to some basis of your choice rather than working with its "raw" expression. In order to do this, we sometimes represent a vector v with coordinates $\alpha_1, \ldots, \alpha_n$ by a column matrix having the coordinates as entries. For example, in the previous example, the vector $(1, 3 + 2i, -2)$ would be represented by the column matrix of coordinates

$$\begin{pmatrix} 1 \\ 3 + 2i \\ -2 \end{pmatrix}$$

with respect to the canonical basis $\{e_1, e_2, e_3\}$.

Important note

It is very important to remember that the column matrix of coordinates of a vector is always defined with respect to a certain basis.

If we considered, for instance, the basis $\{e_1, e_3, e_2\}$, then the coordinates of the aforementioned vector would be

$$\begin{pmatrix} 1 \\ -2 \\ 3 + 2i \end{pmatrix}.$$

And, yes, order matters.

Linear maps and eigenstuff

Now that we know what vector spaces are, it is natural to wonder how we can define transformations $L : V \longrightarrow W$ between some $\mathbb{F}$-vector spaces V and W. In fairness, you could define any such transformation L however you wanted—we are not here to set boundaries on your mathematical freedom. But, if you want L to play nicely with the vector space structure of V and W, you will want it to be linear. That is, you will want to have, for any vectors $v_1, v_2 \in V$ and any scalar $\alpha \in \mathbb{F}$,

$$L(v_1 + v_2) = L(v_1) + L(v_2), \qquad L(\alpha \cdot v_1) = \alpha L(v_1).$$

Keep in mind that the addition and multiplication by scalars on the left-hand side of these expressions is that of V, while the operations on the right-hand side of the expressions are those of W.

Linear maps are wonderful. Not only do they have very nice properties, but they are also very easy to define. If $v_1, \ldots, v_n$ is a basis of V and you want to define a linear map $L : V \longrightarrow W$, all you have to do is give a value—any value—to $L(v_k)$ for every $k = 1, \ldots, n$. Then, by linearity, the function can be extended to all of V as

$$L(\alpha_1 v_1 + \cdots + \alpha_n v_n) = \alpha_1 L(v_1) + \cdots + \alpha_n L(v_n)$$

for any scalars $\alpha_1, \ldots, \alpha_n \in \mathbb{F}$. Furthermore, if we let $\{w_1, \ldots, w_m\}$ be a basis of W and we let $a_{k,l} \in \mathbb{F}$ be the unique scalars such that

$$L(v_k) = a_{1k} w_1 + \cdots + a_{nk} w_n,$$

then the coordinates of $L(v)$ for any $v = \alpha_1 v_1 + \cdots + \alpha_n v_n \in V$ with respect to $\{w_1, \ldots, w_m\}$ will be

$$\begin{pmatrix} a_{11} & \cdots & a_{1n} \\ \vdots & \ddots & \vdots \\ a_{n1} & \cdots & a_{nn} \end{pmatrix} \begin{pmatrix} \alpha_1 \\ \vdots \\ \alpha_n \end{pmatrix}.$$

To put it in perhaps more schematic terms,

$$\begin{pmatrix} | \\ L(v) \\ | \end{pmatrix} = \begin{pmatrix} a_{11} & \cdots & a_{1n} \\ \vdots & \ddots & \vdots \\ a_{n1} & \cdots & a_{nn} \end{pmatrix} \begin{pmatrix} | \\ v \\ | \end{pmatrix},$$

where the column matrices represent the coordinates of the vectors with respect to the bases $\{v_1, \dots, v_n\}$ and $\{w_1, \dots, w_m\}$. We say that the matrix $(a_{kl})_{kl}$ is the **coordinate matrix** of L with respect to these bases. If $V = W$ and we have a map $L : V \longrightarrow V$, we say that L is an **endomorphism** or **operator**[1] and, usually, we consider the same basis everywhere.

As an example, consider the endomorhpism on the Euclidean plane $\mathbb{R}^2$ that rotates all the vectors in the plane by a fixed angle α counter-clockwise (this operation is clearly linear). The vectors $(1, 0)$ and $(0, 1)$, if they suffer such a rotation, become $(\cos\alpha, \sin\alpha)$ and $(-\sin\alpha, \cos\alpha)$. Thus, the coordinate matrix of this rotation operation must be

$$\begin{pmatrix} \cos\alpha & -\sin\alpha \\ \sin\alpha & \cos\alpha \end{pmatrix}.$$

Any two linear operators A and B acting on the same spaces can be added, thus giving rise to the linear operator $A + B$ which brings any vector v to $A(v) + B(v)$; the coordinate matrix of $A + B$ is, clearly, the sum of the matrices A and B. This means that, the matrix entries of $A + B$ are $(A + B)_{jk} = A_{jk} + B_{jk}$:

$$\begin{pmatrix} A_{11} & \cdots & A_{1n} \\ \vdots & \ddots & \vdots \\ A_{n1} & \cdots & A_{nn} \end{pmatrix} + \begin{pmatrix} B_{11} & \cdots & B_{1n} \\ \vdots & \ddots & \vdots \\ B_{n1} & \cdots & B_{nn} \end{pmatrix} = \begin{pmatrix} A_{11} + B_{11} & \cdots & A_{1n} + B_{1n} \\ \vdots & \ddots & \vdots \\ A_{n1} + B_{n1} & \cdots & A_{nn} + B_{nn} \end{pmatrix}.$$

Any two linear functions $A : \mathbb{C}^r \longrightarrow \mathbb{C}^n$ and $B : \mathbb{C}^m \longrightarrow \mathbb{C}^r$ can be composed into the linear function $A \circ B : \mathbb{C}^m \longrightarrow \mathbb{C}^n$. The coordinate matrix of the composition $A \circ B$ is the matrix product AB, whose entries are given by

$$(AB)_{jk} = \sum_{t=1}^{r} A_{jt} B_{tk}.$$

Notice that if the dimensions of A and B are $n \times r$ and $r \times n$, those of AB are $n \times m$.

[1]This name is more ambiguous, as some authors use the term "linear operator" to refer to any linear map.

There is a very special kind of endomorphism that can be defined on any vector space: the **identity**. This is just a function I that takes any vector v to $I(v) = v$. If $L : V \longrightarrow V$ is an endomorphism, we say that a function L^{-1} is the **inverse** of L if both $L \circ L^{-1}$ and $L^{-1} \circ L$ are equal to the identity—actually, checking either of the two conditions is already sufficient when working with endomorphisms on finite-dimensional vector spaces. The coordinate matrix of the inverse of a map with coordinate matrix A is just the usual inverse matrix A^{-1}. What is more, a linear map is invertible if and only if so is its coordinated matrix.

When you have an endomorphism $L : V \longrightarrow V$, there may be some vectors $0 \neq v \in V$ for which there exists a scalar λ such that $L(v) = \lambda v$. These vectors are said to be **eigenvectors** and the corresponding value λ is said to be their **eigenvalue**. In some cases, you will be able to find a **basis of eigenvectors** $v_1, \ldots, v_n$ with some associated eigenvectors $\lambda_1, \ldots, \lambda_n$. With respect to this basis, the coordinate matrix of L would be a diagonal matrix

$$\begin{pmatrix} \lambda_1 & & \\ & \ddots & \\ & & \lambda_n \end{pmatrix}.$$

For all this, we say that an operator for which there exists a basis of eigenvectors is **diagonalizable**. Obviously, by the very definition of basis, if an operator is diagonalizable, then all the vectors in the space must be a linear combination of eigenvectors of the operator.

The eigenvalues of a linear operator A are all the values λ such that $\det(A - \lambda I) = 0$, where I is the identity matrix and det denotes the **determinant**. For any 2×2 coordinate matrix, the determinant can be computed as

$$\det\begin{pmatrix} a & b \\ c & d \end{pmatrix} = ad - bc.$$

This means that the eigenvectors of a matrix like the preceding one will be the values λ such that

$$\det\begin{pmatrix} a-\lambda & b \\ c & d-\lambda \end{pmatrix} = (a-\lambda)(d-\lambda) - bc = 0.$$

Inner products and adjoint operators

On an $\mathbb{F}$-vector space V, we may wish to define an **inner product** $\langle - | - \rangle$. This will be an operation taking any pair of vectors and returning a scalar, that is, a function $V \times V \longrightarrow \mathbb{F}$, satisfying the following properties for any $u, v_1, v_2 \in V$, and $\alpha_1, \alpha_2 \in \mathbb{F}$:

- **Conjugate symmetry**: $\langle v_1 | v_2 \rangle = \langle v_2 | v_1 \rangle^*$. Of course, if the vector space is defined over $\mathbb{R}$, then $\langle v_2 | v_1 \rangle^* = \langle v_2 | v_1 \rangle$, so $\langle v_1 | v_2 \rangle = \langle v_2 | v_1 \rangle$.
- **Linearity**: $\langle u | \alpha_1 v_1 + \alpha_2 v_2 \rangle = \alpha_1 \langle u | v_1 \rangle + \alpha_2 \langle u | v_2 \rangle$.
- **Positive-definiteness**: If $u \neq 0$, $\langle u | u \rangle$ is real and greater than 0.

It is easy to check that the following is an inner product on $\mathbb{C}^n$:

$$\langle (\alpha_1, \dots, \alpha_n) | (\beta_1, \dots, \beta_n) \rangle = \alpha_1^* \beta_1 + \cdots + \alpha_n^* \beta_n.$$

When we have a vector space with an inner product—which is commonly said to be an **inner product space**—two vectors v and w are said to be **orthogonal** if $\langle v | w \rangle = 0$. Moreover, a basis is said to be orthogonal if all its vectors are pairwise orthogonal.

With an inner product, we can define a **norm** on a vector space. We won't get into the details of what norms are but, very vaguely, we can think of them as a way of measuring the length of a vector (don't think about arrows, please, don't think about arrows…). The norm induced by a scalar product $\langle \cdot | \cdot \rangle$ is

$$\|v\| = \sqrt{\langle v | v \rangle}.$$

A vector is said to be a **unit vector** if its norm is one. In addition, we say that a basis is **orthonormal** if, in addition to being orthogonal, all its vectors are unit vectors.

When we are given a matrix $A = (a_{kl})$, we define its **conjugate transpose** to be $A^\dagger = (a^*_{lk})$, that is

$$\begin{pmatrix} a_{11} & \cdots & a_{1n} \\ \vdots & \ddots & \vdots \\ a_{n1} & \cdots & a_{nn} \end{pmatrix}^\dagger = \begin{pmatrix} a^*_{11} & \cdots & a^*_{n1} \\ \vdots & \ddots & \vdots \\ a^*_{1n} & \cdots & a^*_{nn} \end{pmatrix}.$$

The following identities can be easily checked for square matrices and, therefore, for linear maps:

$$(A+B)^\dagger = A^\dagger + B^\dagger, \qquad (AB)^\dagger = B^\dagger A^\dagger.$$

Here, AB denotes the usual matrix multiplication.

If $L : V \longrightarrow V$ is an endomorphism on a finite-dimensional vector space V, we can define its **Hermitian adjoint** as the only linear map $L^\dagger : V \longrightarrow V$ that has, as coordinate basis with respect to some basis, the conjugate transpose of the coordinated matrix of L with respect to that same basis. It can be shown that this notion is well defined, that is, that you always get the same linear map regardless of your choice of basis.

To learn more...

The definition that we have given is, well, not the most rigorous one. Usually, when you have a pair of inner product spaces V and W with inner products $\langle \cdot | \cdot \rangle_V$ and $\langle \cdot | \cdot \rangle_W$, the adjoint of a linear map $L : V \longrightarrow W$ is defined to be the only linear map $L^\dagger : W \longrightarrow V$ such that, for every $v \in V$ and $w \in W$,

$$\langle w | L(v) \rangle_W = \langle L^\dagger(w) \mid v \rangle_V.$$

We invite you to check that, for the particular case that we have considered ($V = W$ finite dimensional), both definitions agree.

We say that an endomorphism L is **self-adjoint** or **Hermitian** if $L = L^\dagger$. And it is a fact of life (again, we encourage you to check your favorite linear algebra textbook) that every Hermitian operator has an orthonormal basis of eigenvectors with real eigenvalues.

Also, we say that an endomorphism U is **unitary** if $U^\dagger U = UU^\dagger = I$, where I denotes the identity matrix. Among the many good properties of unitary operators, they preserve the inner products of vectors: this means that, for any two vectors v and w, if U is unitary, then

$$\langle v|w\rangle = \left\langle v\middle|U^{-1}Uw\right\rangle = \left\langle v\middle|U^\dagger Uw\right\rangle = \langle Uv|Uw\rangle .$$

This means that they also preserve the norms of vectors; therefore, all the eigenvectors of unitary operators must be phases. Indeed, if λ is an eigenvector of a unitary operator U then, for any eigenvector v,

$$\|v\| = \|Uv\| = \|\lambda v\| = \sqrt{\langle \lambda v|\lambda v\rangle} = \sqrt{\lambda\lambda^* \langle v|v\rangle} = |\lambda| \cdot \|v\|,$$

hence $|\lambda|$ must be equal to 1.

A crash course in modular arithmetic

If your watch says it's 15:00 and we ask you the time, you will say that it is three o'clock. But you would be lying, wouldn't you? Your watch says it's 15:00 but you've just said that it is *three* o'clock. What is wrong with you? Well, probably nothing. It turns out that, when you were telling us the time, you were subconsciously working in arithmetic modulo 12.

Vaguely speaking, when you work with numbers modulo n, all you are doing is assuming that n and 0 represent the same number. In this way, when you work in arithmetic modulo 4, for example,

$$0 \equiv 4 \equiv 8 \equiv 12 \equiv 16 \quad \text{mod } 4,$$

$$1 \equiv 5 \equiv 9 \equiv 13 \equiv 17 \quad \text{mod } 4,$$

$$2 \equiv 6 \equiv 10 \equiv 14 \equiv 18 \quad \text{mod } 4,$$

and so on and so forth. Notice how we have written $\equiv$ rather than $=$ to denote that those numbers are not, well, equal on their own, but just that they are equal modulo 4—that's also why we have that cute mod 4 on the right.

In this modular arithmetic setting, you can compute additions and multiplications as usual. For example, when working modulo 4,

$$2 \cdot 3 = 6 \equiv 2 \mod 4.$$

Ha! Look at what we have done! Now you can tell all your friends that 2 times 3 is 2 (you can then silently whisper "modulo 4" and still be technically correct).

We have just seen how, when you work with modular arithmetic, you can multiply numbers just as you can add them. And this begs the question, can you find a **multiplicative inverse** for any number in modular arithmetic? By this, we mean whether for any number x and any number n, there exists a number y such that $xy \equiv 1 \mod n$. For example, the multiplicative inverse of 3 modulo 5 would clearly be 2 as $3 \cdot 2 \equiv 6 \equiv 1 \mod 5$, but can these inverses be found for any number? Turns out that the answer is negative.

As a consequence of a result known as Bézout's identity (which we will neither state nor prove here), a necessary and sufficient condition for a number x to have a multiplicative inverse modulo n is for x and n to be coprime, i.e., for their greatest common divisor to be 1. Now if x has a multiplicative inverse y, then the multiplicative inverse of x^k will be y^k (modulo whatever). Thus, if (and only if) x and n are coprime, x^k will have a multiplicative inverse modulo n for any natural number k.

To learn more...

Can't get enough of modular arithmetic? Dummit and Foote have you covered! Have fun [117].

B

The Bra-Ket Notation and Other Foundational Notions

The bra-ket notation breaks the Law of Conservation of Sticks

— One of the authors of this book

The kets that are used to represent quantum states are simply a fancy way of representing vectors in a complex vector space. In particular, the state of an n-qubit system is just a unit vector in $\mathbb{C}^{2^n}$. Consequently, all the notions that we considered for arbitrary vectors also apply to states. Just as two vectors could be orthogonal, two states can be orthogonal if their scalar product is zero. Just as a bunch of vectors could form a basis, so can a bunch of states, and this basis can be orthonormal if all the vectors within it are pairwise orthogonal. There's not a lot of mystery to this.

In this appendix, we will explore the missing ingredient of the bra-ket notation: bras. And we will also briefly touch upon a very widely-used tool to represent one-qubit states: the Bloch sphere.

Ket's evil sibling: bras

Know those amazing kets we have been working with? We will now introduce a construction that is dual to them and that will allow us to easily refer to the amplitudes of any qubit state.

Consider the canonical basis $|0\rangle, \ldots, |N-1\rangle$ in $\mathbb{C}^N$ and a ket of the form $|\psi\rangle := \alpha_0|0\rangle + \cdots + \alpha_{N-1}|N-1\rangle$. Its associated **bra** is the construction

$$\langle\psi| := \alpha_0^* \langle 0| + \cdots + \alpha_{N-1}^* \langle N-1|,$$

where we use the superscript $*$ to denote complex conjugation (c.f. Appendix A). In terms of matrices, the bras $\langle 0|, \langle 1|, \ldots \langle N-1|$ denote the row matrices

$$\langle 0| = \begin{pmatrix} 1 & 0 & \cdots & 0 \end{pmatrix}, \qquad \langle 1| = \begin{pmatrix} 0 & 1 & 0 \cdots & 0 \end{pmatrix}, \qquad \ldots, \qquad \langle N-1| = \begin{pmatrix} 0 & \cdots & 0 & 1 \end{pmatrix}.$$

This means that, in general,

$$|\psi\rangle = \begin{pmatrix} \alpha_0 \\ \vdots \\ \alpha_{N-1} \end{pmatrix} \implies \langle\psi| = \begin{pmatrix} \alpha_0^* & \cdots & \alpha_{N-1}^* \end{pmatrix}.$$

In short, in order to transform a ket into a bra, matrix-wise, we just have to transpose the matrix and transform all the entries into their conjugates. This is denoted with a superscript $\dagger$; thus, $\langle\psi| = |\psi\rangle^\dagger$, and $|\psi\rangle = \langle\psi|^\dagger$.

Now we get to the real pun of the bra-ket notation. Imagine that you have two kets

$$|\psi\rangle = \begin{pmatrix} \psi_0 \\ \vdots \\ \psi_{N-1} \end{pmatrix}, \qquad |\varphi\rangle = \begin{pmatrix} \varphi_0 \\ \vdots \\ \varphi_{N-1} \end{pmatrix}.$$

What is their scalar product equal to? Well, it is not hard to notice that it will be equal to the product $\langle\psi|\,|\varphi\rangle$. Indeed,

$$\langle\psi|\,|\varphi\rangle = \begin{pmatrix}\psi_0^* & \vdots & \psi_{N-1}^*\end{pmatrix}\begin{pmatrix}\varphi_0 \\ \vdots \\ \varphi_{N-1}\end{pmatrix} = \psi_0^*\varphi_0 + \cdots + \psi_{N-1}^*\varphi_{N-1}.$$

Physicists have a great sense of aesthetics, so—ignoring the Law of Conservation of Sticks—they (and we) write $\langle\psi|\varphi\rangle$ instead of the more awkward looking $\langle\psi|\,|\varphi\rangle$. The construction $\langle\psi|\varphi\rangle$ is known as a **braket**, as it is a bra followed by a ket—and that's the pun of the bra-ket notation!

One of the beautiful things about scalar products is that they allow us to effortlessly refer to the amplitudes of a qubit state. Indeed, given any $|\psi\rangle = \alpha_0|0\rangle + \alpha_1|1\rangle$,

$$\langle 0|\psi\rangle = \alpha_0\langle 0|0\rangle + \alpha_1\cancelto{0}{\langle 0|1\rangle} = \alpha_0, \qquad \langle 1|\psi\rangle = \alpha_1,$$

where we have used the fact that all the states in the computational basis are orthogonal. This means that, given any state of the preceding form, the probabilities of getting an outcome 0 or 1 after a measurement can be computed as

$$P_0 = |\langle 0|\psi\rangle|^2, \qquad P_1 = |\langle 1|\psi\rangle|^2.$$

To learn more...

For those of you who, like us, are mathematical geeks, a few formal clarifications are in order. Kets are just a fancy way of denoting that a certain object is a vector in a complex Hilbert space, and their associated bras are just their Riesz representations in the dual space (see Theorem 16.1 in [118]). Thus, if $|v\rangle \in H$ is a vector in a Hilbert space,

$$\langle v| := \big(x \in H \mapsto \langle x, v\rangle\big) \in H^*,$$

where H^* is the topological dual of H and we use $\langle \cdot, \cdot \rangle$ to denote the inner product of H assuming it to be linear with respect to the second argument (for consistency with the rest of the book).

As a final remark on the computation of bras from kets and kets from bras, if $|\psi\rangle$ is a ket and A is a linear operator, the associated bra to $A|\psi\rangle$ would be $\langle\psi| A^\dagger$. Analogously, the associated ket to a bra $\langle\psi| A$ is $A^\dagger |\psi\rangle$. If this is not intuitively obvious to you, you can convince yourself by writing A and $|\psi\rangle$ in terms of matrices.

The Bloch sphere

The state of a quantum computer can be somewhat difficult to visualize, especially as it involves complex numbers. Nevertheless, at least for one-qubit systems, there is a simple and elegant way of visualizing their states, and it only involves a humble sphere!

Consider an arbitrary one-qubit state of the form $|\psi\rangle = a|0\rangle + b|1\rangle$ and let $r_1 = |a|$, $r_2 = |b|$, $\alpha_1 = \arg(a)$ and $\alpha_2 = \arg(b)$. Clearly, we will have

$$a = r_1 e^{i\alpha_1}, \qquad b = r_2 e^{i\alpha_2}.$$

We know that $r_1^2 + r_2^2 = |a|^2 + |b|^2 = 1$ and, since $0 \leq r_1, r_2 \leq 1$, there must exist an angle θ in $[0, \pi]$ such that $\cos(\theta/2) = r_1$ and $\sin(\theta/2) = r_2$. The reason for considering $\theta/2$ instead of θ in the cosine and sine will be apparent in a moment. Notice that, by now, we have

$$|\psi\rangle = \cos\frac{\theta}{2} e^{i\alpha_1} |0\rangle + \sin\frac{\theta}{2} e^{i\alpha_2} |1\rangle.$$

If we multiply $|\psi\rangle$ by the global phase $e^{-i\alpha_1}$, we can obtain the equivalent representation

$$|\psi\rangle = \cos\frac{\theta}{2} |0\rangle + \sin\frac{\theta}{2} e^{i\varphi} |1\rangle,$$

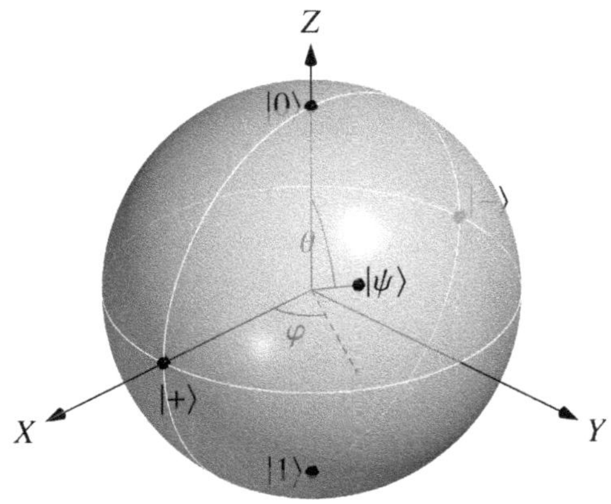

Figure B.1: An arbitrary one-qubit state in the Bloch sphere

where we have defined $\varphi = \alpha_2 - \alpha_1$. Remember that, as we discuss in *Chapter 2*, global phases are computationally negligible.

In this way, we can describe the state of any qubit with just two numbers $\theta \in [0, \pi]$ and $\varphi \in [0, 2\pi]$ that we can interpret as a polar angle and an azimuthal angle, respectively (that is, we are using what are known as **spherical coordinates**). This gives us a three-dimensional point

$$(\sin\theta\cos\varphi, \sin\theta\sin\varphi, \cos\theta)$$

that locates the state of the qubit on the surface of a sphere, called the **Bloch sphere** (see *Figure B.1*).

Notice that θ runs from 0 to π to cover the whole range from the top to the bottom of the sphere. This is why we used $\theta/2$ in the representation of our preceding qubit. We only needed to get up to $\pi/2$ for our angles in the sines and cosines!

In the Bloch sphere, $|0\rangle$ is mapped to the North pole and $|1\rangle$ to the South pole. In general, states that are orthogonal with respect to the inner product are antipodal on the sphere. For instance, $|+\rangle$ and $|-\rangle$ both lie on the equator, but on opposite points of the sphere. As we already know, the X gate takes $|0\rangle$ to $|1\rangle$ and $|1\rangle$ to $|0\rangle$, but leaves $|+\rangle$ and $|-\rangle$ unchanged, at least up to an irrelevant global phase. In fact, this means that the X gate acts like a rotation of π radians around the X axis of the Bloch sphere… so now you know why we use that name for the gate! In the same manner, Z and Y are rotations of π radians around the Z and Y axes, respectively.

We can generalize this behavior to obtain rotations of any angle around any axis of the Bloch sphere. For instance, for the X, Y, and Z axes we may define

$$R_X(\theta) = e^{-i\frac{\theta}{2}X} = \cos\frac{\theta}{2}I - i\sin\frac{\theta}{2}X = \begin{pmatrix} \cos\frac{\theta}{2} & -i\sin\frac{\theta}{2} \\ -i\sin\frac{\theta}{2} & \cos\frac{\theta}{2} \end{pmatrix},$$

$$R_Y(\theta) = e^{-i\frac{\theta}{2}Y} = \cos\frac{\theta}{2}I - i\sin\frac{\theta}{2}Y = \begin{pmatrix} \cos\frac{\theta}{2} & -\sin\frac{\theta}{2} \\ \sin\frac{\theta}{2} & \cos\frac{\theta}{2} \end{pmatrix},$$

$$R_Z(\theta) = e^{-i\frac{\theta}{2}Z} = \cos\frac{\theta}{2}I - i\sin\frac{\theta}{2}Z = \begin{pmatrix} e^{-i\frac{\theta}{2}} & 0 \\ 0 & e^{i\frac{\theta}{2}} \end{pmatrix} \equiv \begin{pmatrix} 1 & 0 \\ 0 & e^{i\theta} \end{pmatrix},$$

where we use the $\equiv$ symbol for equivalent action up to a global phase. Notice that $R_X(\pi) \equiv X$, $R_Y(\pi) \equiv Y$, $R_Z(\pi) \equiv Z$, $R_Z(\frac{\pi}{2}) \equiv S$, and $R_Z(\frac{\pi}{4}) \equiv T$.

In fact, it can be proved (see, for instance, the book by Nielsen and Chuang [13]) that for any one-qubit gate U, there exists a unit vector $r = (r_x, r_y, r_z)$ and an angle θ such that

$$U \equiv \cos\frac{\theta}{2}I - i\sin\frac{\theta}{2}(r_xX + r_yY + r_zZ).$$

For example, choosing $\theta = \pi$ and $r = (1/\sqrt{2}, 0, 1/\sqrt{2})$, we can obtain the Hadamard gate, for it holds that

$$H \equiv -i\frac{1}{\sqrt{2}}(X + Z).$$

Additionally, it can also be proved that, again for any one-qubit gate U, there exist three angles α, β, and γ such that

$$U \equiv R_Z(\alpha)R_Y(\beta)R_Z(\gamma).$$

In fact, you can obtain such a decomposition for any two rotation axes as long as they are not parallel, not just for Y and Z.

To learn more...

In this book, we only use rotation gates every now and then, but they are core in other areas of quantum computing such as quantum machine learning. If that sounds like something you could be interested in, we invite you to read our other book *A Practical Guide to Quantum Machine Learning and Quantum Optimization: Hands-on Approach to Modern Quantum Algorithms* [16].

C

Measuring the Complexity of Algorithms

The art of simplicity is a puzzle of complexity

— Douglas Horton

Computational complexity theory is the branch of theoretical computer science that is concerned with quantifying the resources needed to solve problems with algorithms. It asks questions such as "How much time is needed to multiply two integer numbers of n bits each?", "Do you need more memory space to solve a problem than to check its solution?", or "Is randomness useful in computational tasks?".

In this appendix, we will focus mainly on the mathematical techniques involved in estimating how much time is required to solve certain problems. For a thorough treatment of this and other topics (including space or memory complexity, the role of randomness in computation, approximation algorithms and other advanced matters), you can check standard computational complexity books such as the ones by Sipser [3], Papadimitriou [119], or Arora and Barak [120].

To get started, in order to measure how much time is used by a certain algorithm, we first need to assign a cost to its different instructions. This will depend on the computational model used to describe the algorithm, but usually, simple instructions are assumed to be executed in a unit of time. For instance, when it comes to finding a time metric for quantum circuits, it is common to consider the number of quantum gates (usually, from a fixed universal set of one- and two-qubit gates) needed to solve the problem as a function of the size of the input, all under the assumption that the gates in the set are simple enough to implement.

When analyzing the running time of an algorithm, we are usually more interested in measuring how it grows with the size of the input than we are in finding concrete running-time values for concrete problem instances. For example, we could be interested in knowing whether the time needed for a certain task grows so rapidly that it quickly becomes unfeasible to solve the problem when the input size becomes moderately big.

For this reason, we will define the running time of an algorithm as a function of the length of its inputs, and not as a function of the inputs themselves. Namely, the running time of an algorithm A is a function T that takes as input a non-negative integer n and returns the maximum number of steps (or instructions) that A performs with an input x of n bits before it stops. Notice that this is a worst-case definition of running-time: it is defined in terms of the string that needs the most time in order to be processed.

Important note

The running time of an algorithm A is a function T such that $T(n)$ is the maximum number of steps that A performs when given an input of length n.

In order to compare the running times of different algorithms, it is convenient to perform some additional simplifications. We usually do not care about whether the running time is exactly $T_1(n) = 4321n^2 + 784n + 142$ or, rather, $T_2(n) = n^3 + 3n^2 + 5n + 3$. In fact, we are more interested in whether $T(n)$ grows roughly like n^3 or like n^2, because this implies a qualitative difference: for values of n that are big enough, any polynomial of degree 3 grows more rapidly than any polynomial of degree 2. In the context of computational complexity theory, we would always prefer a $T(n)$ that grows as n^2 over one that grows as n^3, because its behavior for big inputs (its asymptotic growth, in other words) is better.

This intuitive idea is captured by the famous **big O notation**. Given two time functions, $T_1(n)$ and $T_2(n)$, we say that $T_1(n)$ is $O(T_2(n))$ (and we read it is as "$T_1(n)$ is big O of $T_2(n)$") if there exist an integer constant n_0 and a real constant $C > 0$ such that for all $n \geq n_0$, it holds that

$$T_1(n) \leq CT_2(n).$$

For example, you can check that $4321n^2 + 784n + 142$ is $O(n^3 + 3n^2 + 5n + 3)$.

The main idea behind this definition is that if $T_1(n)$ is $O(T_2(n))$, then the growth of T_1 is not worse than that of $T_2(n)$. For example, it is easy to prove that n^a is $O(n^b)$ whenever $a \leq b$ and that n^a is $O(2^n)$ for any a. On the other hand, when $a > b$, n^a is not $O(n^b)$ and 2^n is not $O(n^a)$. See Figure C.1 for an example with linear, quadratic, cubic, and exponential functions. Notice how the exponential function eventually dominates all the others, despite having 10^{-4} as its coefficient.

Important note

Given two non-negative functions $T_1(n)$ and $T_2(n)$, we say that $T_1(n)$ is $O(T_2(n))$ if there exist n_0 and $C > 0$ such that

$$T_1(n) \leq CT_2(n)$$

for every $n \geq n_0$.

Big O notation is extremely useful to estimate the behavior of running times without having to focus on small, cumbersome details. If the running time of an algorithm is $4321n^2 + 784n + 142$, we can just say that it is $O(n^2)$ and forget about the particular

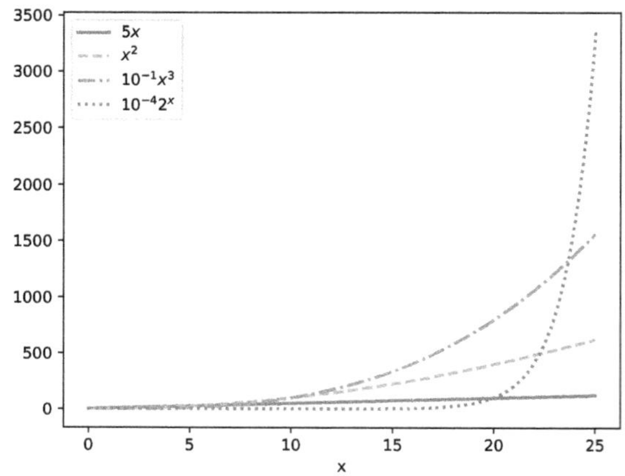

Figure C.1: Growth of linear, quadratic, cubic, and exponential functions

coefficients in the time function. This is also the reason why we can abstractly think about number of steps and not, for example, miliseconds. The particular amount of time that each step takes is a constant that will be "absorved" by the big O notation.

An example of this that is particularly important to us is that of logarithmic functions. It is a well-known fact that logarithms in different bases are related by the following formula:

$$\log_b x = \frac{\log_a x}{\log_a b}.$$

Hence, $\log_a$ and $\log_b$ only differ in a multiplicative constant and they are equivalent as far as the big O notation is concerned.

However, all these simplifications come at a cost. A running time such as $10^{100}n^2$ is certainly $O(n^2)$. But it is not preferable to n^3 unless $n > 10^{100}$, something that will never happen in practical situations, as 10^{100} is much, much bigger than the number of atoms in the visible universe. So use this notation wisely: with big O comes big responsibility.

D

Installing the Tools

We shape our tools and then our tools shape us.

— John M. Culkin

In this appendix, we will give you all the instructions needed to run the code examples provided in the main text. We will start by guiding you through the process of installing the software that we will use, and then we will learn how to access the real quantum computers on which we will run our code.

Getting Python

All the quantum programming libraries that we use in this book are based on Python, so you need to have a working Python distribution. If your operating system is macOS or a Linux distribution, you probably have one already. If your Python version is at least 3.10, then you are ready to go.

However, even if you already have Python installed on your system, we recommend that you consider following one of these two options:

- **Installing Anaconda**: Anaconda is a data science software distribution that includes, among other things, Python and many of its scientific libraries. In addition, it also includes Jupyter, an extremely useful web-based interactive computing platform that allows you to run code, write text and formulas, and visualize graphics, all organized into notebooks. For convenience, we provide all the code of the book in Jupyter notebooks that you can download from the book's Github reporisoty: `https://github.com/PacktPublishing/A-Practical-Guide-to-Quantum-Computing`.

  If you install Anaconda, you will have most of the non-quantum software libraries that we use in the book already on your system, plus some additional ones that you may find convenient for other related projects.

  There is a version of Anaconda called **Anaconda Distribution**, which is free to download from `https://www.anaconda.com/products/distribution`. It is available for Windows, Linux, and macOS. Anaconda Distribution provides a graphical installer, so it is super easy to set up. In case of doubt, you can always check the installation instructions at `https://docs.anaconda.com/anaconda/install/index.html`.

  Once you install Anaconda, we recommend that you launch it and run JupyterLab. This will open an IDE in your web browser that you can use to manage Jupyter notebooks and start running code right away. For a quick introduction to how to use JupyterLab, you can check the overview of its interface included in the JupyterLab documentation: `https://jupyterlab.readthedocs.io/en/stable/user/interface.html`.

- **Using Google Colab**: If you prefer not to install anything on your own computer, we also have an option for you. Google Colab is a web-based environment provided by Google in which you can run Jupyter notebooks with Python code. In fact, its interface its very similar to that of Jupyter and can be used to run all the code in this book (we know because we did it ourselves!) in addition to many other projects, especially those related to machine learning and data science.

The main difference between using Jupyter and Google Colab is that Colab does not run on your computer but is cloud-based: it uses hardware owned by Google. They provide you with a (usually modest) CPU, some amount of RAM, and some disk space, and you also have the chance to request a GPU to accelerate the training of your machine learning models.

The basic version of Google Colab is free to use: you only need a working Google account to start using it at `https://colab.research.google.com/`. And should you ever need more computational power, you can upgrade to a paid version (see more details at `https://colab.research.google.com/signup`).

By the way, the tutorials at `https://colab.research.google.com/` are really helpful, so you will be running your projects in almost no time.

Each of these options has its pros and cons. With Anaconda, you have perfect control over what you install, you get to use your own hardware (which probably is more powerful than the one available at Google Colab, maybe with the exception of those sweet GPUs), and you can work offline. But you need to install everything yourself, keep it up to date, and solve any version conflicts that may arise.

With Google Colab, you can start running code right away from any computer connected to the Internet, without the burden of having to install Python and many other libraries, and you can use quite powerful GPUs for free. However, you need to be online all the time, there are some restrictions on the number of projects that you can run simultaneously (at least, with the free version), and the CPU speed is not that great.

The good thing is that any of these possibilities (or any other that gets you a running Python distribution) works perfectly well for the purpose of running the code in this book. Moreover, they are perfectly compatible with each other, so you can start writing a notebook on Google Colab and complete it with Anaconda, or vice versa. Since both are free, you can try them both and use the one that better suits your needs at any given moment.

Of course, we don't want to be too prescriptive. If you don't feel like relying on Anaconda or on a cloud service, you can use your local machine without any add-ons and everything will work just fine as long as you have the right versions of the packages that we will use.

Installing the libraries

Although both Anaconda and Google Colab come with a lot of data science and visualization libraries already installed by default, they do not yet include any of the quantum computing libraries that we use in this book.

However, getting them set up and running is a breeze with **pip**, a package manager that comes bundled with Python—you don't need to install Anaconda or access Google Colab to use it. In order to install a new library with pip, you just need to run the following instruction on your terminal:

```
pip install name-of-library
```

If you are using a Jupyter notebook to run your code, you can use exactly that same instruction, but you need to write it in a cell of its own, with no additional code. If you need to install several different libraries and you do not want to create a different cell for each pip instruction, then you can put them all together in the same cell but you need to use the escape symbol `!`. So, for instance, you can install three libraries in the same cell of your Jupyter notebook like this:

```
!pip install first-library
!pip install second-library
!pip install last-library
```

Sometimes, you need to install a particular version of a library. This is the case with some of the examples in this book. Don't worry, because pip has your back in this too. You just need to run the following instruction:

```
pip install name-of-library==version-number
```

For example, to install version 2.1 of Qiskit, which is the one that we use in this book, you need to run the following instruction:

```
pip install qiskit==2.1
```

Of course, the same comments that we just made about escape symbols in Jupyter notebooks apply to this case.

Important note

If you run a `pip install` command on a Jupyter notebook to install a different version of a library that was already present on the system, you will probably need to restart the kernel (if you are running a Jupyter notebook on your local machine) or the runtime (in Google Colab) for the changes to take place.

In *Table D.1*, we have collected all the libraries needed for the code in this book in the order they appear in the main text, together with the version that we have used to create the examples. The second column specifies the name of each library in pip, so that is the one that you need to use with the `pip install` command.

Library name	Pip name	Version number
Qiskit	qiskit	2.1
Qiskit Aer	qiskit-aer-gpu	0.17
Qiskit IBM Runtime	qiskit-ibm-runtime	0.40
NumPy	numpy	2.2
Pylatexenc	pylatexenc	2.10
Matplotlib	matplotlib	3.10

Table D.1: Libraries used in the book and their version numbers

You may have noticed that there are a couple of libraries in the list that we never explicitly imported into our code. However, they are used by other packages to be able to plot circuits, so they need them to be present in your system.

Some of the libraries already come with Anaconda and Google Colab. In fact, it is very likely that the code in this book works with whatever version is included in those distributions,

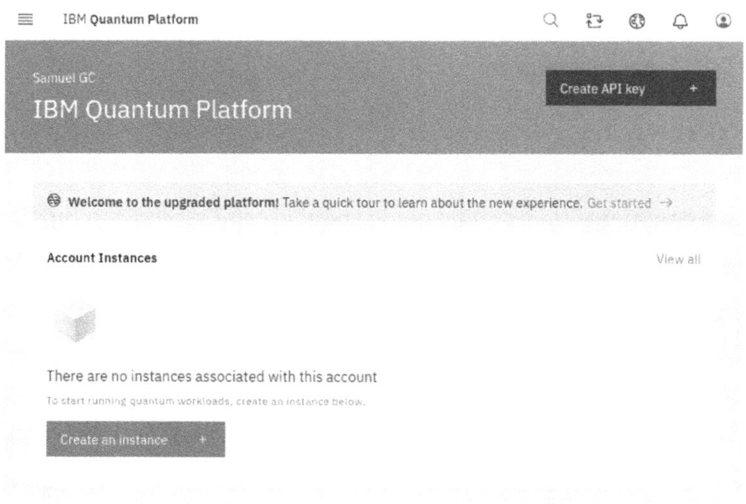

Figure D.1: The IBM Quantum Platform dashboard

so installing the exact version we mention in the table should not be especially important. For the libraries that do not come bundled with Anaconda and Google Colab, it is highly recommended to stick to the versions listed in the table. This is especially important for Qiskit and all its modules, which tend to change their APIs rather frequently.

In any case, for convenience, in the book notebooks that you can download from `https://github.com/PacktPublishing/A-Practical-Guide-to-Quantum-Computing`, we have explicitly included the installation commands of those libraries with the exact version that we have used to create the examples. If you're running the code on a local Python installation, you just need to install these libraries once, so you can remove the `pip install` commands after the first execution. However, if you're using Google Colab, you will need to run those commands every time you create a new runtime, because there is no persistence of data from one session to another.

Accessing IBM's quantum computers

In order to be able to run circuits on IBM's quantum computers from your Python programs, you first need to create an IBM account. This can be done through the IBM Quantum

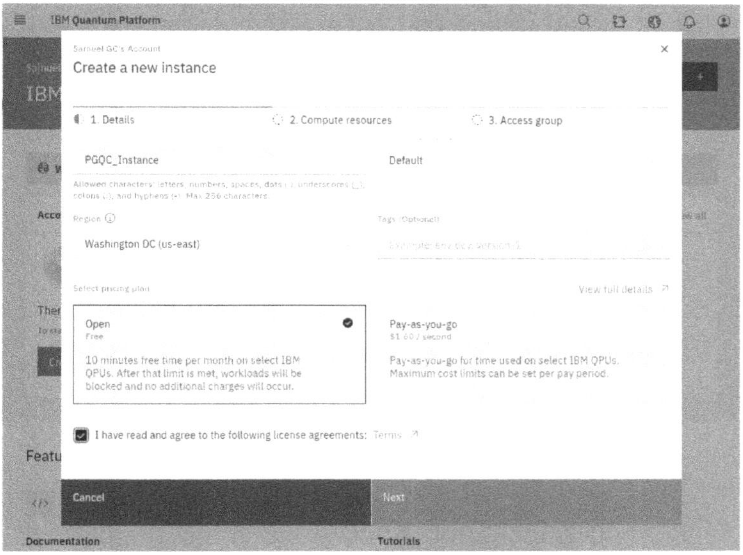

Figure D.2: The setup wizard for instances in the IBM Quantum Platform

Platform login page located at `https://quantum.cloud.ibm.com/`, and it is completely free of charge.

After signing up and loggin in, you will see the page shown in *Figure D.1*. From this page you can get your secret API key, which we use in *Chapter 4*, and you can create an **instance**. Instances are ways of grouping quantum hardware resources and managing their access permissions, but we don't need to get into those details.

When you click the button to create an instance, you will be shown a setup wizard like the one shown in *Figure D.2*—follow its steps and your instance will be set up in no time!

Important note

When creating an instance on the IBM Quantum platform, make sure to choose the Open plan instead of the Pay As You Go plan; otherwise, you might end up paying for additional quantum hardware resources. For the purposes of this book, the Open plan offers more than you need!

With this, you are all set to start coding, and you are ready to run quantum algorithms on real quantum hardware! The adventure continues in *Chapter 4*.

E

Production Notes

I wanna be in the room where it happens.

— Aaron Burr

This book was written in LaTeX by the two of us in two different countries (Spain and the USA) and in a wide variety of places: in offices at Harvard University and the University of Oviedo; in two apartments in Oviedo and an apartment in Cambridge (Massachusetts); on aeroplanes and trains; in airports and train stations; at a hotel in Berlin; on the streets of Madrid; in a sports pavilion; in the waiting rooms of emergency departments of two different hospitals; near the beach; near the mountains; on the backseat of a car; and perhaps in other locations that we can't remember now.

We, Elías and Samuel, have different working styles: one of us is a local-first advocate who won't give up using Vim, and the other is more comfortable on the cloud—you'd be surprised to know who is who. In order to collaborate efficiently, we used Overleaf in conjunction with GitHub, which enabled us to smoothly integrate our workflows.

To help us write formulas, draw circuits, and format code in LaTeX, we used quite a lot of useful packages, such as `quantikz`, `physics`, and `listings`. To write the code examples and run them, we used both Anaconda and Google Colab (as described in *Appendix D*).

All of these are excellent tools that made writing this book a much more pleasant and easy experience.

Bibliography

[1] A. M. Turing, "On computable numbers, with an application to the Entscheidungsproblem," *Proceedings of the London Mathematical Society*, vol. 2, no. 42, pp. 230–265, 1936.

[2] D. Harel, *Computers Ltd: What They Really Can't Do*. Oxford University Press, 2004.

[3] M. Sipser, *Introduction to the Theory of Computation*. Cengage Learning, 2012.

[4] E. Rich, *Automata, Computability and Complexity: Theory and Applications*. Pearson Prentice Hall Upper Saddle River, 2008.

[5] P. Høyer, J. Neerbek, and Y. Shi, "Quantum complexities of ordered searching, sorting, and element distinctness," *Algorithmica*, vol. 34, no. 4, pp. 429–448, 2002.

[6] C. Orzel, *How to Teach Quantum Physics to your Dog*. Simon and Schuster, 2009.

[7] P. Moriarty, *When the Uncertainty Principle Goes to 11: Or How to Explain Quantum Physics with Heavy Metal*. BenBella Books, 2018.

[8] L. Susskind and A. Friedman, *Quantum Mechanics: the Theoretical Minimum*. Basic Books, 2014.

[9] P. Kok, *A First Introduction to Quantum Physics*. Springer, 2018.

[10] D. J. Griffiths and D. F. Schroeter, *Introduction to Quantum Mechanics*. Cambridge University Press, 2018.

[11] K. Young, M. Scese, and A. Ebnenasir, "Simulating quantum computations on classical machines: A survey," *arXiv preprint arXiv:2311.16505*, 2023.

[12] X. Xu, S. Benjamin, J. Sun, X. Yuan, and P. Zhang, "A Herculean task: Classical simulation of quantum computers," *arXiv preprint arXiv:2302.08880*, 2023.

[13] M. A. Nielsen and I. L. Chuang, *Quantum Computation and Quantum Information: 10th Anniversary Edition.* Cambridge University Press, 2011.

[14] M. Howard, J. Wallman, V. Veitch, and J. Emerson, "Contextuality supplies the 'magic' for quantum computation," *Nature*, vol. 510, no. 7505, pp. 351–355, 2014.

[15] J. Preskill, "Quantum computing in the NISQ era and beyond," *Quantum*, vol. 2, p. 79, 2018.

[16] E. F. Combarro and S. González-Castillo, *A Practical Guide to Quantum Machine Learning and Quantum Optimization: Hands-on Approach to Modern Quantum Algorithms.* Packt Publishing Ltd, 2023.

[17] R. Eisberg and R. Resnick, *Quantum Physics of Atoms, Molecules, Solids, Nuclei, and Particles*, 2nd ed. Wiley, 1985.

[18] F. Arute, K. Arya, R. Babbush, *et al.*, "Quantum supremacy using a programmable superconducting processor," *Nature*, vol. 574, no. 7779, pp. 505–510, 2019.

[19] D. C. Marinescu and G. M. Marinescu, *Classical and Quantum Information.* Academic Press, 2011.

[20] P. A. M. Dirac, "A new notation for quantum mechanics," in *Mathematical proceedings of the Cambridge philosophical society*, Cambridge University Press, vol. 35, 1939, pp. 416–418.

[21] R. Schmied, "Quantum state tomography of a single qubit: Comparison of methods," *Journal of Modern Optics*, vol. 63, no. 18, pp. 1744–1758, 2016.

[22] G. Egan, *Quarantine.* Legend, 1992.

[23] J. Bricmont, *Making sense of quantum mechanics.* Springer, 2016, vol. 37.

[24] S. Teufel and D. Dürr, *Bohmian Mechanics.* Springer, 2009.

[25] M.-C. Chen, C. Wang, F.-M. Liu, *et al.*, "Ruling out real-valued standard formalism of quantum theory," *Phys. Rev. Lett.*, vol. 128, p. 040 403, 4 Jan. 2022.

[26] Z.-D. Li, Y.-L. Mao, M. Weilenmann, *et al.*, "Testing real quantum theory in an optical quantum network," *Phys. Rev. Lett.*, vol. 128, p. 040 402, 4 Jan. 2022.

[27] D. Aharonov, "A simple proof that Toffoli and Hadamard are quantum universal," *arXiv preprint arXiv:quant-ph/0301040*, 2003.

[28] S. Wiesner, "Conjugate coding," *ACM Sigact News*, vol. 15, no. 1, pp. 78–88, 1983.

[29] A. Molina, T. Vidick, and J. Watrous, "Optimal counterfeiting attacks and generalizations for Wiesner's quantum money," in *Conference on Quantum Computation, Communication, and Cryptography*, Springer, 2012, pp. 45–64.

[30] J. Katz and Y. Lindell, *Introduction to modern cryptography: principles and protocols. 3rd Edition.* Chapman and hall/CRC, 2020.

[31] S. M. Bellovin, "Frank Miller: Inventor of the one-time pad," *Cryptologia*, vol. 35, no. 3, pp. 203–222, 2011.

[32] C. E. Shannon, "Communication theory of secrecy systems," *The Bell system technical journal*, vol. 28, no. 4, pp. 656–715, 1949.

[33] S. N. Molotkov, "Quantum cryptography and VA Kotel'nikov's one-time key and sampling theorems," *Physics-Uspekhi*, vol. 49, no. 7, p. 750, 2006.

[34] C. H. Bennett and G. Brassard, "Quantum cryptography: Public key distribution and coin tossing," in *Proceedings of IEEE International Conference on Computers, Systems, and Signal Processing*, Bangalore, 1984, p. 175.

[35] C. H. Bennett, G. Brassard, and J.-M. Robert, "Privacy amplification by public discussion," *SIAM journal on Computing*, vol. 17, no. 2, pp. 210–229, 1988.

[36] P. W. Shor and J. Preskill, "Simple proof of security of the BB84 quantum key distribution protocol," *Physical review letters*, vol. 85, no. 2, p. 441, 2000.

[37] R. C. Merkle, "Secure communications over insecure channels," *Communications of the ACM*, vol. 21, no. 4, pp. 294–299, 1978.

[38] W. Diffie and M. E. Hellman, "New directions in cryptography," in *Democratizing Cryptography: The Work of Whitfield Diffie and Martin Hellman*, 2022, pp. 365–390.

[39] D. Bruß, "Optimal eavesdropping in quantum cryptography with six states," *Physical Review Letters*, vol. 81, no. 14, p. 3018, 1998.

[40] H. Bechmann-Pasquinucci and N. Gisin, "Incoherent and coherent eavesdropping in the six-state protocol of quantum cryptography," *Physical Review A*, vol. 59, no. 6, p. 4238, 1999.

[41] V. Scarani, A. Acin, G. Ribordy, and N. Gisin, "Quantum cryptography protocols robust against photon number splitting attacks for weak laser pulse implementations," *Physical review letters*, vol. 92, no. 5, p. 057 901, 2004.

[42] C. Branciard, N. Gisin, B. Kraus, and V. Scarani, "Security of two quantum cryptography protocols using the same four qubit states," *Physical Review A—Atomic, Molecular, and Optical Physics*, vol. 72, no. 3, p. 032 301, 2005.

[43] A. K. Ekert, "Quantum cryptography based on Bell's theorem," *Physical review letters*, vol. 67, no. 6, p. 661, 1991.

[44] A. C. Elitzur and L. Vaidman, "Quantum mechanical interaction-free measurements," *Foundations of physics*, vol. 23, pp. 987–997, 1993.

[45] V. Bergholm, J. Izaac, M. Schuld, *et al.*, "PennyLane: Automatic differentiation of hybrid quantum-classical computations," *arXiv preprint arXiv:1811.04968*, 2022.

[46] C. Gidney, *Quirk: A drag-and-drop quantum circuit simulator*, `https://algassert.com/quirk`, Accessed: 2025-06-01, 2016.

[47] T. C. Developers, *Cirq: A Python framework for creating, editing, and invoking noisy intermediate scale quantum (NISQ) circuits*, `https://github.com/quantumlib/Cirq`, Version 1.5.0, 2024.

[48] T. Jones, A. Brown, I. Bush, and S. C. Benjamin, "QuEST and high performance simulation of quantum computers," *Scientific reports*, vol. 9, no. 1, p. 10 736, 2019.

[49] A. Javadi-Abhari, M. Treinish, K. Krsulich, *et al.*, "Quantum computing with Qiskit," *arXiv preprint arXiv:2405.08810*, 2024.

[50] C. H. Bennett and S. J. Wiesner, "Communication via one-and two-particle operators on Einstein-Podolsky-Rosen states," *Physical review letters*, vol. 69, no. 20, p. 2881, 1992.

[51] J. F. Clauser, M. A. Horne, A. Shimony, and R. A. Holt, "Proposed experiment to test local hidden-variable theories," *Physical review letters*, vol. 23, no. 15, p. 880, 1969.

[52] B. S. Cirel'son, "Quantum generalizations of Bell's inequality," *Letters in Mathematical Physics*, vol. 4, pp. 93–100, 1980.

[53] B. Hensen, H. Bernien, A. E. Dréau, *et al.*, "Loophole-free Bell inequality violation using electron spins separated by 1.3 kilometres," *Nature*, vol. 526, no. 7575, pp. 682–686, 2015.

[54] A. Aspect, P. Grangier, and G. Roger, "Experimental realization of Einstein-Podolsky-Rosen-Bohm gedankenexperiment: A new violation of Bell's inequalities," *Physical review letters*, vol. 49, no. 2, p. 91, 1982.

[55] S. J. Freedman and J. F. Clauser, "Experimental test of local hidden-variable theories," *Physical Review Letters*, vol. 28, no. 14, p. 938, 1972.

[56] D. Bouwmeester, J.-W. Pan, K. Mattle, M. Eibl, H. Weinfurter, and A. Zeilinger, "Experimental quantum teleportation," *Nature*, vol. 390, no. 6660, pp. 575–579, 1997.

[57] D. Deutsch, "Quantum theory, the Church–Turing principle and the universal quantum computer," *Proceedings of the Royal Society of London. A. Mathematical and Physical Sciences*, vol. 400, no. 1818, pp. 97–117, 1985.

[58] D. Deutsch and R. Jozsa, "Rapid solution of problems by quantum computation," *Proceedings of the Royal Society of London. Series A: Mathematical and Physical Sciences*, vol. 439, no. 1907, pp. 553–558, 1992.

[59] C. H. Bennett, G. Brassard, C. Crépeau, R. Jozsa, A. Peres, and W. K. Wootters, "Teleporting an unknown quantum state via dual classical and Einstein-Podolsky-Rosen channels," *Physical review letters*, vol. 70, no. 13, p. 1895, 1993.

[60] J.-G. Ren, P. Xu, H.-L. Yong, *et al.*, "Ground-to-satellite quantum teleportation," *Nature*, vol. 549, no. 7670, pp. 70–73, 2017.

[61] E. Bernstein and U. Vazirani, "Quantum complexity theory," *SIAM Journal on Computing*, vol. 26, no. 5, pp. 1411–1473, 1997.

[62] T. H. Cormen, C. E. Leiserson, R. L. Rivest, and C. Stein, *Introduction to algorithms.* MIT press, 2022.

[63] R. L. Rivest, A. Shamir, and L. Adleman, "A method for obtaining digital signatures and public-key cryptosystems," *Communications of the ACM*, vol. 21, no. 2, pp. 120–126, 1978.

[64] J. P. Buhler, H. W. Lenstra, and C. Pomerance, "Factoring integers with the number field sieve," in *The development of the number field sieve*, Springer, 1993, pp. 50–94.

[65] D. Coppersmith, "Modifications to the number field sieve," *Journal of Cryptology*, vol. 6, pp. 169–180, 1993.

[66] C. Pomerance, "A tale of two sieves," *Notices of the American Mathematical Society*, vol. 43, no. 12, pp. 1473–1485, 1996.

[67] P. W. Shor, "Polynomial-time algorithms for prime factorization and discrete logarithms on a quantum computer," *SIAM Review*, vol. 41, no. 2, pp. 303–332, 1999.

[68] N. FIPS203, "Module-lattice-based key-encapsulation mechanism standard," *Federal Information Processing Standards Publication*, 2023.

[69] N. FIPS204, "Module-lattice-based digital signature standard," *Federal Information Processing Standards Publication*, 2024.

[70] M. Agrawal, N. Kayal, and N. Saxena, "PRIMES is in P," *Annals of mathematics*, pp. 781–793, 2004.

[71] M. O. Rabin, "Probabilistic algorithm for testing primality," *Journal of number theory*, vol. 12, no. 1, pp. 128–138, 1980.

[72] R. Van Meter and K. M. Itoh, "Fast quantum modular exponentiation," *Physical Review A—Atomic, Molecular, and Optical Physics*, vol. 71, no. 5, p. 052 320, 2005.

[73] S. A. Kutin, "Shor's algorithm on a nearest-neighbor machine," *arXiv preprint quant-ph/0609001*, 2006.

[74] I. L. Markov and M. Saeedi, "Constant-optimized quantum circuits for modular multiplication and exponentiation," *arXiv preprint arXiv:1202.6614*, 2012.

[75] C. Gidney, "Windowed quantum arithmetic," *arXiv preprint arXiv:1905.07682*, 2019.

[76] C. Gidney and M. Ekerå, "How to factor 2048 bit RSA integers in 8 hours using 20 million noisy qubits," *Quantum*, vol. 5, p. 433, 2021.

[77] C. Moore and S. Mertens, *The nature of computation.* Oxford University Press, 2011.

[78] G. Brassard, P. Hoyer, M. Mosca, and A. Tapp, "Quantum amplitude amplification and estimation," *Contemporary Mathematics*, vol. 305, pp. 53–74, 2002.

[79] C. H. Bennett, E. Bernstein, G. Brassard, and U. Vazirani, "Strengths and weaknesses of quantum computing," *SIAM Journal on Computing*, vol. 26, no. 5, pp. 1510–1523, 1997.

[80] M. Boyer, G. Brassard, P. Høyer, and A. Tapp, "Tight bounds on quantum searching," *Fortschritte der Physik*, vol. 46, no. 4-5, pp. 493–505, 1998.

[81] S. Ling and C. Xing, *Coding Theory: A First Course.* Cambridge University Press, 2004.

[82] C. Easttom, *Hardware for Quantum Computing.* Springer, 2024.

[83] P. W. Shor, "Scheme for reducing decoherence in quantum computer memory," *Phys. Rev. A*, vol. 52, R2493–R2496, 4 Oct. 1995.

[84] H. Bombín, "Topological codes," in *Quantum Error Correction*, D. A. Lidar and T. A. Brun, Eds. Cambridge University Press, 2013, pp. 455–481.

[85] N. Lacroix, A. Bourassa, F. J. H. Heras, *et al.*, "Scaling and logic in the color code on a superconducting quantum processor," *arXiv preprint arXiv:2412.14256*, 2024.

[86] D. Gottesman and I. L. Chuang, "Demonstrating the viability of universal quantum computation using teleportation and single-qubit operations," *Nature*, vol. 402, no. 6760, pp. 390–393, 1999.

[87] B. W. Reichardt, D. Aasen, R. Chao, *et al.*, "Demonstration of quantum computation and error correction with a tesseract code," *arXiv preprint arXiv:2409.04628*, 2024.

[88] W. van Dam, H. Liu, G. H. Low, *et al.*, "End-to-end quantum simulation of a chemical system," *arXiv preprint arXiv:2409.05835*, 2024.

[89] K. Liu, "Single-bit error," in *Thoughtcrime Experiments*, S. Harihareswara and L. Richardson, Eds., CreateSpace Independent Publishing Platform, 2009.

[90] S. Aaronson and S.-H. Hung, "Certified randomness from quantum supremacy," in *Proceedings of the 55th Annual ACM Symposium on Theory of Computing*, 2023, pp. 933–944.

[91] E. Pednault, J. Gunnels, D. M. Maslov, and J. Gambetta, *On "Quantum Supremacy"*, `https://www.ibm.com/blogs/research/2019/10/on-quantum-supremacy/`.

[92] C. Huang, F. Zhang, M. Newman, *et al.*, "Classical simulation of quantum supremacy circuits," *arXiv preprint arXiv:2005.06787*, 2020.

[93] F. Pan and P. Zhang, "Simulating the Sycamore quantum supremacy circuits," *arXiv preprint arXiv:2103.03074*, 2021.

[94] Y. Liu, X. Liu, F. Li, *et al.*, "Closing the "quantum supremacy" gap: Achieving real-time simulation of a random quantum circuit using a new Sunway supercomputer," in *Proceedings of the International Conference for High Performance Computing, Networking, Storage and Analysis*, 2021, pp. 1–12.

[95] R. Fu, Z. Su, H.-S. Zhong, *et al.*, "Achieving energetic superiority through system-level quantum circuit simulation," *arXiv preprint arXiv:2407.00769*, 2024.

[96] V. Kasirajan, T. Battelle, and B. Wold, "Empowering large scale quantum circuit development: Effective simulation of Sycamore circuits," *arXiv preprint arXiv:2411.12131*, 2024.

[97] Q. Zhu, S. Cao, F. Chen, *et al.*, "Quantum computational advantage via 60-qubit 24-cycle random circuit sampling," *Science bulletin*, vol. 67, no. 3, pp. 240–245, 2022.

[98] M. DeCross, R. Haghshenas, M. Liu, *et al.*, "The computational power of random quantum circuits in arbitrary geometries," *arXiv preprint arXiv:2406.02501*, 2024.

[99] A. Morvan, B. Villalonga, X. Mi, *et al.*, "Phase transitions in random circuit sampling," *Nature*, vol. 634, no. 8033, pp. 328–333, Oct. 2024.

[100] H. Neven, *Meet Willow, our state-of-the-art quantum chip*, `https://blog.google/technology/research/google-willow-quantum-chip/`.

[101] D. Gao, D. Fan, C. Zha, *et al.*, "Establishing a new benchmark in quantum computational advantage with 105-qubit Zuchongzhi 3.0 processor," *arXiv preprint arXiv:2412.11924*, 2024.

[102] A. Bouland, B. Fefferman, C. Nirkhe, and U. Vazirani, "On the complexity and verification of quantum random circuit sampling," *Nature Physics*, vol. 15, no. 2, pp. 159–163, 2019.

[103] F. Mezzadri, "How to generate random matrices from the classical compact groups," *arXiv preprint math-ph/0609050*, 2006.

[104] M. Sohaib Alam and W. Zeng, *Unpacking the quantum supremacy benchmark with Python*, https://medium.com/@sohaib.alam/unpacking-the-quantum-supremacy-benchmark-with-python-67a46709d.

[105] S. Boixo, S. V. Isakov, V. N. Smelyanskiy, *et al.*, "Characterizing quantum supremacy in near-term devices," *Nature Physics*, vol. 14, no. 6, pp. 595–600, 2018.

[106] S. Aaronson and L. Chen, "Complexity-theoretic foundations of quantum supremacy experiments," *arXiv preprint arXiv:1612.05903*, 2016.

[107] S. Aaronson and S. Gunn, "On the classical hardness of spoofing linear cross-entropy benchmarking," *arXiv preprint arXiv:1910.12085*, 2019.

[108] Quantinuum, *Quantinuum's H-Series hits 56 physical qubits that are all-to-all connected, and departs the era of classical simulation*, https://www.quantinuum.com/blog/quantinuums-h-series-hits-56-physical-qubits-that-are-all-to-all-connected-and-departs-the-era-of-classical-simulation.

[109] S. Aaronson and A. Arkhipov, "The computational complexity of linear optics," in *Proceedings of the forty-third annual ACM symposium on Theory of computing*, 2011, pp. 333–342.

[110] H.-S. Zhong, H. Wang, Y.-H. Deng, *et al.*, "Quantum computational advantage using photons," *Science*, vol. 370, no. 6523, pp. 1460–1463, 2020.

[111] L. S. Madsen, F. Laudenbach, M. F. Askarani, *et al.*, "Quantum computational advantage with a programmable photonic processor," *Nature*, vol. 606, no. 7912, pp. 75–81, 2022.

[112] G. Kalai and G. Kindler, "Gaussian noise sensitivity and boson sampling," *arXiv preprint arXiv:1409.3093*, 2014.

[113] J. Renema, V. Shchesnovich, and R. Garcia-Patron, "Classical simulability of noisy boson sampling," *arXiv preprint arXiv:1809.01953*, 2018.

[114] Quantum Computing Report Website, *2024: The year of quantum computing roadmaps*, https://quantumcomputingreport.com/2024-the-year-of-quantum-computing-roadmaps/.

[115] J. Bak and D. J. Newman, *Complex analysis*, 3, Ed. Springer, 2010.

[116] S. Axler, *Linear algebra done right.* Springer, 2015.

[117] D. S. Dummit and R. M. Foote, *Abstract algebra,* 3rd ed. John Wiley and Sons, 2004.

[118] C. Clason, *Introduction to Functional Analysis.* Birkhäuser, 2020.

[119] C. H. Papadimitriou, *Computational complexity.* Addison-Wesley, 1994.

[120] S. Arora and B. Barak, *Computational complexity: a modern approach.* Cambridge University Press, 2009.

Solutions

Chapter 1

(1.1) (a) If n is 6670 or bigger, it is better to use the quantum computer. If n is smaller than 6670, the classical computer will be faster.

(b) If n is 2 000 003 or bigger, it is better to use the quantum computer. If n is smaller than 2 000 003, the classical computer will be faster.

(c) If n is 30 000 000 003 or bigger, it is better to use the quantum computer. If n is smaller than 30 000 000 003, the classical computer will be faster.

Independently of how fast the classical computer is and how slow the quantum computer is, there will always be a value of n from which using the quantum computer will be preferable because $f_q(n) = 10n + 30$ grows asymptotically more slowly than $f_c(n) = n^2 + 1$.

(1.2) We need to prove that there is some n_0 such that, if n is bigger than n_0, then c^n is bigger than n^k. This is equivalent to $k \log n < n \log c$, which is also equivalent to $\frac{\log n}{n} < \frac{\log c}{k}$. But $\log c > 0$ because $c > 1$ and, thus, $\frac{\log c}{k} > 0$. Since, by L'Hôpital's rule, $\frac{\log n}{n}$ tends to 0 when n tends to infinity, there is always some value n_0 such that, if n is bigger than n_0, then $\frac{\log n}{n} < \frac{\log c}{k}$.

Chapter 2

(2.1) (a) It is not a qubit state because it is not normalized ($1^2 + 1^2 = 2 \neq 1$).

(b) It is a qubit state because $4/7 + 3/7 = 1$.

(c) It is not a qubit state because $2^2 \neq 1$.

(d) It is a qubit state as $\left|e^{-i}\right|^2 = 1$.

(e) We could take $x = \sqrt{1 - 1/9} = \sqrt{8}/3$.

(f) The values $x = (1/\sqrt{2})e^{i\theta}$ for any real θ.

(2.2) (a) The probability of obtaining 0 will be $1/2$, and so will the that of obtaining 1.

(b) The measurement will always return 0.

(c) The probability of obtaining 0 is $1/3$. That of obtaining 1 is $2/3$.

(d) The probability of obtaining 0 is p. That of obtaining 1 is $1 - p$.

(2.3) All the probabilities are equal to $1/2$ because $\left|1/\sqrt{2}\right|^2 = \left|-1/\sqrt{2}\right|^2 = \left|i/\sqrt{2}\right|^2 = \left|e^{i\theta}/\sqrt{2}\right|^2 = 1/2$.

(2.4) The conjugate transpose of U_1 is U_1 itself and a simple verification reveals that $U_1^\dagger U_1 = U_1 U_1^\dagger = U_1 U_1$ is equal to the identity matrix. Regarding the other matrices, we have

$$U_2^\dagger = \begin{pmatrix} 1 & 0 \\ 0 & i \end{pmatrix}, \qquad U_3^\dagger = \begin{pmatrix} 0 & -i \\ i & 0 \end{pmatrix}, \qquad U_4^\dagger = \begin{pmatrix} 1-2i & -i \\ 3 & 4 \end{pmatrix}.$$

The matrices U_2 and U_3 are unitary, but U_4 is not because

$$U_4 U_4^\dagger = \begin{pmatrix} 14 & 14-i \\ 14+i & 17 \end{pmatrix},$$

which is not equal to the identity matrix.

(2.5) We have

$$X\left|0\right\rangle = \begin{pmatrix} 0 & 1 \\ 1 & 0 \end{pmatrix}\begin{pmatrix} 1 \\ 0 \end{pmatrix} = \begin{pmatrix} 0 \\ 1 \end{pmatrix}, \qquad X\left|1\right\rangle \begin{pmatrix} 0 & 1 \\ 1 & 0 \end{pmatrix}\begin{pmatrix} 0 \\ 1 \end{pmatrix} = \begin{pmatrix} 1 \\ 0 \end{pmatrix}.$$

(2.6) We already proved Y to be unitary in a previous exercise. Regarding Z, clearly $Z = Z^{\dagger}$ and Z^2 is the identity, so Z is unitary. Moreover,

$$\begin{aligned} Y\left|0\right\rangle &= \begin{pmatrix} 0 & -i \\ i & 0 \end{pmatrix}\begin{pmatrix} 1 \\ 0 \end{pmatrix} = \begin{pmatrix} 0 \\ i \end{pmatrix} = i\left|1\right\rangle, \\ Y\left|1\right\rangle &= \begin{pmatrix} 0 & -i \\ i & 0 \end{pmatrix}\begin{pmatrix} 0 \\ 1 \end{pmatrix} = \begin{pmatrix} -i \\ 0 \end{pmatrix} = -i\left|0\right\rangle, \\ Z\left|0\right\rangle &= \begin{pmatrix} 1 & 0 \\ 0 & -1 \end{pmatrix}\begin{pmatrix} 1 \\ 0 \end{pmatrix} = \begin{pmatrix} 1 \\ 0 \end{pmatrix} = \left|0\right\rangle, \\ Z\left|1\right\rangle &= \begin{pmatrix} 1 & 0 \\ 0 & -1 \end{pmatrix} - \begin{pmatrix} 0 \\ 1 \end{pmatrix} = \begin{pmatrix} -i \\ 0 \end{pmatrix} = -\left|1\right\rangle. \end{aligned}$$

(2.7) We can easily compute that

$$M_x M_x^{\dagger} = \begin{pmatrix} 1 & \frac{x^*}{\sqrt{2}} + \frac{1}{2} \\ \frac{x}{\sqrt{2}} + \frac{1}{2} & xx^* + \frac{1}{2} \end{pmatrix}.$$

The only way for this to be equal to the identity is having $x = -1/\sqrt{2}$. And it can be readily verified that, with that value of x, M_x is unitary.

(2.8) $(UV)(UV)^{\dagger} = UVV^{\dagger}U^{\dagger} = UIU^{\dagger} = UU^{\dagger} = I$, and similarly for $(UV)^{\dagger}(UV)$.

(2.9) The coordinate matrix of HX (with respect to the computational basis) is

$$HX = \frac{1}{\sqrt{2}}\begin{pmatrix} 1 & 1 \\ -1 & 1 \end{pmatrix}.$$

A direct computation reveals that $HX\left|-\right\rangle = -\left|1\right\rangle$.

Chapter 3

(3.1) Indeed, $X\left|0\right\rangle = \left|1\right\rangle$ and $H\left|1\right\rangle = \left|-\right\rangle$.

For a different way of obtaining $|-\rangle$, notice that

$$H|0\rangle = |+\rangle = \frac{1}{\sqrt{2}}(|0\rangle + |1\rangle)$$

and, if we now apply Z, by linearity we obtain

$$Z\left(\frac{1}{\sqrt{2}}(|0\rangle + |1\rangle)\right) = \frac{1}{\sqrt{2}}(Z|0\rangle + Z|1\rangle) = \frac{1}{\sqrt{2}}(|0\rangle - |1\rangle) = |-\rangle,$$

since $Z|0\rangle = |0\rangle$ and $Z|1\rangle = -|1\rangle$. Thus, we can also obtain $|-\rangle$ by first applying H and then Z to a qubit in state $|0\rangle$.

(3.2) If the counterfeiter selects the $|-\rangle$ state, the situation is almost like the one with state $|+\rangle$ presented in the main text, but with cases 1) and 2) interchanged. The probability will then be, again, $1/2$.

For the case in which the selected state is $|0\rangle$ (respectively, $|1\rangle$), the analysis is similar. If the original state is $|0\rangle$ (resp. $|1\rangle$), the test will pass. If it is $|1\rangle$ (resp. $|0\rangle$), it will fail. If the original state is $|+\rangle$ or $|-\rangle$, it will fail with probability $1/2$. Adding everything up, the probability of passing the test is $1/2$.

(3.3) The probability of winning the lottery once is 10^{-6}. The probability of winning it on each of the ten consecutive draws is $(10^{-6})^{10} = 10^{-60}$ (that is 0.0....01 with 59 zeros between the decimal mark and the 1!). The probability of passing the 200 qubit tests is $(\frac{1}{2})^{200} \approx 6.22 \cdot 10^{-61}$, which is less than 10^{-60}.

(3.4) The encrypted message is $100110 \oplus 001101 = 101011$. The original message is $1110011 \oplus 0101101 = 1011110$.

(3.5) They will keep positions 3, 5 and 6 and 7. The final key will be 0110.

(3.6) The conjugate transpose of $R_y(\theta)$ is

$$\begin{pmatrix} \cos\frac{\theta}{2} & \sin\frac{\theta}{2} \\ -\sin\frac{\theta}{2} & \cos\frac{\theta}{2} \end{pmatrix}.$$

A simple computation shows that

$$\begin{pmatrix} \cos\frac{\theta}{2} & -\sin\frac{\theta}{2} \\ \sin\frac{\theta}{2} & \cos\frac{\theta}{2} \end{pmatrix}\begin{pmatrix} \cos\frac{\theta}{2} & \sin\frac{\theta}{2} \\ -\sin\frac{\theta}{2} & \cos\frac{\theta}{2} \end{pmatrix} = \begin{pmatrix} \cos\frac{\theta}{2} & \sin\frac{\theta}{2} \\ -\sin\frac{\theta}{2} & \cos\frac{\theta}{2} \end{pmatrix}\begin{pmatrix} \cos\frac{\theta}{2} & -\sin\frac{\theta}{2} \\ \sin\frac{\theta}{2} & \cos\frac{\theta}{2} \end{pmatrix}$$
$$= \begin{pmatrix} 1 & 0 \\ 0 & 1 \end{pmatrix}.$$

We can prove that $R_Y(\theta)^n = R_Y(n\theta)$ by induction on n. If $n = 1$, the result is trivial. Consider $n > 1$, and assume that $R_Y(\theta)^{n-1} = R_Y((n-1)\theta)$. Then,

$$R_Y(\theta)^n = R_Y(\theta)R_Y(\theta)^{n-1} = \begin{pmatrix} \cos\frac{\theta}{2} & -\sin\frac{\theta}{2} \\ \sin\frac{\theta}{2} & \cos\frac{\theta}{2} \end{pmatrix}\begin{pmatrix} \cos\frac{(n-1)\theta}{2} & -\sin\frac{(n-1)\theta}{2} \\ \sin\frac{(n-1)\theta}{2} & \cos\frac{(n-1)\theta}{2} \end{pmatrix}.$$

The result follows by computing the matrix product and applying the trigonometric identities $\cos\alpha\cos\beta - \sin\alpha\sin\beta = \cos(\alpha+\beta)$ and $\sin\alpha\cos\beta + \cos\alpha\sin\beta = \sin(\alpha+\beta)$.

Chapter 4

(4.1) This circuit can be constructed and represented as follows:

```
circuit_yt = QuantumCircuit(1)
circuit_yt.y(0)
circuit_yt.t(0)
circuit_yt.draw("mpl")
```

(4.2) This is the function that we were asked to implement:

```
def apply_gate(circ, gate):
    if gate == "X":
        circ.x(0)
    elif gate == "Y":
        circ.y(0)
```

```
        elif gate == "Z":
            circ.z(0)
```

(4.3) This piece of code can find the final state of the circuit:

```
state = Statevector(circuit_yt)
print(state)
```

(4.4) This is how the measurement sample can be obtained:

```
sampler = StatevectorSampler()
job = sampler.run([circuit], shots = 4)
print(job.result()[0].data.meas.get_bitstrings())
```

Chapter 5

(5.1) (a) It's valid, because $\left|\frac{1}{\sqrt{3}}\right|^2 + \left|\frac{1}{\sqrt{3}}\right|^2 + \left|\frac{1}{\sqrt{3}}\right|^2 = 1/3 + 1/3 + 1/3 = 1.$

(b) It's valid, because $\left|\frac{1}{2}\right|^2 + \left|\frac{1}{2}\right|^2 + \left|\frac{1}{2}\right|^2 + \left|-\frac{1}{2}\right|^2 = 1/4 + 1/4 + 1/4 + 1/4 = 1.$

(c) It's valid, because $\left|\frac{1}{\sqrt{2}}\right|^2 + \left|-\frac{1}{\sqrt{2}}\right|^2 = 1/2 + 1/2 = 1.$

(d) It's valid, because $\left|\frac{1}{\sqrt{2}}\right|^2 + \left|\frac{i}{\sqrt{2}}\right|^2 = 1/2 + 1/2 = 1.$

(e) It's not valid, because $\left|\frac{2}{\sqrt{3}}\right|^2 + \left|\frac{2}{\sqrt{3}}\right|^2 + \left|-\frac{1}{\sqrt{3}}\right|^2 = 4/3 + 4/3 + 1/3 = 9/3 \neq 1.$

(f) It's valid, because $|i|^2 = 1.$

(g) It's not valid, because $|2|^2 + |-i|^2 = 5 \neq 1.$

(h) It's valid, because $\left|\sqrt{\frac{2}{3}}\right|^2 + \left|-\sqrt{\frac{1}{3}}\right|^2 = 2/3 + 1/3 = 1.$

The values of x that would make $\frac{1}{2}\left|01\right\rangle + x\left|10\right\rangle$ a valid state are those of the form $e^{i\theta}\frac{\sqrt{3}}{2}$, where θ is any real number, because $\left|\frac{1}{2}\right|^2 + \left|e^{i\theta}\frac{\sqrt{3}}{2}\right|^2 = \left|\frac{1}{2}\right|^2 + \left|\frac{\sqrt{3}}{2}\right|^2 = 1/4 + 3/4 = 1.$

(5.2) It is easy to see that

$$|01\rangle = |0\rangle \otimes |1\rangle = \begin{pmatrix} 1 \\ 0 \end{pmatrix} \otimes \begin{pmatrix} 0 \\ 1 \end{pmatrix} = \begin{pmatrix} 1\begin{pmatrix} 0 \\ 1 \end{pmatrix} \\ 0\begin{pmatrix} 0 \\ 1 \end{pmatrix} \end{pmatrix} = \begin{pmatrix} 0 \\ 1 \\ 0 \\ 0 \end{pmatrix}.$$

Also, we have that

$$|10\rangle = |1\rangle \otimes |0\rangle = \begin{pmatrix} 0 \\ 1 \end{pmatrix} \otimes \begin{pmatrix} 1 \\ 0 \end{pmatrix} = \begin{pmatrix} 0\begin{pmatrix} 1 \\ 0 \end{pmatrix} \\ 1\begin{pmatrix} 1 \\ 0 \end{pmatrix} \end{pmatrix} = \begin{pmatrix} 0 \\ 0 \\ 1 \\ 0 \end{pmatrix}.$$

Finally, it holds that

$$|11\rangle = |1\rangle \otimes |1\rangle = \begin{pmatrix} 0 \\ 1 \end{pmatrix} \otimes \begin{pmatrix} 0 \\ 1 \end{pmatrix} = \begin{pmatrix} 0\begin{pmatrix} 0 \\ 1 \end{pmatrix} \\ 1\begin{pmatrix} 0 \\ 1 \end{pmatrix} \end{pmatrix} = \begin{pmatrix} 0 \\ 0 \\ 0 \\ 1 \end{pmatrix}.$$

(5.3) Denote

$$|\psi_1\rangle = \begin{pmatrix} a_1 \\ a_2 \end{pmatrix}, \qquad |\psi_2\rangle = \begin{pmatrix} b_1 \\ b_2 \end{pmatrix}, \qquad |\varphi\rangle = \begin{pmatrix} c_1 \\ c_2 \end{pmatrix}.$$

Then it holds that

$$\alpha_1 |\psi_1\rangle + \alpha_2 |\psi_2\rangle = \begin{pmatrix} \alpha_1 a_1 + \alpha_2 b_1 \\ \alpha_1 a_2 + \alpha_2 b_2 \end{pmatrix}.$$

From this, it follows that

$$(\alpha_1 |\psi_1\rangle + \alpha_2 |\psi_2\rangle) \otimes |\varphi\rangle = \begin{pmatrix} (\alpha_1 a_1 + \alpha_2 b_1)c_1 \\ (\alpha_1 a_1 + \alpha_2 b_1)c_2 \\ (\alpha_1 a_2 + \alpha_2 b_2)c_1 \\ (\alpha_1 a_2 + \alpha_2 b_2)c_2 \end{pmatrix}.$$

We also know that

$$\alpha_1 |\psi_1\rangle \otimes |\varphi\rangle = \begin{pmatrix} \alpha_1 a_1 c_1 \\ \alpha_1 a_1 c_2 \\ \alpha_1 a_2 c_1 \\ \alpha_1 a_2 c_2 \end{pmatrix},$$

and that

$$\alpha_2 |\psi_2\rangle \otimes |\varphi\rangle = \begin{pmatrix} \alpha_2 b_1 c_1 \\ \alpha_2 b_1 c_2 \\ \alpha_2 b_2 c_1 \\ \alpha_2 b_2 c_2 \end{pmatrix}.$$

Adding these two vectors together gives us the first identity. The second identity follows from a similar computation.

(5.4) The results are as follows:

(a) $\frac{1}{\sqrt{2}} |01\rangle - \frac{1}{\sqrt{2}} |11\rangle$

(b) $\frac{1}{2} |00\rangle + \frac{1}{2} |01\rangle + \frac{1}{2} |10\rangle + \frac{1}{2} |11\rangle$

(c) $\frac{1}{2} |00\rangle - \frac{1}{2} |01\rangle + \frac{1}{2} |10\rangle - \frac{1}{2} |11\rangle$

(5.5) (a) Each result has probability $1/4$.

(b) The possible results are 00, 01 and 10, each with probability $1/3$.

(c) The only possible result is 11, and it obviously has probability 1.

(d) We can obtain 01 or 11, each with probability $1/4$, or 10 with probability $1/2$.

(e) We first need to write the state in terms of the computational basis. Since $|+\rangle = \frac{1}{\sqrt{2}}(|0\rangle + |1\rangle)$, it holds that

$$|0\rangle |+\rangle = |0\rangle \frac{1}{\sqrt{2}}(|0\rangle + |1\rangle) = \frac{1}{\sqrt{2}}(|00\rangle + |01\rangle).$$

Hence, the possible results are 00 and 01, each with probability 1/2.

(5.6) (a) We will obtain 0 and 1 each with probability 1/2. The collapsed states will be, respectively, $\frac{1}{\sqrt{2}}(|00\rangle + |01\rangle)$ and $\frac{1}{\sqrt{2}}(|10\rangle - |11\rangle)$.

(b) We will obtain 0 with probability 2/3 and 1 with probability 1/3. The collapsed states will be, respectively, $\frac{1}{\sqrt{2}}(|00\rangle + |10\rangle)$ and $-|01\rangle$.

(c) The result will be 1 with probability 1. The state will still be $\frac{1+i}{\sqrt{2}}|11\rangle$.

(d) The result will be 0 with probability 1/2 and 1 with probability 1/2. The collapsed states will be, respectively, $-|10\rangle$ and $\frac{1}{\sqrt{2}}(|01\rangle + i|11\rangle)$.

(e) If we measure the first qubit, we will obtain 0 with certainty and the state will not change. If we measure the second qubit, we will obtain 0 with probability 1/2 and 1 with probability 1/2. The collapsed states will be, respectively, $|00\rangle$ and $|01\rangle$.

(5.7) Since $|0\rangle \otimes |0\rangle$, $|0\rangle \otimes |1\rangle$, $|1\rangle \otimes |0\rangle$, and $|1\rangle \otimes |1\rangle$ form a basis of the vector space of 4-dimensional column vectors, it is enough to consider the case in which both $|\psi_1\rangle$ and $|\psi_2\rangle$ are taken from $\{|0\rangle, |1\rangle\}$. Assume that

$$U_1 = \begin{pmatrix} a_{11} & a_{12} \\ a_{21} & a_{22} \end{pmatrix},$$

$$U_2 = \begin{pmatrix} b_{11} & b_{12} \\ b_{21} & b_{22} \end{pmatrix}.$$

Then we have that

$$U_1 \otimes U_2 = \begin{pmatrix} a_{11}b_{11} & a_{11}b_{12} & a_{12}b_{11} & a_{12}b_{12} \\ a_{11}b_{21} & a_{11}b_{22} & a_{12}b_{21} & a_{12}b_{22} \\ a_{21}b_{11} & a_{21}b_{12} & a_{22}b_{11} & a_{22}b_{12} \\ a_{21}b_{21} & a_{21}b_{22} & a_{22}b_{21} & a_{22}b_{22} \end{pmatrix}.$$

It follows then that

$$(U_1 \otimes U_2)(|0\rangle \otimes |0\rangle) = \begin{pmatrix} a_{11}b_{11} & a_{11}b_{12} & a_{12}b_{11} & a_{12}b_{12} \\ a_{11}b_{21} & a_{11}b_{22} & a_{12}b_{21} & a_{12}b_{22} \\ a_{21}b_{11} & a_{21}b_{12} & a_{22}b_{11} & a_{22}b_{12} \\ a_{21}b_{21} & a_{21}b_{22} & a_{22}b_{21} & a_{22}b_{22} \end{pmatrix}\begin{pmatrix} 1 \\ 0 \\ 0 \\ 0 \end{pmatrix} = \begin{pmatrix} a_{11}b_{11} \\ a_{11}b_{21} \\ a_{21}b_{11} \\ a_{21}b_{21} \end{pmatrix}.$$

On the other hand, we have that

$$\begin{aligned} (U_1 |0\rangle) \otimes (U_2 |0\rangle) &= \begin{pmatrix} a_{11} & a_{12} \\ a_{21} & a_{22} \end{pmatrix}\begin{pmatrix} 1 \\ 0 \end{pmatrix} \otimes \begin{pmatrix} b_{11} & b_{12} \\ b_{21} & b_{22} \end{pmatrix}\begin{pmatrix} 1 \\ 0 \end{pmatrix} \\ &= \begin{pmatrix} a_{11} \\ a_{21} \end{pmatrix} \otimes \begin{pmatrix} b_{11} \\ b_{21} \end{pmatrix} = \begin{pmatrix} a_{11}b_{11} \\ a_{11}b_{21} \\ a_{21}b_{11} \\ a_{21}b_{21} \end{pmatrix}, \end{aligned}$$

as needed. The computation in the other cases is analogous.

From this, we deduce that for any column vectors $|\psi_1\rangle$ and $|\psi_2\rangle$ of size 2 it holds that

$$\begin{aligned} (U_1 \otimes U_2)(U_3 \otimes U_4)(|\psi_1\rangle \otimes |\psi_2\rangle) &= (U_1 \otimes U_2)((U_3 |\psi_1\rangle) \otimes (U_4 |\psi_2\rangle)) \\ &= (U_1U_3 |\psi_1\rangle) \otimes (U_2U_4 |\psi_2\rangle) \\ &= ((U_1U_3) \otimes (U_2U_4))(|\psi_1\rangle \otimes |\psi_2\rangle). \end{aligned}$$

In particular, this holds for $|\psi_1\rangle \otimes |\psi_2\rangle$ taken from $\{|0\rangle \otimes |0\rangle, |0\rangle \otimes |1\rangle, |1\rangle \otimes |0\rangle, |1\rangle \otimes |1\rangle\}$, which is a basis. Thus, it follows that $(U_1 \otimes U_2)(U_3 \otimes U_4) = (U_1 U_3) \otimes (U_2 U_4)$.

Now, if U_1 and U_2 are unitary, clearly $(U_1 \otimes U_2)^\dagger = U_1^\dagger \otimes U_2^\dagger$ and

$$(U_1 \otimes U_2)(U_1^\dagger \otimes U_2^\dagger) = (U_1^\dagger \otimes U_2^\dagger)(U_1 \otimes U_2) = I_2 \otimes I_2 = I_4,$$

where I_2 is the identity matrix of size 2 and I_4 is the identity matrix of size 4. Thus, $U_1 \otimes U_2$ is unitary.

(5.8) (a)

$$\begin{pmatrix} 0 & 1 \\ 1 & 0 \end{pmatrix} \otimes \begin{pmatrix} \frac{1}{\sqrt{2}} & \frac{1}{\sqrt{2}} \\ \frac{1}{\sqrt{2}} & -\frac{1}{\sqrt{2}} \end{pmatrix} = \begin{pmatrix} 0 & 0 & \frac{1}{\sqrt{2}} & \frac{1}{\sqrt{2}} \\ 0 & 0 & \frac{1}{\sqrt{2}} & -\frac{1}{\sqrt{2}} \\ \frac{1}{\sqrt{2}} & \frac{1}{\sqrt{2}} & 0 & 0 \\ \frac{1}{\sqrt{2}} & -\frac{1}{\sqrt{2}} & 0 & 0 \end{pmatrix}.$$

(b)

$$\begin{pmatrix} 0 & 1 \\ 1 & 0 \end{pmatrix} \otimes \begin{pmatrix} 0 & 1 \\ 1 & 0 \end{pmatrix} = \begin{pmatrix} 0 & 0 & 0 & 1 \\ 0 & 0 & 1 & 0 \\ 0 & 1 & 0 & 0 \\ 1 & 0 & 0 & 0 \end{pmatrix}.$$

(c)

$$\begin{pmatrix} 1 & 0 \\ 0 & 1 \end{pmatrix} \otimes \begin{pmatrix} 1 & 0 \\ 0 & -1 \end{pmatrix} = \begin{pmatrix} 1 & 0 & 0 & 0 \\ 0 & -1 & 0 & 0 \\ 0 & 0 & 1 & 0 \\ 0 & 0 & 0 & -1 \end{pmatrix}.$$

(5.9) It holds that

$$\begin{pmatrix} 1 & 0 & 0 & 0 \\ 0 & 1 & 0 & 0 \\ 0 & 0 & 0 & 1 \\ 0 & 0 & 1 & 0 \end{pmatrix} \begin{pmatrix} \alpha_{00} \\ \alpha_{01} \\ \alpha_{10} \\ \alpha_{11} \end{pmatrix} = \begin{pmatrix} \alpha_{00} \\ \alpha_{01} \\ \alpha_{11} \\ \alpha_{10} \end{pmatrix},$$

which is exactly the action of the $CNOT$ gate.

(5.10) Assume that there exist matrices $A = \begin{pmatrix} a_{11} & a_{12} \\ a_{21} & a_{22} \end{pmatrix}$ and $B = \begin{pmatrix} b_{11} & b_{12} \\ b_{21} & b_{22} \end{pmatrix}$ such that

$$A \otimes B = \begin{pmatrix} a_{11}b_{11} & a_{11}b_{12} & a_{12}b_{11} & a_{12}b_{12} \\ a_{11}b_{21} & a_{11}b_{22} & a_{12}b_{21} & a_{12}b_{22} \\ a_{21}b_{11} & a_{21}b_{12} & a_{22}b_{11} & a_{22}b_{12} \\ a_{21}b_{21} & a_{21}b_{22} & a_{22}b_{21} & a_{22}b_{22} \end{pmatrix} = \begin{pmatrix} 1 & 0 & 0 & 0 \\ 0 & 1 & 0 & 0 \\ 0 & 0 & 0 & 1 \\ 0 & 0 & 1 & 0 \end{pmatrix}.$$

This implies that $a_{11}b_{12} = 0$. Since $a_{11}b_{11} = 1$, a_{11} cannot be 0. Then, b_{12} must be 0. But $a_{22}b_{12} = 1$, which is impossible. As a consequence, there are not matrices A and B such that $A \otimes B = \text{CNOT}$.

(5.11) Let's see how the matrix acts on the computational basis states. It holds that

$$\begin{pmatrix} 1 & 0 & 0 & 0 \\ 0 & 0 & 0 & 1 \\ 0 & 0 & 1 & 0 \\ 0 & 1 & 0 & 0 \end{pmatrix} \begin{pmatrix} 1 \\ 0 \\ 0 \\ 0 \end{pmatrix} = \begin{pmatrix} 1 \\ 0 \\ 0 \\ 0 \end{pmatrix},$$

so it takes $|00\rangle$ to $|00\rangle$. It also holds that

$$\begin{pmatrix} 1 & 0 & 0 & 0 \\ 0 & 0 & 0 & 1 \\ 0 & 0 & 1 & 0 \\ 0 & 1 & 0 & 0 \end{pmatrix} \begin{pmatrix} 0 \\ 1 \\ 0 \\ 0 \end{pmatrix} = \begin{pmatrix} 0 \\ 0 \\ 0 \\ 1 \end{pmatrix},$$

so it takes $|01\rangle$ to $|11\rangle$. Additionally,

$$\begin{pmatrix} 1 & 0 & 0 & 0 \\ 0 & 0 & 0 & 1 \\ 0 & 0 & 1 & 0 \\ 0 & 1 & 0 & 0 \end{pmatrix} \begin{pmatrix} 0 \\ 0 \\ 1 \\ 0 \end{pmatrix} = \begin{pmatrix} 0 \\ 0 \\ 1 \\ 0 \end{pmatrix},$$

so it leaves $|10\rangle$ unchanged. Finally,

$$\begin{pmatrix} 1 & 0 & 0 & 0 \\ 0 & 0 & 0 & 1 \\ 0 & 0 & 1 & 0 \\ 0 & 1 & 0 & 0 \end{pmatrix} \begin{pmatrix} 0 \\ 0 \\ 0 \\ 1 \end{pmatrix} = \begin{pmatrix} 0 \\ 1 \\ 0 \\ 0 \end{pmatrix},$$

and hence $|11\rangle$ is taken to $|01\rangle$. All this coincides exactly with the action of a CNOT gate with control on the second qubit and target on the first one.

(5.12) (a) It is a product state because $|10\rangle = |1\rangle \otimes |0\rangle$.

(b) It is entangled. The proof is almost the same as the one provided in the main text for $\frac{1}{\sqrt{2}}(|00\rangle + |11\rangle)$.

(c) It is a product state. In fact, it is the state $|+\rangle\,|-\rangle$.

(d) It is entangled. Imagine that there are two states $|\psi_1\rangle = a\,|0\rangle + b\,|1\rangle$ and $|\psi_2\rangle = c\,|0\rangle + d\,|1\rangle$ such that $|\psi_1\rangle \otimes |\psi_2\rangle = \frac{1}{2}(|00\rangle + |01\rangle + |10\rangle - |11\rangle)$. Then, we would have

$$\frac{1}{2}(|00\rangle + |01\rangle + |10\rangle - |11\rangle) = (a\,|0\rangle + b\,|1\rangle) \otimes (c\,|0\rangle + d\,|1\rangle)$$
$$= ac\,|00\rangle + ad\,|01\rangle + bc\,|10\rangle + bd\,|11\rangle\,.$$

As a consequence, $ac = ad = bc = 1/2$ and $bd = -1/2$. Then, $acbd = (ac)(bd) = 1/2 \cdot (-1/2) = -1/4$. But we also have that $acbd = (ad)(bc) = 1/2 \cdot 1/2 = 1/4$, which is a contradiction.

(5.13) We have already proved in the main text that $\frac{1}{\sqrt{2}}(|00\rangle + |11\rangle)$ is entangled. You can prove that the rest of the Bell states are also entangled with a similar approach. But we also can follow a more elegant way. Indeed, suppose that $\frac{1}{\sqrt{2}}(|00\rangle - |11\rangle)$ is not entangled. Then, there exist $|\psi_1\rangle$ and $|\psi_2\rangle$ such that

$$\frac{1}{\sqrt{2}}(|00\rangle - |11\rangle) = |\psi_1\rangle \otimes |\psi_2\rangle\,.$$

We know that $\frac{1}{\sqrt{2}}(|00\rangle - |11\rangle) = (Z \otimes I)(\frac{1}{\sqrt{2}}(|00\rangle + |11\rangle))$. Then, since $Z^2 = I^2 = I$, it holds that $(Z \otimes I)(Z \otimes I) = I \otimes I$ and thus

$$\frac{1}{\sqrt{2}}(|00\rangle + |11\rangle) = (Z \otimes I)(Z \otimes I)(\frac{1}{\sqrt{2}}(|00\rangle + |11\rangle)) = (Z \otimes I)(\frac{1}{\sqrt{2}}(|00\rangle - |11\rangle))$$
$$= (Z \otimes I)(|\psi_1\rangle \otimes |\psi_2\rangle) = (Z\,|\psi_1\rangle) \otimes |\psi_2\rangle\,.$$

Then, $\frac{1}{\sqrt{2}}(|00\rangle + |11\rangle)$ should be a product state too, and that is a contradiction. The proof for the rest of the Bell states is similar.

Chapter 6

(6.1) These are two circuits that can create the Bell state that we need:

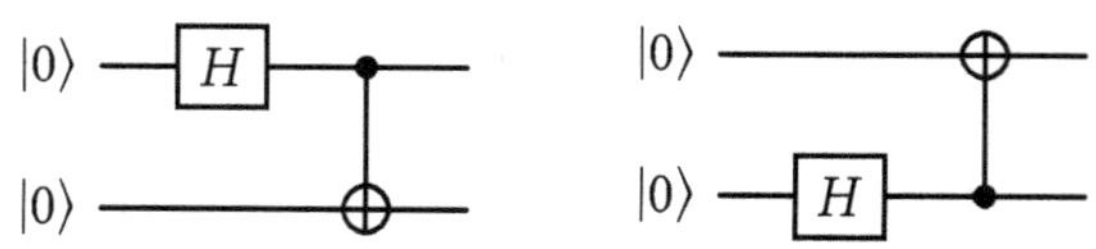

(6.2) We will ignore normalization factors in this solution (just assume there is a global $1/\sqrt{2}$ factor everywhere).

When a Z gate is applied on Alice's qubit, we have

$$(Z \otimes I)(|00\rangle + |11\rangle) = Z(|0\rangle)|0\rangle + Z(|1\rangle)|1\rangle = |00\rangle - |11\rangle.$$

When an X gate is applied, we have

$$(X \otimes I)(|00\rangle + |11\rangle) = X(|0\rangle)|0\rangle + X(|1\rangle)|1\rangle = |10\rangle + |01\rangle.$$

When an XZ gate is applied, we have

$$(XZ \otimes I)(|00\rangle + |11\rangle) = XZ(|0\rangle)|0\rangle + XZ(|1\rangle)|1\rangle = |10\rangle - |01\rangle.$$

(6.3) The case for 00 follows trivially from one of our solutions to Exercise 6.1. For 11, the CNOT gate will transform the state as

$$\frac{1}{\sqrt{2}}(|10\rangle - |01\rangle) \longrightarrow \frac{1}{\sqrt{2}}(|10\rangle - |11\rangle) = |1\rangle \otimes |-\rangle,$$

so the Hadamard matrix will leave the state in $|11\rangle$.

(6.4) It would suffice for her to apply the gates $X^a Z^b X$, thus adding an additional X to bring the state back to the Bell state that we originally considered.

(6.5) Alice and Bob could also agree to always pick 1.

(6.6) Using some basic trigonometry, we have that

$$\begin{aligned}
R_Y(\alpha)R_Y(\beta) &= \begin{pmatrix} \cos(\alpha/2) & -\sin(\alpha/2) \\ \sin(\alpha/2) & \cos(\alpha/2) \end{pmatrix} \begin{pmatrix} \cos(\beta/2) & -\sin(\beta/2) \\ \sin(\beta/2) & \cos(\beta/2) \end{pmatrix} \\
&= \begin{pmatrix} \cos(\alpha/2)\cos(\beta/2) - \sin(\alpha/2)\sin(\beta/2) & -\cos(\alpha/2)\sin(\beta/2) - \sin(\alpha/2)\cos(\beta/2) \\ \sin(\alpha/2)\cos(\beta/2) + \cos(\alpha/2)\sin(\beta/2) & -\sin(\alpha/2)\sin(\beta/2) + \cos(\alpha/2)\cos(\beta/2) \end{pmatrix} \\
&= \begin{pmatrix} \cos((\alpha+\beta)/2) & -\sin((\alpha+\beta)/2) \\ \sin((\alpha+\beta)/2) & \cos((\alpha+\beta)/2) \end{pmatrix} = R_Y(\alpha+\beta)
\end{aligned}$$

(6.7) The expansion of the state is the following:

$$\cos(-\pi/8)\left(\frac{1}{\sqrt{2}}(|00\rangle + |11\rangle)\right) + \sin(-\pi/8)\left(\frac{1}{\sqrt{2}}(|01\rangle - |10\rangle)\right),$$

hence the probability of success is $(\cos -\pi/8)^2 = (\cos \pi/8)^2$.

(6.8) The given state is indeed equal to

$$\frac{1}{\sqrt{2}}(R_Y(3\pi/4) \otimes I) \cdot (R_Y(-\pi/4) \otimes R_Y(-\pi/4)) \cdot (|00\rangle + |11\rangle),$$

which, because of the invariance of Bell states under Y-rotations, is equal to

$$\frac{1}{\sqrt{2}}(R_Y(3\pi/4) \otimes I)(|00\rangle + |11\rangle),$$

and the result follows.

(6.9) It will suffice to check its action on the computational basis states. For $|00\rangle$, the first X gate transforms it into $|10\rangle$, the CNOT gate into $|11\rangle$, and the X gate into $|01\rangle$, as expected. For $|01\rangle$, the sequence is $|11\rangle$, $|10\rangle$, and $|00\rangle$. For $|10\rangle$, it is $|00\rangle$, $|00\rangle$, and $|10\rangle$. And, lastly, for $|11\rangle$, the sequence of states is $|01\rangle$, $|01\rangle$, and $|11\rangle$.

(6.10) An oracle for f_0 is just the identity: no gates at all are needed!

This is an oracle for f_1:

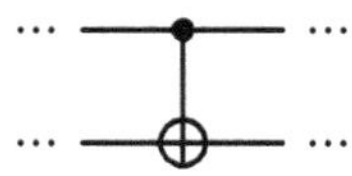

And this is an oracle for f_3:

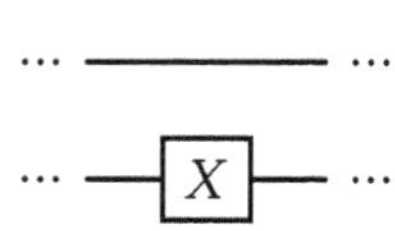

Chapter 7

(7.1) You can use the following piece of code:

(a)
```
state = Statevector(circuit)
print(state)
```

(b)
```
backend = AerSimulator(seed_simulator = 18620123)
sampler = Sampler(backend)

circuit.measure_all()

job = sampler.run([circuit], shots = 8)
result = job.result()[0].data.meas

print(result.get_counts())
```

(c)
```
from qiskit_ibm_runtime import QiskitRuntimeService
token = "YOUR TOKEN GOES HERE" # Replace with your IBM token
service = QiskitRuntimeService(
    token = token, channel = "ibm_quantum"
)

backend = service.least_busy(
    simulator = False, operational = True
)

from qiskit.transpiler.preset_passmanagers import (
    generate_preset_pass_manager
)

pm = generate_preset_pass_manager(
    backend=backend, optimization_level=1
)
```

```
transpiled = pm.run(circuit)

sampler = Sampler(mode=backend)
job = sampler.run([transpiled], shots = 1024)
result = job.result()[0].data.meas
print(result.get_counts())
```

The result for the statevector computation will be as follows:

```
Statevector([0.70710678+0.j, 0.        +0.j, 0.        +0.j,
             0.70710678+0.j],
            dims=(2, 2))
```

When you run the circuit on the simulator, you will obtain 11 five times and 00 three times. This is all consistent with the preparation of the $\frac{1}{\sqrt{2}}(|00\rangle + |11\rangle)$ state that we are doing with our circuit. However, when you use an actual quantum computer, you may obtain some stray 01 and 10 results. This is caused by the noise in the machine.

(7.2) You only need to change the values of `b1` and `b2` to those that Alice wants to send.

(7.3) In order to run the circuits on an actual quantum computer, you first need to create a `Sampler` object based on an IBM backend and preset manager with the following code:

```
from qiskit_ibm_runtime import QiskitRuntimeService
token = "YOUR TOKEN GOES HERE" # Replace with your actual IBM token
service = QiskitRuntimeService(token = token, channel = "ibm_quantum")

backend = service.least_busy(simulator = False, operational = True)

from qiskit.transpiler.preset_passmanagers import (
    generate_preset_pass_manager
)
```

```
pm = generate_preset_pass_manager(
    backend=backend, optimization_level=1
)
sampler = Sampler(mode=backend)
```

Notice that this needs to replace all the instructions based on `AerSimulator` that we show in the main text. Then, you also need to replace the instruction

```
circuit_list.append(circuit)
```

with

```
circuit_list.append(pm.run(circuit))
```

The rest of the code is identical to that used in the main text. When we ran this simulation of the CHSH game on actual quantum hardware, we obtained a win percentage of 84.3%, which is well above the 75% win rate that we would obtain if Nature were classical instead of quantum. We think that this is very, very cool.

(7.4) The oracle for f_0 is the identity function. Thus, you need nothing to implement it!

The oracle for f_1 can be implemented with the following code:

```
circuit.cx(0,1)
```

Finally, the oracle for f_3 can be implemented with the following instruction:

```
circuit.x(1)
```

In the cases of f_0 and f_3, the measurement results are always 0. For f_1, the measurement results are always 1.

Chapter 8

(8.1) To prove that the tensor product is associative, consider column vectors

$$a = \begin{pmatrix} a_1 \\ a_2 \\ \vdots \\ a_k \end{pmatrix}, \qquad a = \begin{pmatrix} b_1 \\ b_2 \\ \vdots \\ b_l \end{pmatrix}, \qquad a = \begin{pmatrix} c_1 \\ c_2 \\ \vdots \\ c_m \end{pmatrix}.$$

We know that

$$a \otimes b = \begin{pmatrix} a_1 b_1 \\ a_1 b_2 \\ \vdots \\ a_k b_l \end{pmatrix}$$

and that

$$b \otimes c = \begin{pmatrix} b_1 c_1 \\ b_1 c_2 \\ \vdots \\ b_l c_m \end{pmatrix}.$$

Thus, we have that

$$(a \otimes b) \otimes c = \begin{pmatrix} a_1 b_1 c_1 \\ a_1 b_1 c_2 \\ \vdots \\ a_1 b_2 c_1 \\ \vdots \\ a_k b_l c_m \end{pmatrix}$$

and also

$$a \otimes (b \otimes c) = \begin{pmatrix} a_1 b_1 c_1 \\ a_1 b_1 c_2 \\ \vdots \\ a_1 b_2 c_1 \\ \vdots \\ a_k b_l c_m \end{pmatrix}.$$

Assume now that we are working with an n-qubit system. We will prove that $|j\rangle$ is a column vector of size 2^n with all zeros but the element on position $j+1$, which is a 1, by induction. For $n = 1$, $n = 2$, and $n = 3$, we have already proved it in the main text. Consider now $n \geq 4$ and assume that the result is valid for $n-1$, and consider a basis state $|\psi\rangle$ of $n-1$ qubits.

If $|\psi\rangle = |0\rangle |\psi'\rangle$, then the column vector for $|\psi\rangle$ will start with the elements of the column vector of $|\psi'\rangle$ and, after that, it will have 2^{n-1} zeros. But the column vector for $|\psi'\rangle$ is, by the induction hypothesis, exactly of the form that we are interested in. It follows that $|\psi\rangle$ also has the desired structure. The case when $|\psi\rangle = |1\rangle |\psi'\rangle$ is analogous.

(8.2) We can prove it by induction. When $n = 1$, it is trivial. Now, consider $n > 1$ and assume that the result is true for $n-1$. Then, we have that

$$\begin{aligned} \bigotimes_{j=1}^{n} |+\rangle &= |+\rangle \otimes \bigotimes_{j=1}^{n-1} |+\rangle = |+\rangle \otimes \left(\frac{1}{\sqrt{2^{n-1}}} \sum_{j=0}^{2^{n-1}-1} |j\rangle \right) \\ &= \left(\frac{1}{\sqrt{2}}(|0\rangle + |1\rangle) \right) \otimes \left(\frac{1}{\sqrt{2^{n-1}}} \sum_{j=0}^{2^{n-1}-1} |j\rangle \right) \\ &= \frac{1}{\sqrt{2^n}} \left(|0\rangle \sum_{j=0}^{2^{n-1}-1} |j\rangle + |1\rangle \sum_{j=0}^{2^{n-1}-1} |j\rangle \right) = \frac{1}{\sqrt{2^n}} \left(\sum_{j=0}^{2^n-1} |j\rangle \right), \end{aligned}$$

as we needed.

(8.3) (a) It is valid, because $\left|\frac{1}{\sqrt{2}}\right|^2 + \left|\frac{1}{\sqrt{2}}\right|^2 = 1/2 + 1/2 = 1$.

(b) It is not valid, because you cannot add $|000\rangle$ (a vector of size 8) and $|11\rangle$ (a vector of size 4).

(c) It is valid, because $\left|\frac{1}{\sqrt{3}}\right|^2 + \left|\frac{1}{\sqrt{3}}\right|^2 + \left|\frac{1}{\sqrt{3}}\right|^2 = 1/3 + 1/3 + 1/3 = 1$.

(d) It is valid, because $\left|\frac{1}{\sqrt{2}}\right|^2 + \left|\frac{i}{2}\right|^2 + \left|\frac{-1}{2}\right|^2 = 1/2 + 1/4 + 1/4 = 1$.

(e) It is not valid, because $\left|\frac{1}{2}\right|^2 + \left|\frac{1}{2}\right|^2 + \left|\frac{1}{2}\right|^2 = 1/4 + 1/4 + 1/4 = 3/4 \neq 1$.

(8.4) Since the GHZ has 3 qubits, it can only be the product of a one-qubit state with a two-qubit state, or of a two-qubit state with a one-qubit state. We will show that the first case is impossible (the other is analogous). Assume that

$$(a|0\rangle + b|1\rangle)(c|00\rangle + d|01\rangle + e|10\rangle + f|11\rangle) = \frac{1}{\sqrt{3}}(|001\rangle + |010\rangle + |100\rangle).$$

Then, $ac = 0$. If $a = 0$, then $ad = 0$, but we know that $ad = \frac{1}{\sqrt{3}}$. Then, we must have $c = 0$. But then $bc = 0$, which is a contradiction.

(8.5) (a) You will obtain 000 or 111, each with probability $1/2$.

(b) You will obtain 00101 with probability $1/2$, 01101 with probability $1/4$, and 10101 with probability $1/4$.

(c) You will obtain 111000 with probability $9/25$ and 000111 with probability $16/25$.

(d) You will obtain 0001, 0010, 0100, or 1000, each with probability $1/4$.

(8.6) (a) You will obtain 0 with probability $2/3$ and the state will then collapse to $\frac{1}{\sqrt{2}}(|001\rangle + |100\rangle)$. You will obtain 1 with probability $1/3$ and the state will then collapse to $|010\rangle$.

(b) You will obtain 0 with probability $7/8$ and the state will then collapse to $\frac{1}{\sqrt{14}}|0001\rangle + \frac{2}{\sqrt{14}}|0010\rangle + \frac{3}{\sqrt{14}}|0100\rangle$. You will obtain 1 with probability $1/8$ and the state will then collapse to $|1000\rangle$.

(c) You will obtain 0 with probability 1/2 and the state will then collapse to $\frac{1}{\sqrt{2}}(|000\rangle - i\,|110\rangle)$. You will obtain 1 with probability 1/2 and the state will then collapse to $\frac{1}{\sqrt{2}}(i\,|011\rangle - |101\rangle)$.

(8.7) Since the CCNOT gate acts on three qubits, it could only be obtained as the product of either a one-qubit gate and a two-qubit gate, or a two-qubit gate and a one-qubit gate. The proof is similar in both cases, so we will only focus on the first situation.

Suppose, then, that there are matrices such that

$$\begin{pmatrix} a_{11} & a_{12} \\ a_{21} & a_{22} \end{pmatrix} \otimes \begin{pmatrix} b_{11} & b_{12} & b_{13} & b_{14} \\ b_{21} & b_{22} & b_{23} & b_{24} \\ b_{31} & b_{32} & b_{33} & b_{34} \\ b_{41} & b_{42} & b_{43} & b_{44} \end{pmatrix} = \begin{pmatrix} 1 & 0 & 0 & 0 & 0 & 0 & 0 & 0 \\ 0 & 1 & 0 & 0 & 0 & 0 & 0 & 0 \\ 0 & 0 & 1 & 0 & 0 & 0 & 0 & 0 \\ 0 & 0 & 0 & 1 & 0 & 0 & 0 & 0 \\ 0 & 0 & 0 & 0 & 1 & 0 & 0 & 0 \\ 0 & 0 & 0 & 0 & 0 & 1 & 0 & 0 \\ 0 & 0 & 0 & 0 & 0 & 0 & 0 & 1 \\ 0 & 0 & 0 & 0 & 0 & 0 & 1 & 0 \end{pmatrix}.$$

Then, $a_{11}b_{11} = 1$, so a_{11} is not 0. Since $a_{11}b_{34} = 0$, it follows that $b_{34} = 0$. But $a_{22}b_{34} = 1$, which is a contradiction.

(8.8) (a) It holds that

$$T^2 = \begin{pmatrix} 1 & 0 \\ 0 & e^{i\pi/4} \end{pmatrix}^2 = \begin{pmatrix} 1 & 0 \\ 0 & e^{i\pi/2} \end{pmatrix} = \begin{pmatrix} 1 & 0 \\ 0 & i \end{pmatrix} = S.$$

(b) It holds that

$$S^2 = \begin{pmatrix} 1 & 0 \\ 0 & i \end{pmatrix}^2 = \begin{pmatrix} 1 & 0 \\ 0 & -1 \end{pmatrix} = Z.$$

(c) It holds that

$$S^4 = Z^2 = \begin{pmatrix} 1 & 0 \\ 0 & -1 \end{pmatrix}^2 = \begin{pmatrix} 1 & 0 \\ 0 & 1 \end{pmatrix}.$$

Hence, $S^3S = SS^3 = I$ and, thus, the inverse of S is S^3. Since S is unitary, then $S^\dagger = S^3$.

(d) It holds that $T^7T = TT^7 = T^8 = S^4 = I$. Hence, T^7 is the inverse of T, which is also $T^\dagger$.

(e) We could obtain the result by simply multiplying the matrices, but we will do it in a different way. Notice that $X|0\rangle = |1\rangle$ and $X|1\rangle = |0\rangle$. Also, $HZH|0\rangle = HZ|+\rangle = H|-\rangle = |1\rangle$. And $HZH|1\rangle = HZ|-\rangle = H|+\rangle = |0\rangle$. Since X and HZH coincide on a basis, they must be the same matrix.

(8.9) It is easy to see that the matrix for CU is

$$\begin{pmatrix} 1 & 0 & 0 & \cdots & 0 \\ 0 & 1 & 0 & \cdots & 0 \\ 0 & 0 & u_{11} & \cdots & u_{12^n} \\ 0 & 0 & u_{21} & \cdots & u_{22^n} \\ \vdots & \vdots & \vdots & \ddots & \vdots \\ 0 & 0 & u_{2^n1} & \cdots & u_{2^n2^n} \end{pmatrix},$$

with

$$U = \begin{pmatrix} u_{11} & \cdots & u_{12^n} \\ u_{21} & \cdots & u_{22^n} \\ \vdots & \ddots & \vdots \\ u_{2^n1} & \cdots & u_{2^n2^n} \end{pmatrix},$$

then, $(CU)^\dagger = C(U^\dagger)$. Multiplying the matrices for CU and $CU^\dagger$, and taking into account that $UU^\dagger = U^\dagger U = I$, the result follows.

Chapter 9

(9.1) (a) It would collapse to $|0\rangle \otimes \frac{1}{\sqrt{2}}(|00\rangle + |11\rangle)$.

(b) It would collapse to $|\psi\rangle \otimes |00\rangle$.

(c) This implies that performing a measurement on the first qubit will have no effect on the two other qubits, and vice versa.

(9.2) Substituting $|\psi\rangle$ for $\alpha|0\rangle + \beta|1\rangle$, we can deduce that the CNOT gate will transform the state as

$$\begin{aligned}(\alpha|0\rangle + \beta|1\rangle) \otimes \frac{1}{\sqrt{2}}(|00\rangle + |11\rangle) &= \alpha|0\rangle \otimes \frac{1}{\sqrt{2}}(|00\rangle + |11\rangle) + \beta|1\rangle \otimes \frac{1}{\sqrt{2}}(|00\rangle + |11\rangle) \\ \longrightarrow\ & \alpha|0\rangle \otimes \frac{1}{\sqrt{2}}(|00\rangle + |11\rangle) + \beta|1\rangle \otimes \frac{1}{\sqrt{2}}(X \otimes I)(|00\rangle + |11\rangle) \\ &= \alpha|0\rangle \otimes \frac{1}{\sqrt{2}}(|00\rangle + |11\rangle) + \beta|1\rangle \otimes \frac{1}{\sqrt{2}}(|10\rangle + |01\rangle).\end{aligned}$$

It is (at least from our view) much easier to take this approach than to multiply 8×8 matrices.

(9.3) If the initial Bell state for the entangled pair is $\frac{1}{\sqrt{2}}(|10\rangle + |01\rangle)$, then the state right after the application of the first CNOT gate is

$$\frac{1}{\sqrt{2}}\left(\alpha(|010\rangle + |001\rangle) + \beta(|100\rangle + |111\rangle)\right),$$

and thus the state after the application of the subsequent Hadamard gate will be

$$\frac{1}{2}\left(\alpha(|010\rangle + |110\rangle + |001\rangle + |101\rangle) + \beta(|000\rangle - |100\rangle + |011\rangle - |111\rangle)\right).$$

If Alice's measurement then yields 00, the state will collapse to $\alpha|001\rangle + \beta|000\rangle$, so Bob will have to apply an X gate on its qubit in order to retrieve $|\psi\rangle$.

(9.4) These are all the two-bit constant functions:

$$c_1(0,0) := 0, \qquad c_1(0,1) := 0, \qquad c_1(1,0) := 0, \qquad c_1(1,1) := 0$$
$$c_2(0,0) := 1, \qquad c_2(0,1) := 1, \qquad c_2(1,0) := 1, \qquad c_2(1,1) := 1$$

And these are all the balanced ones:

$$b_1(0,0) := 0, \qquad b_1(0,1) := 0, \qquad b_1(1,0) := 1, \qquad b_1(1,1) := 1$$
$$b_2(0,0) := 0, \qquad b_2(0,1) := 1, \qquad b_2(1,0) := 0, \qquad b_2(1,1) := 1$$
$$b_3(0,0) := 1, \qquad b_3(0,1) := 0, \qquad b_3(1,0) := 0, \qquad b_3(1,1) := 1$$
$$b_4(0,0) := 0, \qquad b_4(0,1) := 1, \qquad b_4(1,0) := 1, \qquad b_4(1,1) := 0$$
$$b_5(0,0) := 1, \qquad b_5(0,1) := 0, \qquad b_5(1,0) := 1, \qquad b_5(1,1) := 0$$
$$b_6(0,0) := 1, \qquad b_6(0,1) := 1, \qquad b_6(1,0) := 0, \qquad b_6(1,1) := 0$$

(9.5) Using the enumeration of one-bit Boolean gates that we considered in *Section 6.3*, there would only exist such gates for f_1 and f_2. For f_1, G_{f_1} would be the identity (applying no gates), and for f_2, G_{f_2} would be an X gate. Notice that, since f_0 and f_3 are constant, the corresponding "gates" $G_{(\cdot)}$ would not be invertible, hence why they cannot exist.

(9.6) The state of the system right before the measurement is

$$\left(\sum_{x=0}^{2^n-1}\sum_{y=0}^{2^n-1}\frac{(-1)^{f(x)+x\odot y}}{2^n}\left|y\right\rangle\right)\otimes\frac{1}{\sqrt{2}}(\left|0\right\rangle-\left|1\right\rangle)=\left(\sum_{x=0}^{2^n-1}\sum_{y=0}^{2^n-1}\frac{(-1)^{f(x)+x\odot y}}{2^n}\left|y\right\rangle\right)\otimes\left|+\right\rangle,$$

and, since this is a tensor product of two states and $\left|+\right\rangle$ is a (normalized) state, we can simply ignore the last qubit and assume that we have n qubits in the state

$$\sum_{x=0}^{2^n-1}\sum_{y=0}^{2^n-1}\frac{(-1)^{f(x)+x\odot y}}{2^n}\left|y\right\rangle.$$

The amplitude of $|0 \cdots 0\rangle$ is then

$$\sum_{x=0}^{2^n-1} \frac{(-1)^{f(x)+x\odot 0}}{2^n}$$

and the result follows.

Chapter 10

(10.1) We would only have to run the following piece of code:

```
circuit = QuantumCircuit(6)
circuit.mcx([0,2,4,5], 3)
```

(10.2) If we wanted to have certainty that the final state is $|+\rangle$, it would suffice for us to add a Hadamard gate right before the measurement. By doing this, since unitary matrices are invertible, $|+\rangle$ (and only $|+\rangle$) would become $|0\rangle$)—and all other states would have a non-zero amplitude for $|1\rangle$. Therefore, if we run a sufficiently large amount of shots and we always get `0` as an outcome, we can be very certain that the state is $|0\rangle$. Of course, the more shots we run, the more certain we will be!

(10.3) We can run the following instructions:

```
state_1 = QuantumCircuit(1)
state_1.x(0)
result_1 = quantum_teleportation(state_1)
print(result_1.get_counts())
```

We will get that all the outcomes are 1.

(10.4) The following would be one possible alternative:

```
oracle = build_oracle(["010", "111", "011", "100"])
df = DJ(oracle)
job = sampler.run([dj], shots = 1)
```

```
result = job.result()[0].data.c
print(result.get_bitstrings())
```

(10.5) We can try it for $b = 001$, in which case the oracle would be constructed as

```
oracle = build_oracle(["001", "011", "101", "111"])
```

Chapter 11

(11.1) If you select $a = 6, 7, 9, 12, 14, 15, 18$ you will finish at step 2, because you will directly find a non-trivial factor of 21 when computing $\gcd(a, 21)$.

The value $a = 16$ is not valid, because its period is 3, which is odd. For $a = 8, 10, 11, 13, 19$ the periods are, respectively, $r = 2, 6, 6, 2, 6$ and, in these cases, you will be able to factor 21. For $a = 17$, the period is 6, but you will obtain only trivial factors. The same happens with $a = 20$, whose period is 2.

(11.2) If you select a to be one of $3, 5, 6, 9, 10$, or 12, you will find a factor of 15 at step 2. If you select $a = 14$, you will find that its period is 2, but it will give you only trivial factors. For $a = 2, 4, 7, 8, 11, 13$, the periods are $r = 4, 2, 4, 4, 2, 4$ and they will lead you to completely factor 15.

(11.3) If the result for the measurement of the second register is 1, the possible values for the first register are $0, 2, 4$, and 6. The possible differences are, thus, $2, 4$, and 6. The number 8 raised to any of those values is 1 mod 21. If the measurement on the second register gives 8 as a result, the possible values are $1, 3, 5$, and 7, and their differences are again $2, 4$, and 6.

(11.4) The coordinate on row j, column k is $\frac{1}{2}e^{\frac{\pi ijk}{2}}$. By noticing that $e^0 = 1$, $e^{\frac{\pi i}{2}} = i$, $e^{\pi i} = -1$, and $e^{\frac{3\pi i}{2}} = -i$, it is easy to check that the matrix is correct.

(11.5) The first row is composed entirely of the element 1 (which we can consider as the sole first root of unity). The second, fourth, sixth, and eighth rows are the 8th roots of unity, because they correspond to values of k equal to $1, 3, 5$, and 7, which have no common factors with 8. The third and seventh rows are the fourth roots of unity, because they correspond to values of k equal to 2 and 6 and $\gcd(2,8) = \gcd(6,8) = 2$. These roots are repeated twice. Finally, the fifth row consists of four repetitions of 1 and -1, which are the second roots of unity.

(11.6) Consider a computational basis state $|x\rangle |y\rangle$, where $x, y = 0, 1$. Let's see how this is transformed by the first circuit. After the first CNOT gate, the state will be $|x\rangle |y \oplus x\rangle$. After the second, it will be $|x \oplus y \oplus x\rangle |y \oplus x\rangle = |y\rangle |y \oplus x\rangle$. Finally, after the third gate, we will have $|y\rangle |y \oplus x \oplus y\rangle = |y\rangle |x\rangle$, as required.

A similar computation shows that the second circuit also implements the SWAP gate, but you can also reason by symmetry. If you read the first circuit from bottom to top, you get the second one. But exchanging the bottom qubit with the top one is the same as interchanging the top qubit with the bottom one, so this "bottom-up" circuit also implements the SWAP gate.

(11.7) The state $|1\rangle$ is an eigenvector of S with eigenvalue $e^{\frac{i\pi}{2}}$. Thus, $\theta = \frac{1}{4}$. You can determine this phase exactly if you use $m = 2$ qubits on the upper register, because you can express θ exactly as $\frac{1}{2^m} = \frac{1}{4}$.

Since $S^2 = Z$, the circuit would be as follows:

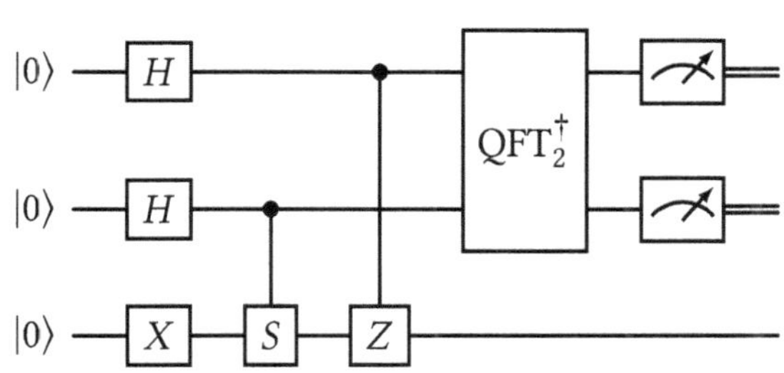

In the case of $|1\rangle$ and T, you would need 3 qubits because $\theta = \frac{1}{8} = \frac{1}{2^3}$. The circuit would be the following:

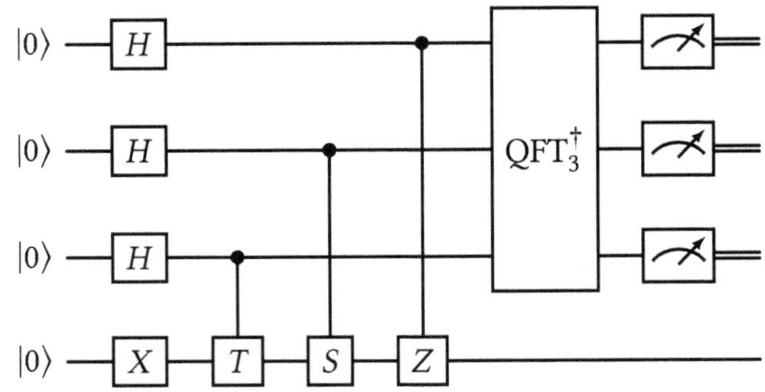

Chapter 12

(12.1) The multi-controlled Z gate will multiply the global state by the phase -1 if the input state is $|1 \cdots 1\rangle$, and it will leave any other computational basis states untouched. Therefore, there is no real distinction between control and target qubits in terms of the way in which the gate behaves. We can then conclude that, as long as there is one target qubit and the remaining qubits are controls, the gate will behave in the same way.

(12.2) Using L'Hôpital's rule,

$$\lim_{x\to 0} \frac{\sin x}{x} = \lim_{x\to 0} \frac{(\sin x)'}{x'} = \lim_{x\to 0} \frac{\cos x}{1} = 1.$$

(12.3) (a) We would need $\lfloor (\pi/4)\sqrt{8192} \rfloor = 71$ iterations.

(b) The probability would be $\left|\sin\left(101 \arcsin\left(1/\sqrt{(2^{13})}\right)\right)\right|^2 \approx 81\%$.

(12.4) The probability of getting s in a measurement is bigger than or equal to $1/2$ if and only if $|\sin \theta_k| \geq 1/\sqrt{2}$. Therefore, if we restrict ourselves to the positive Y half-plane, the probability is bigger than $1/2$ if $\pi/4 < \theta_k < 3\pi/4$. By a reasoning analogous to the one used to compute k_0, these constraints are satisfied by the values of k such that$(\pi/8)\sqrt{N} < k < (3\pi/8)\sqrt{N}$.

(12.5) (a) No, it wouldn't make sense because m is too large. We could just pick elements at random as, having picked k elements, the probability of finding one of the marked elements would be $1 - (1/2)^k$. Therefore, in order for the probability of finding a marked element to be bigger than 90%, we would need to perform $k = 4$ iterations, as $1 - (1/2)^4 \approx 94\%$, yet $1 - (1/2)^3 = 87.5\%$.

On the other hand, if we performed Grover's algorithm, after k iterations, the probability of finding a marked element would be

$$|\sin \theta_k|^2 = \left|\sin\big((2k+1)\arcsin\left(1/\sqrt{2}\right)\big)\right|^2 = \frac{1}{2},$$

regardless of the value of k, as $\arcsin\left(1/\sqrt{2}\right) = \pi/4$ and the sine of any odd integer multiple of $\pi/4$ is $1/\sqrt{2}$.

(b) We would need to perform

$$\frac{\pi}{4}\sqrt{\frac{1024}{5}} \approx 11$$

applications of the Grover operator. After this many iterations, the probability of finding a marked element will be

$$|\sin \theta_{11}|^2 = \left|\sin\left((11 \cdot 2 + 1)\arcsin\sqrt{\frac{5}{1024}}\right)\right|^2 \approx 99.85\%.$$

On the other hand, if we had picked 11 elements at random, the probability of having found a marked element would have been

$$1 - \left(1 - \frac{5}{1024}\right)^{11} \approx 5.24\%,$$

so it is very much worth it to use Grover's algorithm.

(12.6) The characteristic polynomial of this matrix is

$$\det(R(2\omega) - Ix) = x^2 - 2x(\cos 2\omega) + 1,$$

ant its roots are thus given by

$$x = \frac{2\cos(2\omega) \pm \sqrt{4\cos^2(2\omega) - 4}}{2} = \cos(2\omega) \pm i\sin(2\omega) = e^{\pm i(2\omega)}.$$

Since the matrix has two distinct eigenvalues, it is diagonalizable. The only instance in which it as a single eigenvalue is when $\omega = 0$ (or a multiple of π), but in this case the matrix is the identity (and it is already diagonal).

Chapter 13

(13.1) You can use the following piece of code:

```
from qiskit import QuantumCircuit
import numpy as np
qft3 = QuantumCircuit(3)
qft3.h(0)
qft3.cs(1,0)
qft3.cp(np.pi/4,2,0)
qft3.h(1)
qft3.cs(2,1)
qft3.h(2)
qft3.swap(0,2)
qft3.draw("mpl")
```

And you will obtain the following figure:

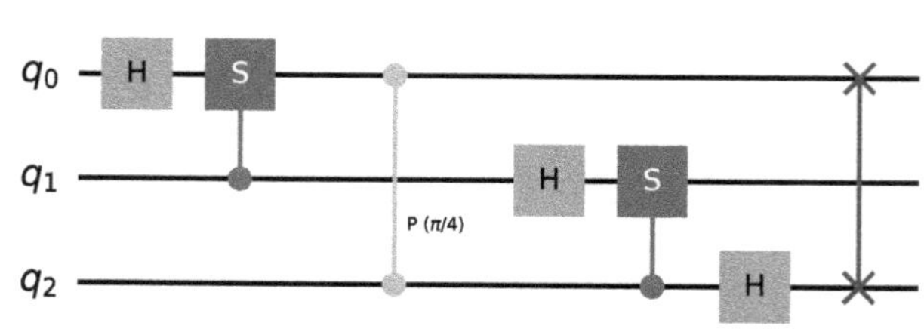

This is exactly the circuit we were looking for, because the controlled phase gate between qubits 0 and 2 is the controlled-T gate that we needed.

(13.2) With $N = 15$ and $a = 7$, you easily recover $r = 4$. With $N = 21$ and $a = 11$, you obtain $r = 6$. In both cases, you can successfully factor the numbers in a way very similar to what we did in the main text. For $a = 2$ and $N = 35$, you need to change the circuit a little bit, because the binary expansion of 35 is 6 bits long. Then, you need to create the circuit with `shor = circuit_shor(2,35,12)`. This is a much wider and longer circuit, so it will take longer to run. Eventually, you will obtain measurement results that, very likely, will include 1024, 1707, 2389, 2731, 3413, or 3755, among other values. Using, for instance, `limit_denominator`, you can discover that $r = 12$ and factor the number.

(13.3) You can find out with the following piece of code:

```
for i in range(11):
    grover = grover_circuit(oracle, i)
    sv = Statevector(grover)
    print("The probability of 111 with", i, "iterations is",
          abs(sv["0111"])**2+abs(sv["1111"])**2)
```

This will give you the following output:

```
The probability of 111 with 0 iterations is 0.12499999999999992
The probability of 111 with 1 iterations is 0.781249999999998
The probability of 111 with 2 iterations is 0.9453124999999958
The probability of 111 with 3 iterations is 0.3300781249999979
The probability of 111 with 4 iterations is 0.01220703124999987
The probability of 111 with 5 iterations is 0.5479736328124941
The probability of 111 with 6 iterations is 0.9997863769531125
The probability of 111 with 7 iterations is 0.5769729614257729
The probability of 111 with 8 iterations is 0.019456863403319972
The probability of 111 with 9 iterations is 0.3028912544250431
```

```
The probability of 111 with 10 iterations is 0.9312659502029226
```

Notice how the probability of success decreases to almost 0 if you set your number of iterations to 4. It is extremely important to correctly set this value in order for Grover's search to work properly!

Chapter 14

(14.1) (a) The probability of correctly identifying whether or not an error has taken place will be $P_{\text{identify}} = 1 - 0.05^3 \approx 99.99\%$, and that of correctly returning the original bit will be $P_{\text{correct}} = 1 - 3(0.05)^2 + 2(0.05)^3 \approx 99.28\%$.

(b) $P_{\text{identify}} = 1 - 0.1^3 = 99.90\%$,
$P_{\text{correct}} = 1 - 3(0.1)^2 + 2(0.1)^3 = 97.20\%$.

(c) $P_{\text{identify}} = 1 - 0.3^3 = 97.30\%$,
$P_{\text{correct}} = 1 - 3(0.3)^2 + 2(0.3)^3 = 78.40\%$.

(d) $P_{\text{identify}} = 1 - 0.5^3 = 87.50\%$,
$P_{\text{correct}} = 1 - 3(0.5)^2 + 2(0.5)^3 = 50.00\%$.

(14.2) The action of the first CNOT gate leaves the state as

$$(\alpha_0 \left|00\right\rangle + \alpha_1 \left|11\right\rangle) \left|0\right\rangle,$$

and the action of the second CNOT gate brings it to $\alpha_0 \left|000\right\rangle + \alpha_1 \left|111\right\rangle$, as claimed.

(14.3) Since $\oplus$ and $\cdot$ are commutative, assume without loss of generality that $x = y$. Clearly, $x = y = xy$. Now, if $z = 0$, we have $xy \oplus yz \oplus xz = xy \oplus 0 \oplus 0 = xy$, whereas, if $z = 1$,

$$xy \oplus yz \oplus xz = xy \oplus y \oplus x = xy \oplus xy \oplus xy = xy,$$

thus proving the result.

Alternatively, you can just evaluate the formula for all possible assignments of 0 and 1 to x, y, and z and check that the result holds.

(14.4) Using the properties of matrix multiplication, it can be checked that

$$HZH = \frac{1}{\sqrt{2}}\begin{pmatrix} 1 & 1 \\ 1 & -1 \end{pmatrix}\begin{pmatrix} 1 & 0 \\ 0 & -1 \end{pmatrix}\frac{1}{\sqrt{2}}\begin{pmatrix} 1 & 1 \\ 1 & -1 \end{pmatrix} = \begin{pmatrix} 0 & 1 \\ 1 & 0 \end{pmatrix} = X.$$

Moreover, it then follows that $H^{-1}XH^{-1} = Z$, but $H = H^{-1}$, hence $Z = HXH$.

(14.5) The application of a Hadamard gate on each of the three qubits in a state $a\left|000\right\rangle + b\left|111\right\rangle$ leaves the state as

$$a\left|+\right\rangle\left|+\right\rangle\left|+\right\rangle + b\left|-\right\rangle\left|-\right\rangle\left|-\right\rangle.$$

Then the application of a Z gate on the first qubit transforms the state into

$$a\left|-\right\rangle\left|+\right\rangle\left|+\right\rangle + b\left|+\right\rangle\left|-\right\rangle\left|-\right\rangle.$$

Finally, applying a Hadamard on each qubit brings the state to

$$a\left|100\right\rangle + b\left|011\right\rangle.$$

Informally speaking, this illustrates how phase-flip errors become bit-flip errors when surrounded by Hadamard gates.

(14.6) We will prove that $\{I, X, Z, XZ\}$ is a basis for the four-dimensional vector space of 2×2 matrices.

In a vector space, we say that some vectors $v_1, \ldots, v_k$ are linearly independent if the only way in which 0 can be written as a linear combination of them is $0 \cdot v_1 + \cdots 0 \cdot v_n$. With this definition in mind, we note that, in a finite-dimensional vector space of dimension n, any set of n linearly independent elements forms a basis (this can be easily proved). Thus, for our purposes, it will suffice to show that I, X, Z, XZ are linearly independent, that is, if $a, b, c \in \mathbb{C}$ are such that $aI + bX + cZ + dXZ = 0$, then $a = b = c = d = 0$.

If $aI + bX + cZ + dXZ = 0$, then

$$aI + bX + cZ + dXZ = \begin{pmatrix} a+c & b-d \\ b+d & a-c \end{pmatrix} = 0.$$

This implies that $a+c = a-c$, hence $c = 0$ and, since $a+c = 0$, then $a = 0$. Similarly, one may deduce that $b = d = 0$.

It can be shown that, if some matrices $A_1, \dots, A_k$ are linearly independent, so is the set of all their tensor products

$$A_1 \otimes A_1, A_1 \otimes A_2, A_1 \otimes A_3, \dots, A_k \otimes A_k.$$

This applies, in particular, to the set E of all possible n tensor products of I, X, Z, ZX, which lie in the space of $2^n \times 2^n$ matrices. Since this space has dimension $2^n \cdot 2^n$, and E has the same number of elements ($4^n = 2^n \cdot 2^n$), we can conclude that E is a basis of the space of of $2^n \times 2^n$ matrices.

Index

R

S

T

U

V

W

Other Books You Might Enjoy

If you enjoyed this book, you may be interested in these other books by Packt:

Learn Quantum Computing with Python and IBM Quantum (Second Edition)

Robert Loredo

ISBN: 978-1-80324-092-3

- Get familiar with the features within the IBM Quantum Platform
- Create and visualize quantum gates and circuits
- Operate quantum gates on qubits using the IBM Quantum Composer
- Install and run your quantum circuits on an IBM Quantum computer
- Discover Qiskit and its many features such as the Qiskit Runtime
- Get to grips with fundamental quantum algorithms and error mitigation techniques to help you get started
- Understand the new era of quantum utility and how this moves us closer towards quantum advantage

Dancing with Qubits (Second Edition)

Robert S. Sutor

ISBN: 978-1-83763-462-0

- Explore the mathematical foundations of quantum computing
- Discover the complex, mind-bending concepts that underpin quantum systems
- Understand the key ideas behind classical and quantum computing
- Refresh and extend your grasp of essential mathematics, computing, and quantum theory
- Examine a detailed overview of qubits and quantum circuits
- Dive into quantum algorithms such as Grover's search, Deutsch-Jozsa, Simon's, and Shor's
- Explore the main applications of quantum computing in the fields of scientific computing, AI, and elsewhere

Packt is searching for authors like you

If you're interested in becoming an author for Packt, please visit `authors.packtpub.com` and apply today. We have worked with thousands of developers and tech professionals, just like you, to help them share their insight with the global tech community. You can make a general application, apply for a specific hot topic that we are recruiting an author for, or submit your own idea.

Share your thoughts

Now you've finished *A Practical Guide to Quantum Computing*, we'd love to hear your thoughts! Please click here to go straight to the Amazon review page for this book and share your feedback or leave a review on the site that you purchased it from.

Your review is important to us and the tech community and will help us make sure we're delivering excellent quality content.

Download a free PDF copy of this book

Thanks for purchasing this book!

Do you like to read on the go but are unable to carry your print books everywhere? Is your eBook purchase not compatible with the device of your choice?

Don't worry, now with every Packt book you get a DRM-free PDF version of that book at no cost.

Read anywhere, any place, on any device. Search, copy, and paste code from your favorite technical books directly into your application.

The perks don't stop there, you can get exclusive access to discounts, newsletters, and great free content in your inbox daily.

Follow these simple steps to get the benefits:

1. Scan the QR code or visit the link below:

   `https://packt.link/free-ebook/9781835885949`

2. Submit your proof of purchase.
3. That's it! We'll send your free PDF and other benefits to your email directly.

Subscribe to Deep Engineering

Join thousands of developers and architects who want to understand how software is changing, deepen their expertise, and build systems that last.

Deep Engineering is a weekly expert-led newsletter for experienced practitioners, featuring original analysis, technical interviews, and curated insights on architecture, system design, and modern programming practice.

Scan the QR or visit the link to subscribe for free.

`https://packt.link/deep-engineering-newsletter`

Join us on Discord!

Read this book alongside other users, developers, experts, and the author himself.

Ask questions, provide solutions to other readers, chat with the authors via Ask Me Anything sessions, and much more. Scan the QR or visit the link to join the community.

https://packt.link/deep-engineering-quantum

www.ingramcontent.com/pod-product-compliance
Ingram Content Group UK Ltd.
Pitfield, Milton Keynes, MK11 3LW, UK
UKHW051957240426
470334UK00010B/310

9 781835 885